AF595081

Natural Sciences in Archaeology and Cultural Heritage

Natural Sciences in Archaeology and Cultural Heritage

Editor

Ioannis Liritzis

MDPI • Basel • Beijing • Wuhan • Barcelona • Belgrade • Manchester • Tokyo • Cluj • Tianjin

Editor
Ioannis Liritzis
Henan University
China

Editorial Office
MDPI
St. Alban-Anlage 66
4052 Basel, Switzerland

This is a reprint of articles from the Special Issue published online in the open access journal *Sustainability* (ISSN 2071-1050) (available at: https://www.mdpi.com/journal/sustainability/special_issues/Archaeology_Cultural_Heritage).

For citation purposes, cite each article independently as indicated on the article page online and as indicated below:

LastName, A.A.; LastName, B.B.; LastName, C.C. Article Title. *Journal Name* **Year**, *Volume Number*, Page Range.

ISBN 978-3-0365-3120-5 (Hbk)
ISBN 978-3-0365-3121-2 (PDF)

© 2022 by the authors. Articles in this book are Open Access and distributed under the Creative Commons Attribution (CC BY) license, which allows users to download, copy and build upon published articles, as long as the author and publisher are properly credited, which ensures maximum dissemination and a wider impact of our publications.
The book as a whole is distributed by MDPI under the terms and conditions of the Creative Commons license CC BY-NC-ND.

Contents

About the Editor

Ioannis Liritzis (www.liritzis.eu) is a professor of archaeometry, paleoenvironment and cultural heritage at the Institute of Capital Civilization and Cultural Heritage at Henan University, China. He specializes in the natural sciences, archaeology and cultural heritage, although, during his career, he has conducted research on a range of multidisciplinary fields. Amongst his innovative research projects are surface luminescence dating, and obsidian hydration dating. He has made contributions are in the interdisciplinary fields of geophysics, astronomy, archaeoastronomy, digital cultural heritage, palaeoclimatic variations, seismic periodicities, historical auroral records, Holocene climates, and chronological methods in the quaternary (U-Th). He is the initiator and Co-PI of the Sino-Hellenic academic project, and is currently developing projects focused on the role of sustainability in archaeometrical results, with case studies in Greek and Yellow River cultural remains. He worked for the Ministry of Culture, and the Academy of Athens, Greece. Ioannis has a BSc from Patras University, and earned his PhD from Edinburgh University (Physics Dept.), with graduate studies at McMaster University..

Preface to "Natural Sciences in Archaeology and Cultural Heritage"

The eleven papers cover a basic range of topics from the interdisciplinary field of Natural Sciences coupled with Archaeology and Cultural Heritage.

All the works deal with material culture and cultural environment properly studies with scientific tools.

Many types of ancient artifact are studied for the investigation of past human ingenuity; thus, the remains of material culture and human artifacts, works of art, and human remains of organic and inorganic origins, are the subject of destructive or mostly preferred non-evasive techniques.

Multiple techniques are can be used for all these investigations and are used to support instrumental analysis including nuclear, spectroscopic, chemical, and electronic devices. The obtained data, after being suitably assessed, fulfil multiple roles: (a) documentation of the preservation of at-risk antiquities from natural, and anthropogenic destruction; (b) reconstruction of a past human cultural environment; (c) available to education and society; (d) development of sustainability for the economy, society and the environment.

The content of the collected papers deal with the aspects and value of legacy concerning ancient organic residues, the sacred landscape of the "Pyramids" of the Han Emperors, a cultural palimpsest in the lower yellow river floodplain at Kaifeng, a mobile augmented reality, archaeometry's impact on cultural heritage, sustainability and development, the emergence of archaeoastronomy, providing a suitable way to link past societies' ingenuity with the skyscape, sustainable governance with archaeometry for Wari brewing traditions in ancient Peru, 3D digital heritage models as sustainable resources, the documentation of white marbles of the Tomb of Christ in Jerusalem, the archaeological chemistry of the ancient consumption of Shellfish purple, and the investigation and documentation of ancient pigments.

These papers delve into the realms of archaeometry and the natural sciences by analyzing artifacts and deciphering the past. The natural sciences, along with archaeometry, strengthen and develop the spirit of interdisciplinarity, delve into the past and strengthen our memory. In the remote past, we meet our future, and enhance the growth of sustainability; the use of the natural sciences in art strengthens intercultural dialogue and this also fosters ecumenical values.

Ioannis Liritzis
Editor

Article

Ancient Organic Residues as Cultural and Environmental Proxies: The Value of Legacy Objects

Andrew J. Koh [1,*] and Kathleen J. Birney [2]

1 Center for Materials Research in Archaeology and Ethnology, Massachusetts Institute of Technology, Cambridge, MA 02139-4307, USA
2 Department of Classical Studies and Archaeology Program, Wesleyan University, Middletown, CT 06459-3207, USA; kbirney@wesleyan.edu
* Correspondence: akoh@mit.edu

Received: 23 December 2018; Accepted: 24 January 2019; Published: 27 January 2019

Abstract: Often treated as an accessory science, organic residue analysis (ORA) has the capacity to illuminate otherwise hidden aspects of ancient technology, culture, and economy, and therein can play a central role in archaeological inquiry. Through ORA, both the intact vessel freshly excavated from a tomb and the sherd tucked away in a museum storage closet can offer insights into their contents, their histories, and the cultures that created them—provided the results can be carefully calibrated to account for their treatment during and after excavation. The case study below presents ORA data obtained from a range of artifacts from Late Bronze Age Crete, setting results from freshly-excavated and legacy objects alongside one another. Although legacy objects do tend to yield diminished results from both a quantitative and qualitative perspective, our comparative work has demonstrated both their value and untapped potential when their object biographies are carefully considered. It also sheds light on biomarker degradation processes, which have implications for methodologies of extraction and interpretation of legacy objects. Comparative studies such as these broaden the pool of viable ORA candidates, and therein amplify ORA's ability to reveal patterns of consumption as well as ecological and environmental change. They also highlight the role and value of data-sharing in collaborative environments such as the OpenARCHEM archaeometric database.

Keywords: organic residue analysis (ORA); archaeochemistry; phytochemistry; ethnobotany; ethnohistory; paleoenvironment; paleoecology; legacy artifacts; perfumed oils; Minoan Crete; OpenARCHEM

1. Introduction

Although often viewed as a scientific "sidecar", organic residue analysis (ORA) is gradually becoming a core component of archaeological research design, as its capacity to illuminate otherwise invisible data becomes increasingly evident. Since 2003, the ARCHEM project has collected and curated thousands of ORA field samples taken from ancient artifacts at sites across the eastern Mediterranean and western Asia, including Greece, Israel, Turkey, and Egypt. These include not only samples extracted in the field from freshly-excavated objects (typically, though not exclusively, ceramic), but also samples taken from "legacy" objects excavated up to a century prior, and stored in varying conditions in museums or excavation storage facilities. Although legacy objects do tend to yield diminished results from both a quantitative and qualitative perspective (as their compounds are subject to greater degradation), our work has demonstrated both their value and untapped potential as proxies for cultural, commercial, or environmental change when object parameters and biographies are carefully considered [1]. This is best illustrated through comparison between ORA results from freshly-excavated and legacy objects, an approach that offers valuable insight into the processes of chemical, environmental, and anthropogenically-induced degradation that affect them; it also has significant implications for future methodologies of sample collection and interpretation.

In the comparative case study below, we present, in parallel, ORA results taken from artifacts excavated at Mochlos and Tourloti, both Late Bronze Age (LBA) sites in East Crete (Figure 1). We offer some preliminary comments on their ramifications for the variable expression of compounds, the significance of vessel topography, and the overarching value of legacy data in ORA studies as a whole. This is the second in a series of short studies undertaken with the ARCHEM library of samples integrated into the OpenARCHEM archaeometric database, a new open-access repository, resource, and publication outlet for archaeometric data [2].

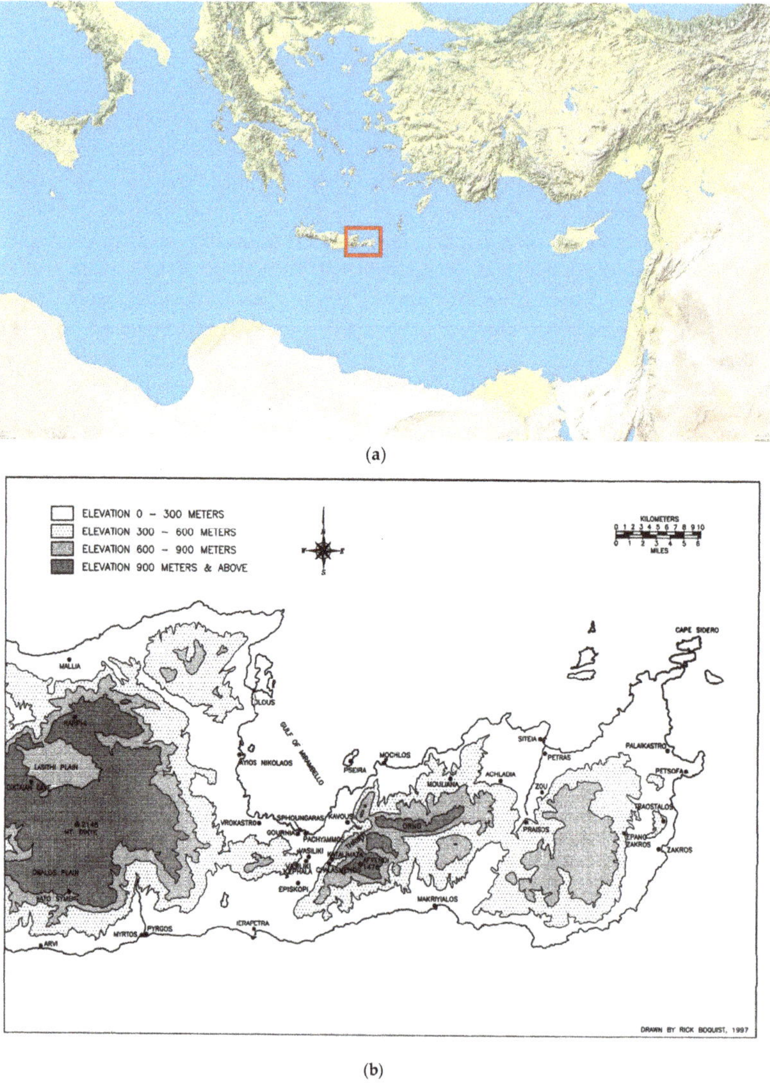

Figure 1. (**a**) Map of the eastern Mediterranean. Map data © 2019 Google Maps. (**b**) Map of East Crete. Base map by R. Boquist. Courtesy of J. G. Younger.

The Tale of Two Perfumes

In 2004, excavations at the coastal site of Mochlos in East Crete, co-directed by C. Davaras and J. S. Soles, revealed a Late Minoan (LM) IB (ca. 1500 B.C.) perfume workshop, which functioned at the height of the Minoan palaces only to be abandoned after a major LM IB earthquake [3,4]. The perfume manufacturing installation was found largely in situ, and its centerpiece was a complete circular ceramic vat (P6267), ca. 50 cm in diameter, with a specialized, detachable spout created from a conical vessel ("rhyton"); a fragmentary, undecorated oval-mouthed amphora (P6313A) was found nearby [4]. The amphora was pierced through the bottom in antiquity and may have served as a filter or volumetric standard for bottling the final aromatic product in finer, decorated amphoras. These objects were selected as good candidates for ORA given their likely role in a specialized production process, although its nature was unclear at the time. Once removed from the soil with gloves (typically nitrile) or placed directly into aluminum foil without direct contact, the artifacts were left uncleaned, packaged (some in plastic bags), and transported to the nearby Institute for Aegean Prehistory Study Center in Pacheia Ammos, where organic residues were extracted within two weeks of excavation. The goal was to minimize time between excavation and extraction, and, therefore, to reduce potential for harmful exposure. These precautions, coupled with the careful avoidance of cleaning and conservation prior to sampling, have all been shown to reduce the risk of anthropogenic and environmental contamination, alteration, and degradation while maximizing the potential for significant results [4–6]. Samples were taken according to an early version of the ARCHEM non-destructive extraction protocols described in the methodology section below.

Figure 2. Late Minoan IIIC Early stirrup jar from Tourloti.

In 1905–1906, during excavations of a LM III (ca. 1400–1050 B.C.) cemetery at Tourloti-Plakalona, approximately 4 km to the southeast of Mochlos, Richard B. Seager [7] uncovered a LM IIIC tomb (ca. 1175 B.C), whose contents included a locally-made stirrup jar (Penn Museum MS4494, Figure 2). This is a small globular, spouted vessel whose twin handles run from its shoulder to attach to a short "false neck" at its top, giving the appearance of stirrups. It is painted in the so-called Close style, and depicts a stylized octopus, an image descended from marine motifs popular in Minoan palatial traditions several centuries prior, but painted in a manner consistent with stylized trends typical for the early 12th century, or LM IIIC Early [1,8]. After excavation, the Tourloti jar was included by Seager in a group of local East Cretan artifacts gifted by the Candia Museum (now the Herakleion Archaeological Museum) to the Free Museum of Science and Art (now the University of Pennsylvania Museum of Archaeology and Anthropology), where it has been in storage since 1906 [9], most recently in modern, enclosed archival cabinets for the past several decades. MS4494 is almost completely intact and completely enclosed with only a ca. 5 mm opening providing access to its interior, and—based on our inspection and according to museum records—has never been displayed, handled excessively, or cleaned aggressively (e.g., scrubbed, acid washed, immersed in water). With a permit from the Penn Museum Scientific Testing Committee, ORA samples were taken from this vessel in 2015, over a century after its original excavation, employing the most recent version of the ARCHEM extraction protocol first applied to the Mochlos objects in 2003 [1,4–6].

2. Materials and Methods

OpenARCHEM prioritizes in-field and on-site extractions, for which non-destructive sampling methods are best suited. The value and viability of these approaches have been demonstrated in numerous studies [1,4–6,10–13]. Ancient organic compounds pose a challenge to the archaeological scientist in that, even under ideal conditions, they are prone to volatilization, degradation, and decomposition over time. As such, the diagnostic components of ancient compounds register more weakly than in comparable modern samples. Beyond these naturally occurring processes, archaeological samples are also susceptible to anthropogenic or environmental contamination in their journey from excavation to storage and exhibition, processes that have the potential to further obscure, dilute, or alter chemical signatures. Extraction and instrumentation protocols, if not carefully calibrated, can actually exacerbate these challenges. As such the ARCHEM project has developed and tested methodologies to maximize diagnostic output while minimizing undue risk to the already fragile contents as well as the artifacts themselves.

2.1. Non-Destructive Extraction Methodologies

The vessels from both Mochlos and Tourloti were sampled according to ARCHEM extraction protocols [4–6], which were developed utilizing non-destructive precedents [10,11] with field extractions in mind and subsequently refined through years of continuous field-testing [13]. While they share the same fundamental approach, the protocol used for Tourloti [1], being more recent, reflects the latest methodology. This is the product of an additional decade and a half of experimental refinement, which has shown that attention to the preservation, history, and therein potential contamination of an object is essential to the acquisition of qualitatively and quantitatively significant samples. With proper consideration of these features (i.e., a pre-experimental "diagnosis"), sampling strategies can be adapted to improve the collection and contamination and degradation risks likewise factored into interpretive frameworks. Object biography—including excavation conditions in the field – thus was and remains one of the key considerations in our methodology.

The ARCHEM extraction protocols are of two types: "swishing" and "boiling", both of which are non-destructive processes typically utilizing heated absolute, analytical-grade ethanol. Given that some mineral tempers in clay (e.g., calcites) are lipophilic [14], preferentially adsorbing certain fatty organic residues, the introduction of heated solvents induces dissolution of organics from the ceramic matrix. Ethanol is the ideal ORA solvent for field sampling, as it is both the least toxic of all alcohols, mitigating the need for a fume hood, and readily procurable worldwide, even in remote locations. In the early years, ARCHEM employed a two-step extraction process utilizing non-polar and polar solvents in rapid succession (e.g., ethyl acetate and methanol) with the goal of minimizing compounds to separate in each chromatogram [4,10]. However, over a decade of research in the field has empirically demonstrated that the downsides of a two-solvent extraction process in the field (e.g., working with highly toxic solvents) typically outweigh any marginal benefits that the more unwieldy solvent systems might provide in certain limited circumstances with ancient organic residues [5,6]. Beyond its ready availability and low toxicity, ethanol's solvent properties obviate the need for a two-step extraction process [5], as its hydroxyl group attracts polar and ionic molecules, while its ethyl group attracts non-polar ones. These qualities facilitate the rapid and effective extraction of a broad range of organic residues while leaving the artifact undamaged, making it ideally suited to field extraction. On-site extraction not only allows the ORA specialist to adapt extraction methods to the immediate archaeological environment and take regular soil samples as future controls, but also results in a comprehensive library of samples with detailed provenance. This is information that can offer uncommon insight into bigger issues such as the spatial distribution of a particular ancient organic commodity, or help to identify large scale contamination events across a site [13], which are almost always anthropogenically induced.

Boiling extractions are best suited for vessel fragments that have a clear chain of custody. This includes those that have a clean excavation and sampling history (e.g., handled with gloves, isolated in aluminum foil, unwashed) and subsequently have a low probability of contamination on any surface. The isolated fragments are placed in suitably sized Griffin beakers and are immersed in heated ethanol for at least several minutes. The type of pottery fragment makes a difference: For boiling extractions, vessel fragments are optimally chosen from the base, followed by the rim, especially for cookpots [15] or pouring vessels, and then the body, in that order. This is predicated on the assumption that the interiors of bases have the greatest exposure to organics, which pool at the bottom of a vessel or its lowest point at deposition, while certain diagnostic compounds float to the top during its history of use. Care must be taken to ensure that samples extracted from multiple fragments originate from a single object (i.e., that they all join). In cases of uncertainty, or where there are constraints upon resources (time or materials), which necessitate more limited sampling, then a single fragment chosen from the base at a location where it joins the body is usually preferable.

Swishing extractions are better suited for complete and near-complete vessels, or vessel fragments such as large bases, which, given their location on the vessel (as noted above), have typically had more prolonged exposure to organic contents, and therein greater opportunity for extant residues to permeate the matrix of the clay fabric. For the swishing of large vessel fragments, solvents are heated in a beaker as usual (to just below boiling), but they are then poured in small volumes onto the fragment and gently swished and/or stirred for approximately one minute on an interior surface, where the original organic contents would have come into prolonged contact for adsorption. For swishing of largely intact vessels, hot solvent is directly pipetted or poured into vessels, and is immediately swished and/or stirred around the object for up to a minute, and subsequently filtered directly into 20 mL master scintillation vials. Multiple swish extractions are typically executed.

Here again, object biography makes a difference. For objects with a clear chain of custody, and for which there are no indications of contaminants, all swished extractions can be collected into a single master sample prior to instrumentation. However, as our earlier work has indicated, the longer an object has been unearthed, the more likely the first extraction—those taken from the exposed surface—will include organics suffering from some level of contamination and degradation. Therefore, for those objects whose condition, or biography, is less certain, the swishing process is repeated and each resulting solution sample isolated, in order to moderate the removal of potential contaminants in early stages and facilitate compound separation and identification in multiple chromatograms. This can be a successful method to filter out plasticizers such as phthalates, which tend to appear in the first extraction and diminish or disappear in subsequent ones, as the results from plastic-packaged vessels from other areas of Mochlos and other sites have shown [4].

As freshly excavated artifacts, the Mochlos vessels—which included both complete and fragmentary vessels—were considered at lower risk for environmental degradation and anthropogenic contamination. Four samples were taken from the Mochlos vat and two from the nearby undecorated amphora, for a total of six samples from Mochlos and three from Tourloti (Table 1). Being smaller and more fragmentary, sherds from the vat's rhyton-spout were boiled using the two extraction solvents in sequence, ethyl acetate first followed by ethanol, which produced two samples (Figure 3). The large pieces of the main body of the vat, being too large for beakers, were swished using the two extraction solvents in sequence, which produced two more samples (Figure 4). The undecorated amphora (P6313A) was fragmented, but as the full profile was preserved and the sherds from the base, body, and rim were clearly from the same vessel, these sherds were boiled together first in ethyl acetate followed by ethanol to produce two samples (Figure 5).

Table 1. Gas Chromatography-Mass Spectrometry data from Tourloti and Mochlos vessels.

Vessel	Figure	Sample ID §	Solvent	Maximum Peak Height	Maximum Peak AA *	Maximum Peak Concentration (%) †	Cholesterol RA (%) ‡	Oleic Acid RA (%) ‡	Linoleic Acid RA (%) ‡	Azelaic Acid RA (%) ‡	Cinnamic Acid RA (%) ‡	Linalool RA (%) ‡	Manoyl Oxide RA (%) ‡	Docosane RA (%) ‡	Camphor RA (%) ‡
Mochlos Vat spout (P6267)	3a	Run 9.06	ethyl acetate	12224	25516	37.49	trace	trace	7.12	0	6.50	trace	44.30	0	7.62
Mochlos Vat spout (P6267)	3b	Run 2.17	ethanol/methanol	20793	41916	21.06	7.40	25.20	18.10	0	3.13	trace	trace	0	20.52
Mochlos Vat Base (P6267)	4a	Run 9.05	ethyl acetate	9353	20609	32.08	0	24.30	trace	0	15.03	trace	49.70	8.64	trace
Mochlos Vat Base (P6267)	4b	Run 3.10	ethanol/methanol	11574	23998	47.95	0	62.50	18.00	0	12.20	trace	trace	trace	8.96
Mochlos amphora (P6313A)	5a	Run 9.16	ethyl acetate	7989	14218	10.70	0	20.90	22.80	0	22.10	16.60	trace	4.66	42.40
Mochlos amphora (P6313A)	5b	Run 6.14	ethanol/methanol	19014	39287	33.73	0	#####	53.10	0	45.80	0	trace	trace	52.00
Tourloti Jar (MS4944) 1	6a	ARCHEM 4426	ethanol	7586	11674	41.02	0	18.70	n/a	100	0	14.90	19.46	0	0
Tourloti Jar (MS4944) 2	6b	ARCHEM 4427	ethanol	6673	10907	56.60	0	0	n/a	100	43.54	3.28	0	0	4.78
Tourloti Jar (MS4944) 3	6c	ARCHEM 4428	ethanol	10231	16473	37.22	0	0	n/a	100	0	2.64	3.34	0	0

trace = 1–3%, §Run # (as assigned by Koh 2008) or standard ARCHEM #, * Absolute Abundance, or peak area determined by integration in chromatograms, † Percentage of the sum of all peaks in a chromatogram, ‡ Relative Abundance, or peak area as a percentage relative to the maximum peak.

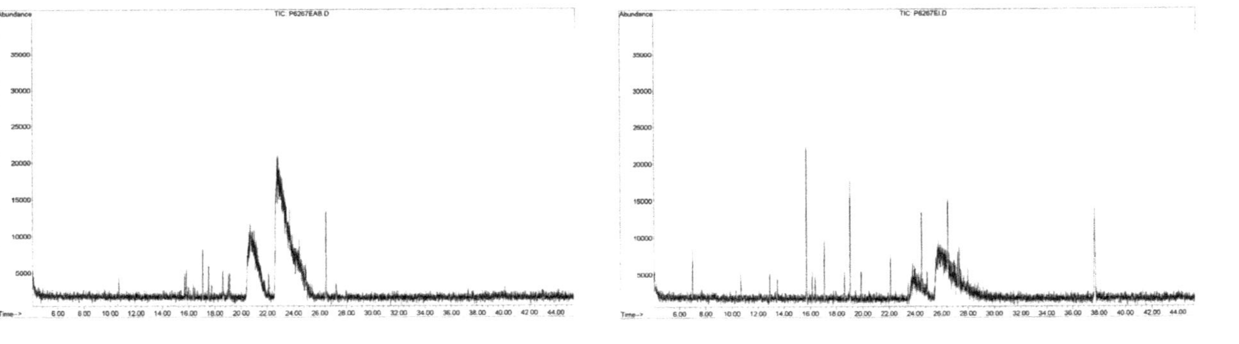

Figure 3. Total Ion Chromatogram from: (**a**) ethyl acetate extraction of Mochlos vat spout; (**b**) ethanol extraction of Mochlos vat spout.

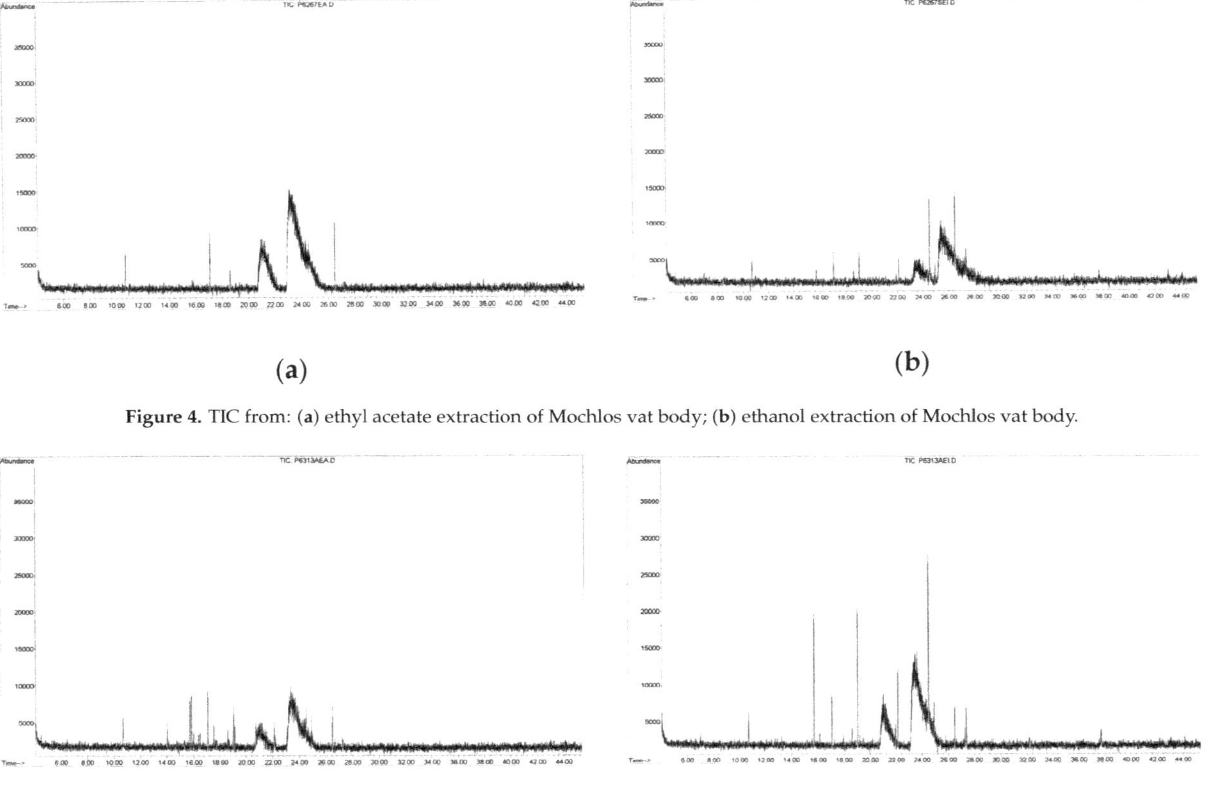

(**a**) (**b**)

Figure 4. TIC from: (**a**) ethyl acetate extraction of Mochlos vat body; (**b**) ethanol extraction of Mochlos vat body.

(**a**) (**b**)

Figure 5. TIC from: (**a**) ethyl acetate extraction of Mochlos amphora; (**b**) ethanol extraction of Mochlos amphora.

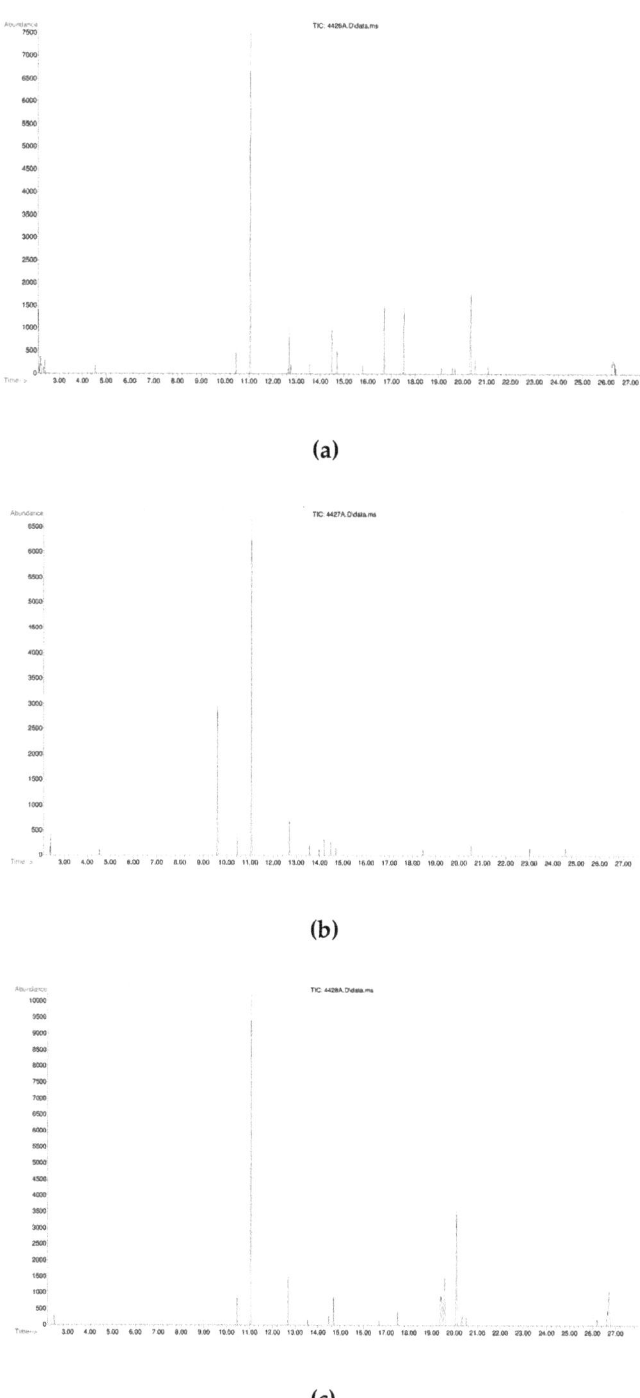

Figure 6. TIC from: (**a**) first extraction of MS4494; (**b**) second extraction of MS4494; **(c)** third extraction of MS4494.

The Tourloti jar, by contrast, was sampled over a century after its excavation. The top-flight archaeological museum in which it currently resides now boasts climate-controlled storage; however, the jar certainly experienced more environmentally variable conditions prior to our sampling. Although neither the museum records nor visual inspection showed any signs of cleaning or conservation, it was unclear whether any viable compounds could be extracted from a jar so long exposed after excavation. As an experimental precaution, additional measures were employed to account for the potential buildup of dust or other contaminants on the interior of the vessel, and to maximize the likelihood that residues trapped within the fabric could be chemically extracted: Three separate hot solvent washes (i.e., swishes) were performed in rapid succession despite using the same solvent (i.e., ethanol), resulting in three separate filtered solution samples (Figure 6).

2.2. Instrumentation Methodology

The low concentrations of organic compounds preserved in ancient samples demand analytical instruments of high sensitivity, which allow researchers to separate and characterize organic compounds, both qualitatively and quantitatively. Instruments of choice have been chromatographic, especially liquid (LC-MS) or gas chromatography (GC-MS) coupled with mass spectrometry [16–18], with the protocols for each refined for the best expression of fragile and low concentration compounds.

The extracted Mochlos ORA samples, evaporated to dryness, were held in climate-controlled storage until export permissions from the Hellenic Ministry of Culture were granted, after which they were sent at first opportunity to the University of Pennsylvania Department of Chemistry in spring of 2005 [4]. Once safely in Philadelphia, preparations were made to inject the samples into the department's most advanced GC-MS housed in its Mass Spectrometry Facility. These preparations were made at the Roy and Diana Vagelos Laboratories at the invitation of G. Palladino, executive director, and G. A. Molander, the Hirschmann-Makineni Professor of Chemistry. The Mochlos ORA samples were processed and reintroduced into solution using ethyl acetate. Before injection into the Mass Spec Facility GC-MS, general conditions were determined using a HP 5890A Gas Chromatograph interfaced with a HP 3395 Integrator, housed in the Vagelos research laboratories of the Molander Group. This protocol was then further optimized in collaboration with the Molander Group over the course of several weeks using the actual GC-MS at the Mass Spec Facility, an Agilent Technologies 6890N Network 6C System used in series with an Agilent Technologies 5973N Network Mass Selective. Samples were automatically injected with an Agilent Technologies 7683 Series Injector with 10 μL syringe, which allowed automated overnight runs and the possibility to analyze hundreds of samples in a compressed time frame. Before the injection of each 2 μL sample, the computer automatically cleaned the syringe with 10 μL of methylene chloride twice and also primed it twice with the ORA sample pre-injection. Post-injection, the syringe was cleaned four times using 10 μL of methylene chloride. The GC-MS had a capillary column (30.0 m × 250 mm × 0.15 mm) set to constant flow mode and front inlet. The carrier gas flow (helium) was set to 0.5 mL/min with an average velocity of 41 cm/s. The EPC Split-Splitless Inlet was ultimately set to splitless as experimentation demonstrated that the small quantities of ancient compounds present in the circumstances at hand were better expressed in such fashion. The GC oven temperature had a set point of 110 °C held for two minutes, and increased by 6 °C/minute until it hit 250 °C, and then held for 20 min, with a total run time of 45.33 min. Solvent blanks were utilized before and after the set to verify that no contaminants existed from previous runs or that components were lost in the column.

The Mass Spec Facility GC-MS was interfaced with a HP Vectra VL420 DT computer (P4 2.20 GHz, 256 mb RAM) running Microsoft Windows 2000 Professional SP 2 (5.00.2195) and Agilent Technologies MSD Enhanced Chemstation (G1701DA version D.00.00.38). The compounds were initially identified using the National Institute of Standards and Technology Mass Spectral Database (NIST 02) with its Mass Spectral Search Program (1.7a) before being manually checked for final peak assignation. In order to help ensure the survival of the molecular ions in sample sizes that were anticipated to be relatively minute, the mass spectrometer was eventually set up for "soft" chemical ionization (CI), instead of the harsher electron ionization (EI) more prevalent in labs. This approach made it more laborious to interpret mass spectra since most reference spectra and the computer library searches are EI-based, but the results proved this approach worthwhile in the context of ancient organic residue analysis.

The Tourloti ORA solution samples were transported back to the Brandeis University Department of Chemistry immediately after extraction in August 2015 and processed for instrumentation the following day [1]. The samples were concentrated to solid by rotary evaporator and then redissolved in uninhibited tetrahydrofuran (THF) to produce ~300 μL GC-MS analytes, which supplied 4 μL for auto injection into an Agilent 7890A GC with a HP-5MS column and a 5975C VL MSD Triple Axis Detector. Before the injection of each sample, the computer automatically cleaned the syringe with 10 μL of THF twice, and also primed it twice with the ORA sample pre-injection. Post-injection, the syringe was cleaned two times using 10 μL of THF. The pulsed split injector and interface were both set to 250 °C. The initial oven temperature was set to 100 °C and held for two minutes before reaching 250 °C at a rate of 10 °C/minute, at which time it was held for an additional 11 min, giving the total program time of 28 min/sample. As at Penn Chemistry, solvent blanks were utilized before and after the set to verify that no contaminants existed from previous runs or that components were lost in the column. Within the first sample (ARCHEM 4426), any anticipated dust and dirt were removed by hot filtration during the extraction process. The compounds were initially identified using the NIST Mass Spectral Database (NIST 11) with its Mass Spectral Search Program (2.0 g) before being manually checked for final peak assignation.

3. Results

Table 1 presents the comparative results and quantities of the diagnostic compounds recovered from each extraction of the Mochlos and Tourloti vessels. Note that palmitic and stearic acid, although usually represented in all samples, have been omitted from the list of compounds, as they are so ubiquitous in organic samples as to be non-diagnostic. In the Mochlos samples, oleic acid, cinnamic acid, linalool, manoyl oxide, docosane, and camphor were identified after initial peak assignation using the NIST Mass Spectral Database, NIST 02 (Figures 3–5). Each of these compounds can be connected to a high degree of certainty with known botanical sources local to Crete and the eastern Mediterranean. In the Tourloti jar, azelaic acid, oleic acid, linalool, and manoyl oxide were tentatively identified in the first extraction (ARCHEM 4426, Figure 6a) after initial peak assignation using NIST 11 with the understanding that they could be byproducts or remains of modern contamination until confirmed by subsequent samples. The second extraction (ARCHEM 4427, Figure 6b) clearly produced azelaic acid, cinnamic acid, camphor, and linalool from the ceramic matrix. In the third extraction (ARCHEM 4428, Figure 6c), azelaic acid, linalool, and manoyl oxide were present again confirming their presence as components of the original ancient residue. Mass spectrum fragments for oleanolic acid also clearly came out at 20.07 min from this third extraction. These identifications were confirmed by running chemical reference samples for each individual constituent compound, and also through comparison with the chromatographic signatures from ethnobotanical samples extracted from *Artemisia*, *Cistus*, and *Tilia* sourced from Crete, and from ethnographic samples extracted from modern 18th century pithoi housed in the Museum of Cretan Ethnology Research Centre, which oral history and visual inspection documented had been used to store olive oil. More data can be seen in the Supplementary Materials.

The slate of ingredients suggests that both the Mochlos and Tourloti jars held complex perfumed oils. The combination of botanicals employed suggests a surprisingly early understanding of perfume design through the creation of balanced tripartite blend, as each jar contained constituents that appear to have functioned in the perfumes as lower ("base"), middle ("heart"), and top ("head") notes, respectively. This fragrance framework of base, heart, and head notes is a governing principle in the craftsmanship of perfumes throughout history [19] and its importance was apparently recognized millennia ago.

3.1. *Oil from Olea Europaea (Olive)*

Azelaic Acid ($C_9H_{16}O_4$, 188 MW, 11.05 min at Brandeis, 12.90 min at Penn)
Oleic Acid ($C_{18}H_{34}O_2$, 282 MW, 16.70 min at Brandeis, 24.50 min at Penn)

Oleic acid is a monounsaturated fatty acid (18:1) that is a pale to brownish-yellow liquid at room temperature. It is the most abundant unsaturated fatty acid found in nature, but occurs in high concentrations (55–85%) only in olive oil from *Olea europaea*. Oleic acid is rarely found in animals, so it serves as a good biomarker for olive when detected in sufficient quantity in the proper context. Azelaic acid in sufficient quantity can also be a secondary biomarker for olive oil, in certain archaeological contexts. After a decade of ORA research, the project has verified that legacy objects suspected of containing olive oil in antiquity often contain large quantities of azelaic acid, a saturated dicarboxylic acid, in lieu of oleic acid. This is because oleic acid degrades to azelaic acid through oxidation scission processes that can occur once artifacts are exposed to the open environment [20–22] and discussion below. *O. europaea* is abundant on the island of Crete and was cultivated by 3000 B.C., or Early Minoan I [5]. In addition to being a widely available product of the Mediterranean region, textual sources from the Bronze Age and later classical periods demonstrate that olive oil was frequently used as a liquid base for perfumes [23], likely preferred in antiquity for its ability to be stored at ambient temperatures for prolonged periods without evaporating or spoiling.

3.2. *Storax Balsam from Liquidambar Orientalis (Sweetgum)*

Cinnamic Acid ($C_9H_8O_2$, 148 MW, 9.61 min at Brandeis, 10.68 min at Penn)
Oleanolic Acid ($C_{30}H_{48}O_3$ 456 MW, 20.07 min at Brandeis)

Cinnamic acid is a carboxylic acid with strong fixative, antioxidant, and antimicrobial properties [24]. In the initial Penn Chemistry study, the compound had been noted in the Mochlos chromatograms but was not recognized at the time as cinnamic acid, instead having been categorized among the repeatedly occurring unknowns, and was retroactively identified. Cinnamic acid is a white, crystalline compound with a pleasant, leathery odor. In both the Tourloti and Mochlos perfumes, the cinnamic acid was probably obtained from the balsamic oleoresin exuded from the inner bark of *Liquidambar orientalis* [25], a species that contains up to 150,000 ppm cinnamic acid [26]. While cinnamic acid occurs up to 90,000 ppm in the fruit of *Laurus nobilis* (laurel), the complete absence of cineole, the prevailing diagnostic compound in *L. nobilis* (occurring up to 283,700 ppm in its fruit), disqualifies laurel as the source botanical. Moreover, oleanolic acid, another biomarker of *L. orientalis* according to a GC-MS characterization of aromatic resins carried out by Modugno et al. [27], was identified in sample 4428 (i.e., the third extraction). This is one of the major biomarkers that helped determine that the extant cinnamic acid was derived from storax balsam sourced from *L. orientalis*, rather than storax benzoin (or benzoe) sourced from *Styrax officinalis* [28].

The Roman historian, Pliny the Elder, suggests that *L. orientalis* grew on Crete (*Naturalis Historia* 12.55), although it is unclear whether the species was present on the island 1200 years earlier during the LBA. Well-watered locales, such as the Kephalo Vrisi Valley (Mesa Mouliana) or the Richtis Gorge (Exo Mouliana), both nearby to Tourloti and Mochlos, would have been suitable environments judging by the regular occurrence today of *Platanus orientalis* (plane), which is known to grow in similar riparian ecosystems [1]. Storax balsam from *L. orientalis* could also have been readily sourced from the Dodecanese or southwest Anatolia (roughly 200 km from East Crete), where the tree grows to this day on the island of Rhodes and in the Muğla Province of the Turkish mainland, directly across from the island. Used today to anchor scents in Bourjois' "Soir de Paris" and Hermès' "Bel Ami" [19], in both the Tourloti and Mochlos perfumed oils, storax likely had the same effect of lending the final aromatic product a balsam quality to its lower and middle body (i.e., the base and heart notes). In antiquity, its antimicrobial and antioxidant qualities would have also held great value as a kind of preservative, delaying spoilage of the olive oil-based perfume.

3.3. *Camphor from Artemisia absinthium (Wormwood)*

Camphor ($C_{10}H_{16}O$, 152 MW, 14.19 min at Brandeis, 15.7 min at Penn)

Camphor is a bicyclic terpenoid ketone that is a white (or clear) waxy crystal with a strong, pungent aromatic odor. While camphor appears in a number of species in trace amounts, the significant quantities recovered direct us to the *Artemisia* genus (e.g., wormwood) as the likely botanical source, where it appears in quantities upwards of 210,000 ppm [26]. While camphor also appears in particularly large quantities in *Cinnamomum camphora* (i.e., camphor tree), at upwards of 220,000 ppm, the latter originates in East Asia and is unlikely to occur in a Mediterranean context at this early date. *Artemisia* plants are hardy shrubs that grow in dry or semi-dry climates, such as the Mediterranean, and are known for their volatile oils. Due to its extremely bitter taste, plants in this genus have traditionally been used for medicinal and aromatic purposes (Dioscorides, *De Materia Medica* III.23.1-6), rather than flavoring [19,29]. Camphor has traditionally been used as a pest repellent and antimicrobial agent as well as the initiating top scent, or head note, in spicy perfumes such as Yves St. Laurent's "Jazz." It remained one of the most popular perfume ingredients in the early Arab world [19], affirming its long history of use in aromatics.

3.4. *Essential Oil from Coriandrum sativum (Coriander)*

Linalool ($C_{10}H_{18}O$, 154 MW, 14.50 min at Brandeis, 14.16 min at Penn)

Linalool is a terpene alcohol found in plants with a pleasant, floral scent, hence its regular use in soaps and perfumes. Researchers have also highlighted its ability to elicit physiological effects from inducing calmness to improving sleep [30]. Notably, it is the main constituent of the essential oil of *Coriandrum sativum* (coriander), where it occurs in concentrations upwards of 742,300 ppm [26]. κορίανδρον, or ko-ri-ja-do-no in Linear B tablet PY Un 267, has been connected with the LBA perfumed oil industries both on Crete [4] and the Greek mainland [23]. As an essential oil with some highly volatile compounds, it could have been valued as a head note in ancient perfumed oils, a role it plays today in "Gucci No. 1", Yves St. Laurent's "Jazz", and Max Factor's "Le Jardin d'Amour" [19]; however, its primary value in antiquity likely stemmed from its astringent properties, and therein its role in stypsis, a key step in oil-based perfume manufacture to which the Bronze Age and subsequent ancient sources make reference (see discussion below).

3.5. *Labdanum Amber from Cistus creticus (Cretan Rockrose)*

Manoyl Oxide ($C_{20}H_{34}O$, 290 MW, 17.51 min at Brandeis, 23.71 min at Penn)

Manoyl oxide is a diterpenoid ether found in the Cistaceae (rockrose) and Pinaceae (pine) families, and could arguably represent either. Cistaceae has a long history, however—in the Mediterranean, in general, and Crete, in particular (Herodotus, *The Histories* 3.112)—as a traditional ingredient in perfumes and, moreover, was a species known to and used by the LBA Minoans themselves [4,31]. Manoyl oxide is found (in higher quantities than in Pinaceae) in labdanum, a highly viscous resinous material obtained from *Cistus creticus*, or Cretan rockrose, a plant that is depicted on an LM IA (ca. 1600 B.C.) fresco from Akrotiri, Thera [31–33]. *C. creticus* is native to the island of Crete and has several well-known cultivars including one termed "Lasithi", named after the general region in which Tourloti resides. This Lasithi cultivar seems likely to be the source of the manoyl oxide used in both the Tourloti and Mochlos preparations, as it grows abundantly in the western Siteia foothills to this day.

Ancient Egyptian and Greek sources enumerate labdanum's many medicinal and aromatic uses [34], but it is best known in the present day as "amber" in perfumes such as Taittinger's "Baccarat", Yves St. Laurent's "Jazz", and Calvin Klein's "One" [19]. It is one of the main substitutes for ambergris in the modern perfume industry as a foundational lower, or base, note, and serves also a fixative to help retain scents, as it undoubtedly did for the perfumed oils of LBA East Crete. Some of labdanum's compounds can also serve as a middle, or heart, note just as in the case with storax balsam, and this versatility may explain its use in nearly a third of all modern perfumes [19]. It has filled this role for the past century as an original main ingredient for the so-called chypre (i.e., Cyprus) family of perfumes, such as Hermès' "Bel Ami", which coincidentally uses storax as a complementary base note in a similar fashion to our ancient perfumed oils.

3.6. Tilleul from Tilia (Linden)

Docosane ($C_{22}H_{46}$, 310 MW, 20.70 at Penn)

Docosane is an alkane hydrocarbon found in high concentrations in flower extracts, notably from the *Tilia* genus, or linden, which according to palynological studies grew on Crete at least during the earlier LBA [35–37]. Its identification is based not only on the solitary presence of the biomarker docosane, but also by careful consideration of the presence or absence of companion diagnostic compounds (e.g., isopulegol, a monoterpene alcohol that when appearing concurrently with docosane would instead suggest *Lilium*, or the lily plant [38]). These holistic phytochemical considerations make *Tilia* the most likely candidate to explain the concentration of docosane, a conclusion verified by the matching chromatographic signature produced from an ethnobotanical *Tilia* sample collected on Crete. ORA evidence for *Tilia* was recovered only from the Mochlos samples and absent from Tourloti.

In addition to its use as a treatment for atherosclerosis, *Tilia* also has antimicrobial properties [39], but the double-flowered blossom from this genus is best known for its aromatic compounds used to make perfumes [19,40]. Its fragrance, known primarily today by the industry as the heart-note infusion "tilleul" and used in perfumes such as Bourjois' "Soir de Paris" and Jesus Del Pozo's "Duende", is often described as reminiscent of citrus or honeysuckle, and as such could have also served as a head note. It is notable that Bourjois' "Soir de Paris" employs both storax and tilleul as complementary base and heart notes, respectively [19], echoing the pairing used in the Mochlos perfumed oil [4].

3.7. Summary of ORA Results

The collective chemical composition of the residues from the Mochlos and Tourloti objects suggests that both contained (or were used in the production of) a sophisticated olive oil-based perfume infused with *Cistus* labdanum amber (base and heart notes), *Liquidambar* storax balsam (base and heart notes), *Artemisia* camphor (a head note), and *Coriandrum* essential oil (a head note); all are botanicals that would have been locally available in East Crete or nearby regions such as Rhodes. The consistency between the two perfume "recipes"—both in regard to their ingredients and the techniques of their manufacture—can also be said to be a meaningful indicator of cultural continuity over the course of several centuries, where luxury items offered one means by which to negotiate status both during and after the demise of the Cretan Bronze Age palatial centers [1,41].

The results highlight the exciting potential for organic residue studies to illustrate both products and processes, here particularly related to Bronze Age perfume manufacture. The use of *Coriandrum* in both the Mochlos and Tourloti vessels is particularly meaningful in that it likely reflects the practice of stypsis, the treatment of the base oil with an astringent in a two-stage process to make it more receptive to infusion with aromatic compounds [23] (with the added bonus of a complementary head note). This technique was employed in the perfumeries of the Late Bronze Mycenaean palaces, as documented at Pylos in Linear B tablets Un 267 and Vn 130 [23,42], and is now archaeometrically confirmed [1,4]. In this vein, it is significant that the linalool is present in the completed product—the perfumed oil present in both the Mochlos amphora and the Tourloti stirrup jar—but not in the Mochlos vat. Its absence might simply reflect the rapid volatilization of the compound as it was exposed to air in the open vat, but this seems unlikely given the high boiling point of this compound—thus its lower relative volatility—and furthermore given that the Tourloti jar had similarly been exposed to air for over a century, and yet contained a relative abundance (RA) of linalool nearly identical to the Mochlos amphora. Linalool's absence from the vat may instead show that the vat was not employed for stypsis, but instead used in some other aspect of the perfume manufacturing process, perhaps the infusion of the primary constituent aromatics. The comparable quantities of linalool in the Mochlos and Tourloti jars—in the neighborhood of 15% RA for both—may reflect the absorption potential of olive oil in relation to linalool, or infer something about the proportions of the ancient recipe. While the practice of stypsis needs to be explored more fully, it may also be that biomarkers for coriander could be used to distinguish perfumed oils from other lipid-based organic commodities in the Bronze Age Aegean world and beyond. The comprehensive ORA results, thus, offer unique insight into Bronze Age *chaîne opératoire*, elements of which could have implications for further ORA studies.

Additionally of note is the fact that the earlier Mochlos vessels contained one additional ingredient, tilleul from *Tilia* blossoms, the presence of which was marked by significant quantities of the compound docosane. Additional vessels containing docosane, including cooking pots, were found immediately outside the Vat Room in the same Mochlos workshop and were probably used for maceration, or heated steeping, of the *Tilia* blossoms [4]. The absence of tilleul in the later Tourloti perfume may reflect a shift in the botanical profile of the region from LM IB to IIIC (ca. 1500–1175 B.C.), and in particular a move away from preferential utilization of the plant in the wake of political and social change in East Crete in the intervening centuries. However, it is significant that the LM III-EIA pollen cores from the island echo the same pattern as the perfumes; *Tilia* disappears at the end of the LBA. The disappearance of *Tilia* from the palynological record was, in fact, specifically connected with climatic shifts at the end of the Bronze Age that diminished the tree's natural habitat [36]. It is, therefore, possible that the absence of tilleul from the Tourloti perfume could be a direct reflection of climate change (e.g., aridification) that occurred at the end of the Late Bronze Age [43–48].

The results of our ORA study, thus, shed new light on elite behavior, manufacturing processes, and cultural patterns at the end of the Bronze Age, while also hinting at ORA's exciting and underdeveloped potential as a tool for paleoecological reconstruction.

4. Discussion

The presence of complex perfumed oils with multiple scent layers, or notes, in the LM IB Mochlos and LM IIIC Tourloti vessels offers a window into the structural and market forces that shaped demand and production of these expensive, value-added commodities during the Bronze Age [1,4]. From an archaeometric perspective, however, the comparative Tourloti and Mochlos results are in many ways even more exciting. They are particularly important in that they demonstrate the value and viability of ORA samples taken from legacy objects, even those for which the "chain of custody" cannot be completely known. They moreover illuminate patterns in degradation processes and sample variation that in turn have potentially significant implications for future extractive and interpretive methodologies, upon which we offer a few preliminary comments.

4.1. Compound Expressions in Legacy and Newly Excavated Objects

Diagnostic legacy compounds, which are typically lipids, suffer from long-term autoxidation processes and therein degradation, recognition of which is essential for successful interpretation [49,50]. Oleic acid, for example, the major diagnostic and prevalent fatty acid of olive oil, degrades into azelaic acid and nonanoic acid under oxidizing conditions, a process noted in several artifact residue studies [1,20,21]. Our Tourloti case study demonstrates this lipid oxidation in stark terms: The residues from the freshly excavated Mochlos objects preserved substantial quantities of oleic acid and no azelaic acid. By contrast, in the legacy Tourloti jar the degradation is virtually total, and azelaic acid constitutes the largest peak across all chromatograms.

This same phenomenon was detected in early ARCHEM data from a sixth century B.C. olive press installation from Azoria in East Crete. This installation was discovered in the larger west room of a two-room structure (D300) from the Service Building, excavated in 2006 under the direction of D. Haggis and M. Mook [51]. D300 contained overwhelming archaeological evidence for the processing and pressing of olives: a press bench, press beam sockets, numerous weights, and a basin, found together with substantial quantities of olive pits—including seeds from press-cake. The site was eventually abandoned and destroyed, although apparently not before valuables had been removed [52]. Whereas ORA samples taken from modern olive oil jars and presses on Crete have oleic acid as their predominant compound, the ORA samples taken from the basin, querns, and surrounding soil from the olive press installation at Azoria were dominated by overwhelming evidence for azelaic acid. Indeed, the pressing process during the lifetime of the room would have presumably ensured that the incidental detritus and remnant oils from olives were subjected to constant, unmitigated exposure to air, offering continuous opportunity for natural aerobic processes such as oxidation and degradation.

As a monounsaturated fatty acid, the three-phase overall mechanism (i.e., 1.) initiation, or formation of free radicals; 2.) propagation, or free-radical chain reactions; and 3.) termination, or formation of nonradical products [49,53]) of oleic acid oxidation and therein the rate of degradation, is affected by successive reactions, of which azelaic acid is a dominant terminal nonradical product [1,20]. The OpenARCHEM project is currently conducting a series of fatty acid degradation studies at the Wesleyan University Department of Chemistry in order to better characterize these mechanisms, the nature and timing of which have yet to be comprehensively documented. As we come to understand the processes, which facilitate and catalyze oleic acid degradation, it is possible that the product distribution ratios between oleic and azelaic, and the peroxide value of various intermediary hydroperoxides [53], could eventually serve as meaningful indicators of the length of time for which an object was exposed in antiquity and therein shed light on site use or abandonment processes. It may be possible, for example, that the better preservation of oleic acid in the Mochlos perfume paraphernalia may be due in part to the fact that the space was quickly sealed by the unexpected collapse of the building during the powerful LM IB (ca. 1500 B.C.) earthquake that hit the town [3,4]. These mechanisms clearly demand systematic study, but may eventually also be a useful means by which one could determine something about the history of legacy vessels when the object biography is otherwise opaque.

4.2. Location, Location, Location: Considering Vessel Topography

As we move to refine our methodologies, it is worth noting—and attempting to better understand—the nature of certain variabilities exhibited among extractions from the same vessel and their possible relationship to the location at which sampling occurred [1,14]. One of the more striking results from within the Mochlos group, for example, was the variable evidence for *Tilia*. Docosane was present in quantities of 8.64% RA in samples taken from the lowest point of the Mochlos vat (Table 1), but was absent, or appeared only in trace quantities, from other areas of the vat such as the vessel spout, despite the use of the same solvent. These distinctions are arguably meaningful reflections of the vessel use, as the greatest abundance was found in the vat base where the contents would have most likely pooled and therein had greater opportunity to permeate the clay, as opposed to the sides, which would have had more sporadic exposure to the contents. Similar disparities were noted in relation to the presence of cholesterol, which was detected only in the vertical, conical spout of the vat, and which points to the use of wool as a filter to strain macroscopic impurities [23]. The Tourloti jar, which being a complete closed form was sampled using the swish method, also showed some variability among the three extractions taken, as indeed the total ion chromatograms (TIC) demonstrate. Note, for example the absence of cinnamic acid and camphor in extractions one and three (Figure 6, Table 1). The qualitative differences between the Tourloti samples are not yet fully understood, but do not appear to be related to degradation processes. It may be that, as with the Mochlos objects, these discrepancies are related to the site of sampling, and perhaps to the variable properties and depositional densities and locations of the compounds within, which might differentially permeate the base, side, and shoulders of the jar. Such variability suggests the importance of considering the shape, use, fabric, and function of vessels when sampling organic residues, and the importance of multiple extractions that target different areas of a vessel to ensure the broadest possible coverage and therein greater accuracy.

4.3. Rehabilitating Legacy Objects

Above all, the case study above demonstrates the viability and value of legacy objects, provided proper consideration is afforded to object biography and potential contaminants, whether natural or anthropogenic. It thus offers an important corrective to past criticisms of ORA that have dismissed the utility of such artifacts. The companion volumes *Minoans and Mycenaeans: Flavours of Their Times* [54] and *Archaeology Meets Science: Biomolecular Investigations in Bronze Age Greece* [55], two of the most prominent ORA publications in Mediterranean studies, were particularly influential in conveying this message. These collaborative volumes, which assembled contributions from seven different laboratory groups utilizing wildly varying extraction methodologies and instrumentation (with no evident coordination or standardization), allowed for easy criticism of ORA, and especially of ORA results from legacy objects because of their greater risk of exposure to anthropogenic and environmental contaminants [56,57]. The volumes shed welcome light on Bronze Age organic commodities and drew necessary attention to potentially problematic contaminants, but included unqualified critiques that essentially encouraged a "baby-with-the-bathwater" dismissal of ORA, casting doubt on the viability of ORA studies in general and legacy objects in particular.

It is certainly clear that better ORA results are usually obtained from artifacts recovered from ongoing, or at least recent, excavations [5,6,13,18,58], if only because sampling conditions and object biographies can be better controlled and understood. Yet the case study presented herein showcases a legacy object that in many respects reproduces a result obtained from a freshly-excavated object, thus demonstrating the inherent value of this class. The compounds in legacy objects may appear in diminished quantities—both in number and peak strength—relative to modern standards or freshly-excavated examples, but if properly scaled the peaks are discernible and appear to offer meaningful indicators of the presence and proportions of compounds, as the RA of linalool demonstrates (see above).

Insofar as contaminants are concerned, although there is no foolproof methodology that can guarantee that results from a legacy vessel—or indeed any object—will be free from contamination, our work does point towards a few steps that can be taken to better ensure a sample's reliability and viability, even in circumstances where the full object biography is unknown. Foremost among these is an ability to predict, identify, and screen for common contaminants, such as plasticizers (e.g., phthalates, which are not naturally occurring and can reflect the storage of an object in a plastic bag), as was the case with portions of the Mochlos vat spout. In addition, it may be possible to screen for impurities that are most likely to be present on the outermost oxidized surface of the organic residues, through the use of multiple extractions [1], and by consideration of the geological profile of the clay itself [14]. This latter, too, is a process presently under study by the OpenARCHEM project at the Massachusetts Institute of Technology Center for Materials Research in Archaeology and Ethnology. Most important, it is essential that the condition, quality, and storage of any sample, whether taken from a newly excavated or legacy object, be fully described and acknowledged in any publication. With transparency and proper consideration of object biographies and degradation processes, legacy objects can indeed yield valuable and verifiable results.

Such results are most meaningful when they can be placed in conversation with new samples, or considered as constituents within a larger ecosystem of chemically comparable or contemporary samples. This is often difficult: ORA studies tend to be siloed, published singly in various specialized outlets or buried in archaeological site reports, preventing aggregate study. It is clear that the field would benefit from a common repository that might best facilitate the comparison and sharing of ORA results. The OpenARCHEM archaeometric database [59] was created specifically in response to this challenge. Using the original ARCHEM project library of chromatographic data as its core, which includes over 5000 ORA samples, the OpenARCHEM database is designed to be a platform by which scholars of antiquity across many disciplines and regions can review, compare, and integrate ORA into their own work. Each entry presents the results from a single extraction, and includes not only diagnostic compounds with their GC-MS chromatograms and links to NIST reference chromatograms, but also information about the type of artifact sampled, its quality, condition, and storage history, as well as links to ethnohistorical textual sources that discuss the botanical species identified.

The inclusion of these different categories of information offers routes by which the scientist, the social scientist, and the humanist can ask different kinds of questions of organic data—the scientist searching by chemical compound, for example, or the humanist by ancient author. It also allows for focused comparisons, whether between samples from similar jar types or by those with comparable object histories, thus aiding in further refinement of archaeometric methodologies. Through such data-sharing, chemical signatures that went undetected or unrecognized in early studies may become identifiable (as was the case with our cinnamic acid at Mochlos), and larger patterns, such as connections between contents and their containers, can become visible. Collaborative repositories and archaeometric communities such as these tap into the promise of big data [59,60], and are a resource by which ORA can be engaged in service to larger archaeological questions.

5. Conclusions

ORA is evolving far beyond its traditional lab-bound role to become an integrated element of field methodology and a fundamental part of archaeological research design. It moreover stands ready to harness the power of big data to address larger questions of economy, technology, ecology, and environment. The comparative case study above demonstrates the value of legacy objects as candidates for ORA, and highlights new questions such data can be called upon to ask. The development of responsible and nuanced methods for the sampling of legacy objects, when paired with careful consideration of degradation pathways, can substantially broaden the range of viable artifacts for study and therein the field of inquiry. Yet whether from new or legacy artifacts, ORA results must be considered within a comprehensive ecosystem of archaeological, archaeometric, and ethnographic data.

Supplementary Materials: The following are available online at www.openarchem.org, additional chromatographic, ethnobotanical, and ethnohistorical data.

Author Contributions: The analyses and interpretation of the results were coordinated by A.J.K. and carried out together with K.J.B. A.J.K. and K.J.B. wrote and revised the manuscript drafts. Both authors read and approved the final manuscript.

Funding: This research was supported by the University of Pennsylvania (Penn Museum, Department of Chemistry, and School of Arts and Sciences), Wesleyan University (Department of Chemistry, Archaeology Program), Brandeis University (Office of the Dean of Arts and Sciences, Department of Chemistry, Department of Classical Studies), Institute for Aegean Prehistory, and Archaeological Institute of America.

Acknowledgments: We thank the University of Pennsylvania Department of Chemistry for incubating the analytical facets of the ARCHEM project at its start in 2003. Without the guidance of G. Molander and G. Palladino, the project would not have materialized. At the Penn Museum, C. B. Rose, P. P. Betancourt, and L. Makowsky of the Mediterranean Section and J. Houser Wegner, J. W. Wegner, and D. Silverman of the Egyptian Section have been instrumental in supporting the project. L. Grant, M.-C. Boileau, and the Penn Museum Scientific Testing Committee have offered us valuable feedback and the use of the Conservation Department facilities. E. H. Cline, L. Hitchcock, C. R. Floyd, and A. M. Berlin have offered steady advice during the startup of the OpenARCHEM project and its archaeometric database. A. K. Krohn and A. M. Crandall, as senior staff members of the OpenARCHEM Project, continue to offer their unfailing assistance. P. Der Manuelian and J. A. Greene supported this study by arranging our appointments as research associates of the Harvard Semitic Museum. H. N. Lechtman facilitated access to MIT resources through its Center for Materials Research in Archaeology and Ethnology. We benefited greatly from conversations with A. Kanta, J. Moody, and A. Sarpaki concerning the rich flora of Crete, A. Yasur-Landau and P. Waiman-Barak concerning the same in the Southern Levant, and I. Liritzis concerning the ecology of Southern Phokis. J. G. Younger generously shared his map of East Crete. Colleagues at our institutions—W. Gilstrap, J. Meanwell, C. Parslow, D. Charles, A. O. Koloski-Ostrow, L. Muellner, B. Snider, C. Thomas, C. Wade, M. Sheehy, and I. Roy—have been our pillars of support. We especially thank T. D. Westmoreland and A. Roberts at Wesleyan University Chemistry for their support of OpenARCHEM degradation studies. In Crete, AJK received support from the INSTAP Study Center for East Crete, in collaboration with the 24th Ephorate of Prehistoric and Classical Antiquities and the American School of Classical Studies at Athens, and the Museum of Cretan Ethnology Research Centre, under the directorship of C. Vallianos.

Conflicts of Interest: The authors declare no conflicts of interest.

Abbreviations

GC-MS—gas chromatography-mass spectrometry; ORA—organic residue analysis; EI—electron ionization; CI—chemical ionization; TIC—total ion chromatogram; RA—relative abundance; AA—absolute abundance; LBA—Late Bronze Age; LM—Late Minoan; EIA—Early Iron Age.

References

1. Koh, A.J.; Birney, K.J. Organic Compounds and Cultural Continuity: The Penn Museum Late Minoan IIIC Stirrup Jar from Tourloti. *Mediterr. Archaeol. Archaeom.* **2017**, *17*, 19–33. [CrossRef]
2. OpenARCHEM Project. Available online: http://openarchem.com (accessed on 19 December 2018).
3. Soles, J.S.; Mochlos, I.A. *Period III. Neopalatial Settlement on the Coast: The Artisans' Quarter and the Farmhouse at Chalinomouri.*; INSTAP Academic Press: Philadelphia, PA, USA, 2003.
4. Koh, A.J. Wreathed in a Fragrant Cloud: Reconstructing a Late Bronze Age Aegean Workshop of Aromata. Ph.D. Dissertation, University of Pennsylvania, Philadelphia, PA, USA, 2006.
5. Koh, A.J.; Betancourt, P.P. Wine and Olive Oil from an Early Minoan Hilltop Fort. *Mediterr. Archaeol. Archaeom.* **2010**, *10*, 15–23.
6. Koh, A.J.; Yasur-Landau, A.; Cline, E.H. Characterizing a Middle Bronze Palatial Wine Cellar from Tel Kabri, Israel. *PLoS ONE* **2014**, *9*, e106406. [CrossRef] [PubMed]
7. Seager, R.B. Excavations on the Island of Mochlos, Crete in 1908. *Am. J. Archaeol.* **1909**, *13*, 271–303. [CrossRef]
8. Betancourt, P.P. *Minoan Objects Excavated from Vasilike, Pseira, Sphoungaras, Priniatikos Pyrgos, and Other Sites;* University Museum, University of Pennsylvania: Philadelphia, PA, USA, 1983.
9. Becker, M.J.; Betancourt, P.P. *Richard Berry Seager: Archaeologist and Proper Gentleman;* University of Pennsylvania Museum of Archaeology and Anthropology: Philadelphia, PA, USA, 1997.
10. Gerhardt, K.O.; Searles, S.; Biers, W.R. Corinthian Figure Vases: Non-destructive Extraction and Gas Chromatography-Mass Spectrometry. In *Organic Contents of Ancient Vessels: Materials Analysis and Archaeological Investigation;* Biers, W.R., McGovern, P., Eds.; MASCA Research Papers in Science and Archaeology 7; University of Pennsylvania Museum of Archaeology and Anthropology: Philadelphia, PA, USA, 1990; pp. 41–50.
11. Biers, W.R.; Gerhardt, K.O.; Braniff, R.A. *Lost Scents: Investigations of Corinthian "Plastic" Vases by Gas Chromatography-Mass Spectrometry;* MASCA Research Papers in Science and Archaeology; MASCA, University of Pennsylvania Museum of Archaeology and Anthropology: Philadelphia, PA, USA, 1994; Volume 11.
12. Evershed, R.P.; Vaughan, S.J.; Dudd, S.N.; Soles, J.S. Fuel for Thought? Beeswax in Lamps and Conical Cups from Late Minoan Crete. *Antiquity* **1997**, *71*, 979–985. [CrossRef]
13. Yasur-Landau, A.; Cline, E.H.; Koh, A.J.; Ratzlaff, A.; Goshen, N.; Susnow, M.; Waiman-Barak, P.; Crandall, A.M. The Wine Storage Complexes at the Middle Bronze II Palace of Tel Kabri: Results of the 2013 and 2015 Seasons. *Am. J. Archaeol.* **2018**, *122*, 309–338. [CrossRef]
14. Goldenberg, L.; Neumann, R.; Weiner, S. Microscale Distribution and Concentration of Preserved Organic Molecules with Carbon-Carbon Double Bonds in Archaeological Ceramics. *J. Archaeol. Sci.* **2014**, *42*, 509–518. [CrossRef]
15. Evershed, R.P. Experimental Approaches to the Interpretation of Absorbed Organic Residues in Archaeological Ceramics. *World Archaeol.* **2008**, *40*, 26–47. [CrossRef]
16. Mills, J.S.; White, R. *The Organic Chemistry of Museum Objects*, 2nd ed.; Butterworth-Heinemann: Oxford, UK; Woburn, MA, USA, 1999.
17. Colombini, M.P.; Modugno, F.; Ribechini, E. GC/MS in the Characterization of Lipids. In *Organic Mass Spectrometry in Art and Archaeology;* Colombini, M.P., Modugno, F., Eds.; Wiley: Chichester, UK, 2009; pp. 191–214.
18. Koh, A.J. Methods and Approaches: Organic Residue Analysis. In *The Blackwell Companion to Aegean Art and Architecture;* Hitchcock, L., Davis, B., Eds.; Blackwell: Oxford, UK, in press.
19. Groom, N. *New Perfume Handbook*, 2nd ed.; Blackie Academic and Professional: London, UK, 1997.
20. Regert, M.; Bland, H.A.; Dudd, S.N.; Van Bergen, P.F.; Evershed, R.P. Free and Bound Fatty Acid Oxidation Products in Archaeological Ceramic Vessels. *Proc. R. Soc. Lond.* **1998**, *265*, 2027–2032. [CrossRef]

21. Ribechini, E.; Modugno, R.; Baraldi, C.; Baraldi, P.; Colombini, M.P. An Integrated Analytical Approach for Characterizing an Organic Residue from an Archaeological Glass Bottle Recovered in Pompeii (Naples, Italy). *Talanta* **2008**, *74*, 555–561. [CrossRef]
22. Godard, A.; De Caro, P.; Thiebaud-Roux, S.; Vedrenne, E.; Mouloungui, Z. New Environmentally Friendly Oxidative Scission of Oleic Acid into Azelaic Acid and Pelargonic Acid. *J. Am. Oil Chem. Soc.* **2013**, *90*, 133–140. [CrossRef]
23. Shelmerdine, C.W. *The Perfume Industry of Mycenaean Pylos*; Paul Åströms Förlag: Göteborg, Sweden, 1985.
24. Sova, M. Antioxidant and Antimicrobial Activities of Cinnamic Acid Derivatives. *Mini Rev. Med. Chem.* **2012**, *12*, 749–767. [CrossRef] [PubMed]
25. Feliks, Y. The Incense of the Tabernacle. In *Pomegranates and Golden Bells: Studies in Biblical, Jewish, and Near Eastern Ritual, Law, and Literature in Honor of Jacob Milgrom*; Wright, D.P., Freedman, D.N., Hurvitz, A., Eds.; Eisenbrauns: Winona Lake, IN, USA, 1995; pp. 125–149.
26. Dr. Duke's Phytochemical and Ethnobotanical Databases. U.S. Department of Agriculture, Agricultural Research Service 1992–2016. Available online: http://phytochem.nal.usda.gov (accessed on 19 December 2018).
27. Modugno, F.; Ribechini, E.; Colombini, M.P. Aromatic Resin Characterisation by Gas Chromatography–Mass Spectrometry: Raw and Archaeological Materials. *J. Chromatogr. A* **2006**, *1134*, 298–304. [CrossRef] [PubMed]
28. Birney, K.J.; Koh, A.J. Adventures in Storax. Full Spectrum: Micro-Publications from OpenARCHEM. 2018. Available online: https://openarchem.com/2018/10/16/adventures-in-storax (accessed on 19 December 2018).
29. Nezhadali, A.; Akbarpour, M.; Zarrabi, S.B. Chemical Composition of the Essential Oil from the Aerial Parts of Artemisia Herba. *J. Chem.* **2008**, *5*, 557–561. [CrossRef]
30. Elsharif, S.A.; Banerjee, A.; Buettner, A. Structure-odor relationships of linalool, linalyl acetate and their corresponding oxygenated derivatives. *Front. Chem.* **2015**, *3*, 57. [CrossRef] [PubMed]
31. Vlachopoulos, A. The Wall Paintings from the Xeste 3 Building at Akrotiri: Towards an Interpretation of the Iconographic Programme. In *Horizon: A Colloquium on the Prehistory of the Cyclades*; Brodie, N., Doole, J., Gavalas, G., Renfrew, C., Eds.; McDonald Institute for Archaeological Research: Cambridge, UK, 2008; pp. 451–465.
32. Marinatos, S. *Excavations at Thera VII 1973 Season*; Archaeological Society at Athens: Athens, GR, USA, 1976.
33. Chapin, A. The Lady of the Landscape: An Investigation of Aegean Costuming and the Xeste 3 Frescoes. In *Reading a Dynamic Canvas: Adornment in the Ancient Mediterranean World*; Colburn, C.S., Heyn, M.K., Eds.; Cambridge Scholars Publishing: Newcastle, UK, 2008; pp. 48–83.
34. Oller-López, J.L.; Rodriguez, R.; Cuerva, J.M.; Oltra, J.E. Composition of the Essential Oils of Cistus ladaniferus and, C. monspeliensis from Morocco. *J. Essent. Oil Res.* **2005**, *8*, 455.
35. Moody, J.; Rackham, O.; Rapp, G. Environmental Archaeology of prehistoric NW Crete. *J. Field Archaeol.* **1996**, *23*, 273–297.
36. Moody, J. Hinterlands and Hinterseas: Resources and Production Zones in Bronze Age and Iron Age Crete. *Br. Sch. Athens Stud.* **2012**, *20*, 233–271.
37. Moody, J. Prehistoric and Historic Climate and Weather in (East) Crete. In *A Cretan Landscape through Time: Priniatikos Pyrgos and Environs*; Molloy, B., Duckworth, C., Eds.; British Archaeological Reports; Archaeopress: Oxford, UK, 2014; pp. 23–30.
38. Farsam, H.; Amanlou, M.; Amin, G.; Nezamivand-Chegini, G.; Salehi-Surmaghi, M.-H.; Abbas, S. Anatomical and Phytochemical Study of Lilium ledebourii (Baker) Boiss.; a Rare Endemic Species in Iran. *Daru J. Pharm. Sci.* **2003**, *11*, 164–170.
39. Gönül, S.; Karapinar, M. Inhibitory effect of linden flower (Tilia flower) on the growth of foodborne pathogens. *Food Microbiol.* **1987**, *4*, 97–100. [CrossRef]
40. Vidal, J.P.; Richard, H. Characterization of Volatile Compounds in Linden Blossoms Tilia cordata Mill. *Flavours Fragr.* **1986**, *1*, 7–62.
41. Hitchcock, L.A.; Maeir, A.M. Lost in Translation: Urbanism in Post-Palatial Crete at the End of the Bronze Age. In *From Static Data to Dynamic Processes: New Perspectives on Minoan Architecture and Urbanism*; Knappett, C., Letesson, Q., Eds.; Oxford University: Oxford, UK, 2017; pp. 289–333.

42. Palaima, T. Pylos Tablet Vn 130 and the Pylos Perfume Industry. In *KE-RA-ME-JA: Studies Presented to Cynthia, W. Shelmerdine*; Nakassis, D., Gulizio, J., James, S.A., Eds.; INSTAP Academic Press: Philadelphia, PA, USA, 2014; pp. 83–90.
43. Kaniewski, D.; Paulissen, E.; Van Campo, E.; Weiss, H.; Otto, T.; Bretschneider, J.; Van Lerberghe, K. Late Second–Early First Millennium BC Abrupt Climate Changes in Coastal Syria and Their Possible Significance for the History of the Eastern Mediterranean. *Quat. Res.* **2010**, *74*, 207–215. [CrossRef]
44. Drake, B.L. The Influence of Climatic Change on the Late Bronze Age Collapse and the Greek Dark Ages. *J. Archaeol. Sci.* **2012**, *39*, 1862–1870. [CrossRef]
45. Kaniewski, D.; Van Campo, E.; Guiot, J.; Le Burel, S.; Otto, T.; Baeteman, C. Environmental Roots of the Late Bronze Age Crisis. *PLoS ONE* **2013**, *8*, e71004. [CrossRef]
46. Cline, E.H. *1177 B.C.: The Year Civilization Collapsed*; Turning Points in Ancient History; Princeton University: Princeton, NJ, USA, 2014.
47. Langgut, D.; Finkelstein, I.; Litt, T.; Harald Neumann, F.; Stein, M. Vegetation and Climate Changes during the Bronze and Iron Ages (~3600–600 BCE) in the Southern Levant Based on Palynological Records. *Radiocarbon* **2015**, *57*, 217–235. [CrossRef]
48. Knapp, A.B.; Manning, S.W. Crisis in Context: The End of the Late Bronze Age in the Eastern Mediterranean. *Am. J. Archaeol.* **2016**, *120*, 99–149. [CrossRef]
49. Simic, M.G. Free radical mechanisms in autoxidation processes. *J. Chem. Educ.* **1981**, *58*, 125. [CrossRef]
50. Porter, N.A.; Caldwell, S.E.; Mills, K.A. Mechanisms of Free Radical Oxidation of Unsaturated Lipids. *Lipids* **1995**, *30*, 277–290. [CrossRef]
51. Haggis, D.C.; Mook, M.S.; Fitzsimons, R.D.; Scarry, C.M.; Snyder, L.M.; West, W.C. Excavations in the Archaic Civic Buildings at Azoria in 2005–2006. *Hesperia* **2011**, *80*, 1–70. [CrossRef]
52. Haggis, D.C.; Mook, M.S.; Fitzsimons, R.D.; Scarry, C.M.; Snyder, L.M. The Excavation of Archaic Houses at Azoria in 2005–2006. *Hesperia* **2011**, *80*, 431–489. [CrossRef]
53. Porter, N.A.; Mills, K.A.; Carter, R.L. A Mechanistic Study of Oleate Autoxidation: Competing Peroxyl H-Atom Abstraction and Rearrangement. *J. Am. Chem. Soc* **1994**, *116*, 6690–6696. [CrossRef]
54. Tzedakis, Y. Martlew, H. Minoans and Mycenaeans: Flavours of Their Times. In *Catalogue of the Exhibition*; Kapon: Athens, Greece, 2002.
55. Tzedakis, Y.; Martlew, H.; Jones, M.K. (Eds.) *Archaeology Meets Science: Biomolecular Investigations in Bronze Age Greece*; Oxbow Books: Oxford, UK, 2008.
56. Beeston, R.F.; Palatinus, J.; Beck, C.W. Organic Residue Analysis: Pseira. In *Archaeology Meets Science: Biomolecular Investigations in Bronze Age Greece*; Tzedakis, Y., Martlew, H., Jones, M.K., Eds.; Oxbow Books: Oxford, UK, 2008; pp. 74–86.
57. Barnard, H.; Ambrose, S.A.; Beehr, D.E.; Forster, M.F.; Lanehart, R.E.; Malainey, M.E.; Parr, R.E.; Rider, M.; Solazzo, C.; Yohe, R.M. Mixed results of seven methods for organic residue analysis applied to one vessel with the residue of a known foodstuff. *J. Archaeol. Sci.* **2007**, *34*, 28–37. [CrossRef]
58. Perruchini, E.; Glatz, C.; Hald, M.M.; Casana, J.; Toney, J.L. Revealing Invisible Brews: A New Approach to the Chemical Identification of Ancient Beer. *J. Archaeol. Sci.* **2018**, *100*, 176–190. [CrossRef]
59. Koh, A.J.; Birney, K.J. (Eds.) OpenARCHEM Archaeometric Database. Available online: http://openarchem.org (accessed on 19 December 2018).
60. Berlin, A.M. (Ed.) Levantine Ceramics Project. Available online: http://levantineceramics.org (accessed on 19 December 2018).

© 2019 by the authors. Licensee MDPI, Basel, Switzerland. This article is an open access article distributed under the terms and conditions of the Creative Commons Attribution (CC BY) license (http://creativecommons.org/licenses/by/4.0/).

Article

The Sacred Landscape of the "Pyramids" of the Han Emperors: A Cognitive Approach to Sustainability

Giulio Magli

School of Architecture, Urban Planning and Construction Engineering, Politecnico di Milano, Piazza Leonardo da Vinci, 32, 20133 Milano MI, Italy; Giulio.Magli@Polimi.it; Tel.: +39-02-2399-4568

Received: 20 December 2018; Accepted: 24 January 2019; Published: 2 February 2019

Abstract: The so-called "Chinese pyramids" are huge burial mounds covering the tombs of the Emperors of the Western Han dynasty. If we include also the mounds of the members of the royal families, these monuments sum up to more than 40, scattered throughout the western and the southern outskirts of modern Xi'an. They are mostly unexcavated and poorly known, although taken together, they form a fascinating sacred landscape, which was conceived as a perennial witness of one of the most magnificent Chinese dynasties. This sacred landscape is today encroached by the frenetic urban development of the Xi'an urban area. We discuss and elaborate here some of the results of a recent, new satellite-imagery survey of these monuments, highlighting the aspects which may contribute to solutions for sustainable and compatible development within this important ancient landscape.

Keywords: Chinese Pyramids; Han Dynasty; Feng Shui; protection of ancient landscapes

1. Introduction

The Western Han dynasty of ancient China (202 BC–9 AD) marked important political, economic, and scientific developments. The Han rulers followed the custom initiated by the first emperor of Qin—whose mausoleum is world-famous due to the terracotta army guarding its eastern side—and chose to be buried in tombs located under huge square mounds of rammed earth, today known popularly as "Chinese pyramids" [1,2].

These monuments are located in the outskirts of Xi'an (see Figure 1 for a general map of the area). A main group of them, composed by 9 emperor's tombs and 21 satellite tombs of members of the royal families, is located along the northern bank of the river Wei, close to the airport and subjected in recent times to high population growth and urbanization within the so-called "Xi-xian new area" projects. A second group of monuments is located to the south of Xi'an, not far from other high-impact areas related to high tech and tourism development zones.

The Han mausoleums are thus a fascinating, almost "alien" presence in the rapidly developing landscape of modern China. They have been poorly studied and only two have been (partly) excavated: those of Emperors Jing and Wu. It is rather difficult today, onsite, to have an idea of the ways in which this funerary landscape was conceived, and especially of the visual relationships the monuments bear to each other. This is due to various factors, but chiefly to pollution—which drastically reduces horizon visibility—and to the sheer difficulty of reaching some of the monuments and/or their tops. For these reasons, although I have personally visited many of them, I decided to carry out a new, complete survey using satellite imagery tools. The results of this campaign have been recently fully published in a specialized archaeological Journal [1]. In the present paper, I discuss and expand some of the results obtained, focusing on the problem of the sustainable development of this area so rich in cultural, almost unexplored relics.

Figure 1. General map of the burial mounds of the emperors of the Western Han Dynasty. 1—Gauzu (Chanling), 2—Hui (Anling), 3—Jing (Yanling), 4—Wu (Maoling), 5—Zhao (Pingling), 6—Xuan (Duling), 7—Yuan (Weiling), 8—Cheng (Yangling), 9—Ai (Yiling), 10—Ping (Kanling). The mountain tomb (Baling) of emperor Wen is denoted by **B**. (Image courtesy of Google Earth, editing by the author).

The aim of the paper is as follows. Besides the sheer necessity of understanding the minimal dimensions of buffer areas where future excavations might reasonably be carried out, we shall discuss broader aspects based on a cognitive approach to archaeological relics [3–6] (see Section 2.3 for the methodology used in this paper). As we shall see, this approach can contribute to carrying innovation in linking conservation and sustainable development, not only in the direction of protecting heritage, but also in enhancing present and future senses of place and quality of life.

2. Materials and Methods

2.1. Study Area

In ancient China, the emperors chose for their tombs suitable areas in the proximity of the city they elected, or founded, as the capital. This led to the establishment of royal Necropolises, which modelled and rationalized the landscape as sacred landscapes of power, devoted to transmitting the rights to eternal life of the deceased rulers and to establish the rights to the throne of their successors as well. The study area examined in the present paper is the first (in chronological order) of such sacred landscapes: that of the emperors of the Western Han dynasty. The Han capital was in Chang'an (today an area of present Xi'an), and the Necropolis was developed along the northern bank of the river Wei to the northwest, with an addition located to the southeast, which was due to the will of one emperor, Wen, to be buried under a natural mountain and not under a burial mound. The main study area thus extends for about 34 km in the densely populated, rapidly developing territory nicknamed "Xi-xian", which extends from Xi'an's western suburbs to the Xianyang urban area and airport.

2.2. Survey Methods

The monuments have been surveyed onsite whenever possible (some are of difficult or no access). However, they are in any case very difficult to measure due to vegetation, fences, and to the fact that the sides are not always clean. Therefore, we took advantage of the fact that the area is well covered by satellite imagery (both on Google Earth and Bing), with a resolution which is more than sufficient to measure the average sides and average azimuths of the mounds. Another problem that satellite imagery is helpful in solving is that of the ancient horizon. Indeed, the horizon visibility today is very poor—due to pollution—even when the horizon itself is clear of buildings; moreover, sometimes, modern buildings are actually present. Using satellite tools, it is instead possible to establish whether monuments had mutual intervisibility in the past or not. All in all, the available images—in many cases, the historical archive of Google Earth contains more than one image with a sufficient resolution—were extracted and measured with AutoCAD for length of the sides, orientation, and directions of visibility towards other monuments. The results were mediated in the presence of more than one image. Errors have to be expected, of course, but on account of the high quality of the images and of the low projection error associated with them, the intrinsic error expected from this kind of measurement is quite low, so Google Earth is a quite useful tool for this kind of investigation [7,8]. Unfortunately, however, the original heights of the mounds (certainly greater than the current ones) were impossible to determine because the summits have deteriorated considerably (see, e.g., Figure 2).

Figure 2. The huge burial mound of Maoling, the tomb of emperor Wu of Han, looks like a smoothed, natural hill. It is, instead, fully artificial: a pyramid made of hard rammed earth. (Photograph by the author.)

2.3. Methodology

The approach presented here is that of cognitive archaeology, as developed in the last 30 years or so [3–6]. In a nutshell, it is an approach to the material relics of human past which aims at describing them as objects that had their primary cultural existence as "percepts" in topological relation to one another. They thus fitted within the cognitive schemes of their creators, and can be fully understood—that is, not only "functionally" understood—only if such schemes are also studied and, as far as possible, understood as well.

A particularly interesting case in such a context is that of sacred landscapes; that is, natural landscapes in which monuments were built in accordance with rigorous criteria, usually connected with power and religion (for a general approach to the relationship between traditional built environments and perceived meaning, see [9–12]). In the case of such landscapes, indeed, the cognitive approach turns out to be extremely effective; for instance, in the study of ancient Egyptian Necropolises [13–16]. In such studies, the use of satellite imagery reveals itself as being particularly useful. In fact, it allows to establish the mutual relationships between monuments also in the presence of evolving dynamics of the built features of the landscape.

3. Results

As a first result, the satellite imagery analysis has shown the presence of two different patterns of orientation of the square basis of the pyramids [1]. This result is only of side interest here and will therefore be only briefly recalled.

A first group of monuments is precisely oriented to the cardinal points, with errors not exceeding $\pm 1^\circ$, while a second group has errors of several degrees in relation to the geographic north. These errors are not random: they are always to the west of north and exhibit a tendency to decrease in time from a maximum of 14° to a minimum of 8°. The mounds of the first group were oriented by determining the cardinal directions, while the skewed orientations of the other group can be explained as pointing to the maximal western elongation of Polaris, which was at those times, due to precession, not coinciding with the celestial pole, which was located in a dark region. These orientations are symbolic, as both the North celestial pole and the circumpolar region were of paramount importance for the Chinese: the function of the pole as the "pivot" of the sky was, in fact, equated with the centrality of the imperial power on Earth, and the whole polar region of the sky was identified as a celestial image of the Emperor's palace, the "Purple Enclosure" [17]. These cosmological concepts were reflected in architecture; for instance, in the planning of cities [18]. The different choices made by different emperors (cardinal orientation, or to Polaris) have still to be analyzed in details in historical terms; they are certainly related to their conception of the mandate of the heaven and to the role of Confucianism in the royal court.

The abovementioned results exclude the use of the magnetic compass (first invented in China precisely during the Han) for the orientation and placement of the pyramids, and therefore the use of the traditional Chinese pseudo-scientific doctrine of "geomancy" (Feng Shui) with compass measurements. Similarly, also the standard canons of Form Feng Shui, based on the presence of a mountain to the north of the site, smooth hills to the east and west, and water and a protective hill to the south, which were applied in the necropolises of several subsequent dynasties [19,20], are here clearly excluded, since the monuments lie in the flat plain. We arrive in this way to the results of [1], which are of main relevance for our discussion here, and will be further expanded and elaborated: these monuments were planned one after the other, but taking into account the already existing ones, in order to form a sacred landscape where *mutual placement* and *intervisibility* played a key role.

Let us consider, first of all, the distribution of the monuments in the western necropolis. The tombs are located along a direction of azimuth of about 72 degrees, roughly following the river with an average distance of 4.5 km from today's banks.

One would expect the easternmost tomb to have been built first, because it is the closest to the Han capital Chang'an, and the others to have been built in succession from east to west. However, it is not so: they were not built in a linear succession, and many "jumps" back and forth occurred. The jumps can be explained taking into account that historical sources, such as the Book of the Han [21], mention a doctrine called *Zhaomu*, which was used for choosing the location of emperors' tombs. The doctrine states that left/right (east/west) have to be alternately selected, so that when looking at a tomb, the first successive one will be found to the left (west) and the second to the right (east). From satellite imagery, it is clear that this alternate distribution was followed for two triplets: the tombs of Gaozu, Hui, and Jing (omitting the choice of a natural mountain made by Wen in between) and those of Yuan, his son Cheng, and his grandson Ai. Of special interest are the latter three, since their centers are connected by an almost perfectly straight line (Figure 3).

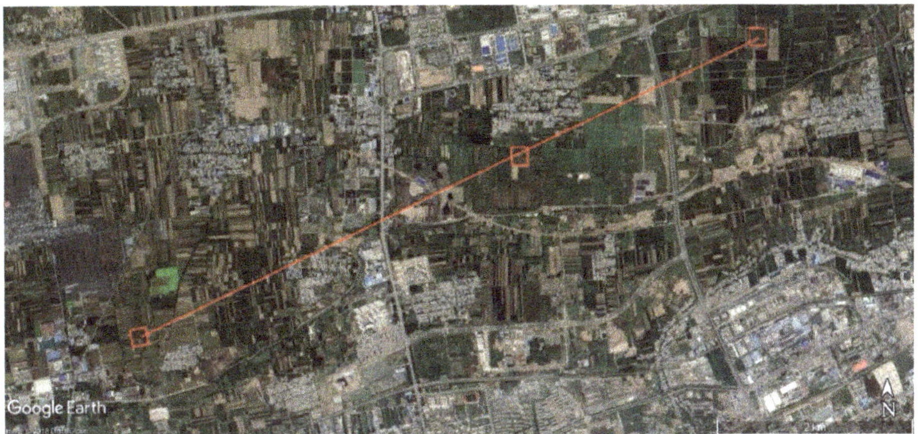

Figure 3. A satellite image showing the burial mounds (red boxes) of the emperors Yuan (center), his son Cheng (right), and his grandson Ai (left). The Zhao-mu doctrine is here self-evident; furthermore, the monuments were fully inter-visible. The view is, still today, almost miraculously preserved (red line), as can be seen from the distribution of modern buildings. (Image courtesy Google Earth, editing by the author.)

This observation finally leads us to our main issue: intervisibility. Indeed, were dynastic connections and topographical connections only intended in the projects, or instead, were they explicitly visible each day when looking at the monuments? To calculate the (theoretical) visibility of an object from a fixed position located at a height h, a simple formula can be used (essentially an application of Pythagoras' theorem). This "horizon formula" takes into account the Earth's curvature and states that the visibility distance in kilometers equals the square root of the product of the number 13 times h, with h measured in meters. So, for instance, a person two meters tall can look as far as the square root of 26 in kilometers; that is, a bit more than five kilometers. Of course, however, for tall objects, the horizon visibility increases considerably as the heights sum up. As mentioned, it is difficult today to assess the original height of the burial mounds; however, even considering only the heights they reach today, it is readily seen that they were practically all intervisible to each other from their summits. For instance, the distance "as the crow flies" (that is, the length of a straight line) from the westernmost emperor's monument, Maoling, and the easternmost, Yangling, is about 35 km. Still today, the highest point of Maoling is about 47 meters above the ground, and that of Yangling is about 25 m, giving a horizon visibility which is comparable to this distance. It follows that with all probability, they were fully intervisible at the time of their construction. With these two pyramids being at the two extrema of the area where all these monuments are located, we can conclude that these magnificent monuments were *all* placed in such a way so as to "speak" to each other along the visibility lines from their summits. What is more, many of them were simultaneously visible from the ground. Still today, in spite of haze and pollution, it is possible to appreciate from each mound the presence of at least the closest of the other monuments (see, e.g., Figure 4).

Figure 4. Pingling, the tomb of emperor Zhao of Han, seen from the Maoling group. (Photograph by the author.)

The "skyline" was made even more fascinating by the presence of the satellite mounds. In [1], we noticed, without having an explanation, the curious fact that the sides of the satellite mounds never align with the sides of the corresponding main mound. This looks odd, because as a consequence, the orientation procedure must have been repeated for each one of them separately, while construction in alignment would simplify the planning. However, it is now clear that in this way, these mounds contribute to the skyline in a significant manner, a thing they would hardly do in the case of parallel alignment with their principal counterparts.

Let us consider now the tombs located in the area to the southeast of Xian. Here, the situation is more complex, as the first monument constructed, Baling, the tomb of emperor Wen, is not an artificial mound. Indeed, Wen is the only Han ruler who selected a mountain for his tomb, and the funerary chambers were hollowed out of the rock. Wen's funerary landscape comprises two satellite burials: those of his wife Empress Dowager Bo and of his daughter Empress Dou. These are huge, almost identical, rectangular mounds located in the plain to the southwest of Baling. The tomb of Bo was orientated towards the Baling peak, which is visible when looking from the summit along the projection of the longest sides of the mound, at a distance of some 3.6 km (Figure 5); the tomb of Dou was a replica of that of Empress Bo, located further south. Later, another Emperor chose the same area for his tomb, Xuan. The diagonal of Xuan's mound passes quite neatly the apex of Baling, which was (barely) visible on the horizon, some 11.5 km away.

All in all, we can conclude that the tombs of the southern group were also conceived according to the idea of creating a sacred landscape of mutually visible monuments.

Figure 5. A satellite image showing the burial mound of Empress Dowager Bo. The monument (highlighted by the red box) is aligned (red line) towards the natural peak which signals the presence of the tomb of her husband, Emperor Wen. The alignment is perfectly preserved, but the Xian southeast development area is expanding from the north (image courtesy Google Earth, editing by the author),

4. Discussion

According to official reports, in the last 20 years, thousands of Chinese cultural sites have disappeared to make room for building projects. The number of endangered or cancelled sites is even higher than the average in the Shaanxi province. This should come as no surprise, as this area was the heartland of the Chinese civilization for so many centuries. It suffices to think that Tang's Xi'an in the years around 750 AD was the most inhabited city in the world, with an estimated population reaching 2 million people. Besides the sheer number of archaeological sites, another cause of this tragedy is that in Shaanxi, rural areas are being rapidly transformed, with population density already higher than 230 people per square km, and increasing. Intense building activity, with the construction of compounds of five to ten high towers each, occurs in the suburbs of Xian, dangerously close to the area of the Han dynasty mausoleums.

The recent history of Xi'an helps in explaining this urbanization phenomenon. During the 1950s, Xi'an was at the centre of a process of industrialization which brought to light many new industries. This process restarted after the Cultural Revolution and during the 1970s, also due to the remoteness of the territory, which offered warranties for industries of military interest, such as aerospace. Finally, in the last two decades of the last century, urbanization and development were triggered, especially by the projects called "New Special Development Zones".

Sustainability of urban development in China is of course an important, much-debated issue [22,23]. However, at least as far as the present author is aware, the problem of compatibility between development and cultural heritage has hardly been brought to attention as a key problem. Of course, there exists in China accurately preserved, magnificent sacred landscapes which are safe from any danger: it suffices to mention the UNESCO site of the 13th tombs of the Ming dynasty. Alas, the same cannot be said of the tombs of the Han emperors, which are, therefore, also an occasion to investigate regarding sustainability with respect to cultural heritage in general, and on an unprecedented scale. Rather few examples may indeed be cited in this connection; one of them is the World Heritage Site of Champaner–Pavagadh in the state of Gujarat, India [24].

This site has been unfortunately subjected to several threats during the violent religious conflict that exploded in the region in 2002. However, before those tragic events, a policy of conservation and sustainable development was established in accordance with a landscape approach that embraced the complex pressures of pilgrimage, tourism, economic development, forest management, and archaeological protection. Among the landscape approach guiding principles, there was the key idea of extending consideration beyond monuments and sites to the rich topographic sense of place. Further, key attention was given to relationships between environmental and social processes and to the copresence of multiple historical layers. Consequently, strategies were developed from landscape analysis to generate solutions (for instance, for designing routes) in order to "harmoniously reconnect people and places" [24]. To attempt a similar approach here is (hopefully) not only an interesting exercise, taking into account that protection of the Han cultural relics is explicitly mentioned in the Government's master plan for the Xi-xian urban development area.

First of all, it has to be observed that excavations at Yanling—where thousands of miniature terracotta statues of warriors and animals have been unearthed—have shown that the pits of the funerary equipment of the emperors were disposed in a radial manner directly near the four sides of the burial mound. From this point of view, the first emperor's burial, with the tremendously huge pits of life-size statues lying kilometers away from the mound, must surely be considered as a unique case. Thus, a minimal, urgent intervention must be the institution of a buffer zone enclosing each of the mounds by all sides for a few hundred meters to assure the possibility of future excavations. The buffer zone should be established also for the satellite burials, as several additional tombs and burial pits have to be expected in their areas [25,26]. In addition, a very recent discovery (press news of November 2018, yet unpublished) has shown that a miniature terracotta army could also accompany the burial areas of members of the Han royal family: they have been discovered in what are probably the annexes of the tomb of Liu Hong, son of Emperor Wu of Han (141–87 BC) in Linzi.

Further to this, another aspect should be taken into account. Indeed, we have shown here that the royal mausoleums have to be considered as an ensemble, which stands as an imposing icon not only of each divine ruler separately, but actually of the Han dynasty as a whole. One might suppose this landscape of power to be related to the traditional Chinese "geomantic" doctrine, but we have shown that is not so: the magnetic compass was not used for orientations and the typical auspicious features of the territory are absent. Actually, although some elements of the tradition must be very old, the first written records about Feng Shui appear later, with the Zang Shu (Book of Burial) by Guo Pu (276–324 CE). It follows that trying to apply Feng Shui canons in future constructions, as sometimes proposed (see, e.g., [27]), would have nothing to do with the way of thinking of the Han builders. For them, it was rather the intervisibility and the imposing presence of the "pyramids" that played the key role. The Han mounds are "mountains where there are no mountains": the floodplain on the northern bank of the river way is flat and each visible "hill" is actually a pyramid. It seems to me, therefore, that a feasible proposal is to respect the sacred landscape in which these imposing monuments were placed involves avoiding construction of compounds along their intervisibility lines. Inspection shows that many of these lines are still preserved; for instance, the alignments between the tombs of Yuan, Cheng, and Ai and those between the tomb of Empress Dowager Bo and the Baling mountain (see again Figures 3 and 5). Intervisibility is also preserved, for instance, between the Wu and Zhao groups (Figures 4 and 6).

Figure 6. A satellite image showing the burial mounds of the emperors Wu and Zhao (highlighted with red boxes). The line of sight between the two is preserved (red line), but in danger. (Image courtesy of Google Earth, editing by the author.)

Unfortunately, instead, the intervisibility between the first two groups in chronological order, those of emperors Gauzu and Hui, is lost forever due to the presence of the huge Weihe Power Station (Figure 7).

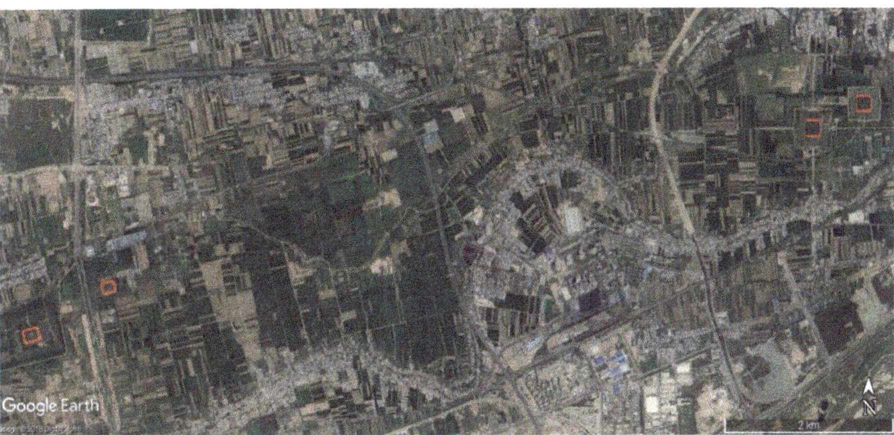

Figure 7. A satellite image showing the burial mounds groups of the emperors Gauzu and Hui. The line of sight between the two groups (highlighted by red boxes) is interrupted by the huge settlement of the Weihe Power Station. (Image courtesy of Google Earth, editing by the author.)

To conclude, it looks desirable and also relatively feasible to plan new building compounds in such a way as to allow the intervisibility from each pyramid to at least the nearest imperial monuments to the east and to the west, so that the Zhaomu tradition remains effective and a glimpse of the ancient global planning of this wonderful landscape, conceived more than 2000 years ago, can still be taken.

A harmonic development in this direction could also act as a laboratory in linking conservation and sustainable development for future projects in China, as well as assuring future possibilities for archaeological research.

Funding: This research received no external funding.

Conflicts of Interest: The author declares no conflict of interest.

References

1. Magli, G. Royal Mausoleums of the Western Han and of the Song Chinese Dynasties: A Satellite Imagery Analysis. *Archaeol. Res. Asia* **2018**, *15*, 45–54. [CrossRef]
2. Zhewen, L. *China's Imperial Tombs and Mausoleums*; Foreign Languages Press: Beijing, China, 1993.
3. Segal, E.M. Archaeology and cognitive science. In *The Ancient Mind. Elements of Cognitive Archaeology*; Renfrew, C., Zubrow, E.B.W., Eds.; Cambridge University Press: Cambridge, UK, 2009.
4. Zubrow, E.B. Knowledge representation and archaeology: A cognitive example using GIS. In *The Ancient Mind. Elements of Cognitive Archaeology*; Renfrew, C., Zubrow, E.B.W., Eds.; Cambridge University Press: Cambridge, UK, 1994.
5. Marcus, J.; Flannery, K. *Ancient Zapotec Ritual and Religion: An Application of the Direct Historical Approach*; Renfrew, C., Zubrow, E.B.W., Eds.; Cambridge University Press: Cambridge, UK, 1994.
6. Flannery, K.; Marcus, J. Cognitive Archaeology. In *Contemporary Archaeology in Theory: A Reader (Social Archaeology)*; Preucel, R.W., Hodder, I., Eds.; Wiley-Blackwell: New York, NY, USA, 1996.
7. Potere, D. Horizontal positional accuracy of Google Earth's high-resolution imagery archive. *Sensors* **2008**, *8*, 7973–7981. [CrossRef] [PubMed]
8. Luo, L.; Wang, X.; Guo, H.; Lasaponara, R.; Shi, P.; Bachagha, N.; Yao, Y.; Masini, N.; Chen, F.; Ji, W.; et al. Google Earth as a Powerful Tool for Archaeological and Cultural Heritage Applications: A Review. *Remote Sens.* **2018**, *10*, 1558. [CrossRef]
9. Smith, J.Z. *To Take Place. Towards Theory in Ritual (Originally, 1987)*; University of Chicago Press: Chicago, IL, USA, 1992.
10. Rapoport, A. *Human Aspects of Urban Form. Towards a Man—Environment Approach to Urban Form and Design*; Pergamon Press: London, UK, 1977.
11. Rapoport, A. *The Meaning of the Built Environment. A Nonverbal Communication Approach*; University of Arizona Press: Tucson, AZ, USA, 1982.
12. Rapoport, A. *The Mutual Interaction of People and Their Built Environment*; De Gruyter: London, UK, 1976.
13. Lehner, M. A contextual approach to the Giza pyramids. *Archiv fur Orientforschung* **1985**, *31*, 136–158.
14. Magli, G. The Giza 'written' landscape and the double project of King Khufu. *Time Mind* **2016**, *9*, 57–74. [CrossRef]
15. Magli, G. From Abydos to the Valley of the Kings and Amarna: the conception of royal funerary landscapes in the New Kingdom. *Mediter. Archaeol. Archaeom.* **2011**, *11*, 23–36.
16. Belmonte, J.; Magli, G. Astronomy, Architecture, and Symbolism: The Global Project of Sneferu at Dahshur. *J. Hist. Astr.* **2016**, *46*, 173–205. [CrossRef]
17. Pankenier, D. *Astrology and Cosmology in Early China: Conforming Earth to Heaven*; Cambridge University Press: Cambridge, UK, 2013.
18. Wheatley, P. *The Pivot of the Four Quarters: A Preliminary Enquiry into the Origins and Character of the Ancient Chinese City*; Edinburgh University Press: Edinburgh, UK, 1971.
19. Paludan, A. *The Imperial Ming Tombs*; Yale University Press: Yale, CT, USA, 1981.
20. Magli, G. Astronomy and Feng Shui in the projects of the Tang, Ming and Qing royal mausoleums: A satellite imagery approach. *Archaeol. Res. Asia* **2018**, in press. [CrossRef]
21. Wilkinson, E. *Chinese History: A Manual*; Harvard University Press: Cambridge, MA, USA, 2000.
22. Liu, H.; Zhou, G.; Wennersten, R.; Frostell, B. Analysis of sustainable urban development approaches in China. *Habit. Int.* **2014**, *41*, 24–32. [CrossRef]
23. Hald, M. *Sustainable Urban Development and the Chinese Eco-City Concepts, Strategies, Policies and Assessments*; Nansen Institute: Lysaker, Norway, 2009.
24. Wescoat, J.L. The Indo-Islamic Garden: Conflict, conservation, and conciliation in Gujurat, India. In *Cultural Heritage and Human Rights*; Silverman, H., Fairchild Ruggles, D., Eds.; Springer: London, UK, 2008; pp. 53–78.
25. Forte, M. Western Han landscape and remote sensing applications at Xi'an (China). In *Space, Time, Place: Third International Conference on Remote Sensing in Archaeology*; Campana, S., Forte, M., Liuzza, C., Eds.; Archaeopress: Oxford, UK, 2010.

26. Kenderdine, S.; Forte, M.; Camporesi, C. Rhizome of Western Han: An Omnispatial Theatre for Archaeology. In *Revive the Past. Computer Applications and Quantitative Methods in Archaeology (CAA), Proceedings of the 39th International Conference, Beijing, April 12–16*; Zhou, M., Romanowska, I., Wu, Z., Xu, P., Verhagen, P., Eds.; Pallas Publications: Amsterdam, The Netherlands, 2012; pp. 141–158.
27. Weller, A.W. Pyramids and the City: Urban Encroachment on Chinese Heritage in Xi'an B.S. Ph.D. Thesis, The Ohio State University, Columbus, OH, USA, 2008.

© 2019 by the author. Licensee MDPI, Basel, Switzerland. This article is an open access article distributed under the terms and conditions of the Creative Commons Attribution (CC BY) license (http://creativecommons.org/licenses/by/4.0/).

Article

A Geoarchaeological Reading of the City-Overlap-City Phenomenon in the Lower Yellow River Floodplain: A Case Study of Kaifeng City, China

Pengfei Wu [1,2], Dexin Liu [2,*], Jianhua Ma [1,2,*], Changhong Miao [1,2], Lingling Chen [3], Lei Gu [2] and Jiahuan Tong [1]

1 Key Research Institute of Yellow River Civilization and Sustainable Development & Collaborative Innovation Center on Yellow River Civilization jointly built by Henan Province and Ministry of Education, Henan University, Kaifeng 475001, China; wupengfei@henu.edu.cn (P.W.); chhmiao@henu.edu.cn (C.M.); 18337678233@163.com (J.T.)

2 College of Environment and Planning, Henan University, Kaifeng 475004, China; hedagulei1981@vip.henu.edu.cn

3 Department of Civil Engineering, Lushan College of Guangxi University of Science and Technology, Liuzhou 545616, Guangxi, China; lingl_chen@163.com

* Correspondence: dxliu@vip.henu.edu.cn (D.L.); vyhorse@163.com (J.M.)

Received: 27 December 2018; Accepted: 13 February 2019; Published: 16 February 2019

Abstract: The unique urban form on the ground and the "city overlap city" phenomenon occurring underground at Kaifeng city, on the Yellow River floodplain, is investigated. Archaeological data and historical geographical analysis were used to study the form of surface remains. Primary data were collected from four 25 m long drill cores which were obtained from different locations at Kaifeng city and the sedimentary cycles were quantitatively divided-out and dated. The results show that the evolution of Kaifeng's surface urban form mainly occurred over four periods, the first of which was before 225 BC; the second took place between 225 BC and 956 AD; the third between 956 AD and 1219 AD; and the fourth between 1219 AD and 1907 AD. The results support the view that the city wall of today has undergone continuous reconstruction on the basis of previous city walls and thus forming the special landscape sequence of overlapped walls as a result of the 1642 AD and 1841 AD floods. The results also substantiate the "city overlap city" phenomenon at Kaifeng city where there are at least "three and a half ancient cities" located underground today, and suggests the "city overlap city" landscape is a harmonious production comprising both natural and human heritages that are of worldwide significance in terms of authenticity and integrity. Our results contribute to understanding the effects of Yellow River flooding on shaping adaptive landscapes and human beings, and suggest that Kaifeng city as well as other lower Yellow River sites become World Heritage sites.

Keywords: Geoarchaeology; city overlap city; the Yellow River floodplain; Kaifeng city; cultural heritage

1. Introduction

Cities are important compositions of human civilization and, arguably, the main space formations of human society. The study of cities, therefore, is of value to society, not only concerning contemporary issues but also in relation to how they developed previously. This is especially the case for cities, or rather urban sites, which have been impacted by historical natural disasters such as earthquakes, tsunamis, and floods that resulted in destruction or the disruption of urban civilization. The work of excavating and researching such urban sites through archaeological methods has, therefore, become a hot topic in academia [1–5]. Well known urban archaeological sites including Pompeii, Volubilis, and the archaeological ensemble of Tarraco, have become important assets of international cultural heritage, and have received much attention in western countries. Among them, Pompeii was built in the 6th century BC but was subsequently destroyed as a result of the Mount Vesuvius eruption in 79 AD, where streets and houses were buried by volcanic ash. Remarkably, these houses have been preserved relatively completely, and archaeological excavations have continued since 1748 AD, providing important information enhancing our understanding of social life, culture and the art of ancient Rome [6,7]. Consequently, the city of Pompeii has been designated as a world cultural and natural heritage site by the United Nations Educational, Scientific and Cultural Organization (UNESCO).

In China, there are also many unique urban areas, similar to "Pompeii", forming what is known as a "city overlap city" landscape because in different historical periods, cities have been built successively in the same area under the influence of natural and social factors, and there is an obvious vertical overlapping phenomenon. Several cities including Kaifeng, Shangqiu, Xuzhou, and Xiapi, which are located in the Yellow River floodplain, characteristically exhibit this special urban landscape phenomenon which was essentially produced through flooding of the Yellow River in the historical period [8–11]. Among these cities, Kaifeng city, the only provincial city in the Yellow River floodplain, has suffered the most from Yellow River floods; for instance, it has been destroyed several times throughout history. Although these cities and their relationship to the Yellow River can be considered to be highly relevant in research on the impact of environmental change, especially concerning human coping mechanisms, research in this area has hitherto been limited to mere "landscape descriptions" and, therefore, little is known about the region's stratigraphic characteristics. This limitation is generally regarded as being related to the fact that the floodplain in the lower reaches of the Yellow River is in an economically underdeveloped region, and consequently, the government, as well as most other sectors of society, is not sufficiently aware of the natural and human heritage intrinsic to these unique cities and the region.

Therefore, this study takes Kaifeng as a case in point, using archaeological data and historical geographical analysis to study the form of surface remains. Primary data were collected from four drill cores which were obtained from different locations at Kaifeng city. The sedimentary cycles were quantitatively divided-out and dated, and the form of Kaifeng's underground city heritage was thus revealed, at least, from the perspective of stratigraphy. The purpose of this study is not only to explain the shape of this "city overlap city" landscape as a harmonious production, made by human beings and nature in the context of the Yellow River floods, but also, in terms of a representation of the "wisdom of the ancients" which has an important heritage value. It is also intended that this study can be used as a reference for other such studies in the Yellow River floodplain.

2. Study Area

2.1. Cities of the Lower Yellow River Floodplain

The Yellow River, known as "China's sorrow" because of the misery caused by its periodic flooding, traverses the lower Yellow River floodplain. During the Holocene, in the lower reaches of the Yellow River, the course of its flow fluctuated frequently over the North China Plain, with flows into the Bohai Sea in the Northeast, or into the Yellow Sea in the Southeast. The area covered by the river basin, which is bounded by Tianjin to the north and Jianghuai to the south, comprises some 250,000 km^2 and had a very important impact on the social and economic development of the region [12,13]. The lower Yellow River in history has been characterized as frequently shifting course with overflows leading to floods. According to records in historical documents, catastrophic levee failure occurred 1593 times and major shifts of the channel occurred 26 times during the past 3000 years [14–18] (Figure 1). These changes brought unimaginable catastrophes to villages, cities, and their peoples, the survival circumstances of which, thus, merits detailed study for future benefit.

Figure 1. Locations and frequencies of flood events in the lower Yellow River. The base map is from the Harvard database https://doi.org/10.7910/DVN/Q9VOF5. The data in the figure is from References [14,16].

Furthermore, the current status of the ancient cities in the Yellow River floodplain were divided into three categories, as shown in Table 1, depending on the location of water: 1) lakes/ponds within the city wall, 2) lakes/ponds encircling the city wall, and, 3) the old city turned into lakes (Table 1).

Table 1. Remaining form of City in the Yellow River floodplain [19–22].

Region	Lakes/Ponds within the Wall Type	Lakes/Ponds Encircling the City Type	The Old City Turned into Lakes Type
Eastern Henan	Kaifeng, Chenliu, Taikang, Ningling, Luyi, Fengqiu, Changyuan	Shangqiu, Huaiyang, Xiayi, Yucheng, Qi county	Sui county, Zhecheng
Southwest Shandong	Heze, Caoxian, Dingtao, Gaotang county, Yuncheng, Yuye	Liaocheng, Chengwu, Shan county	
West Anhui		Dangshan, Bozhou	
Northern Jiangsu	Xuzhou, Feng county		

2.2. *Kaifeng City*

Kaifeng city, extending from 34°11′N to 35°01′N and 113°52′E to 115°15′E in Henan province, is situated on the southern bank of the Yellow River, about 70 km east of the provincial capital, Zhengzhou, and nearly 650 km southwest of Beijing. (What is now Kaifeng has had a number of different names. For example, it was Yiyi in the Spring and Autumn Period, Daliang in the Warring and States Period, Bianzhou in the Sui and Tang dynasties, Dongjing in the North Song dynasty, Bianjing in the Jin dynasty, Bianliang in the Yuan dynasty, and Kaifeng in the Ming and Qing dynasties.) As an ancient capital and one of the most important cities in ancient China, Kaifeng city is often referred to as "the Capital of the Eight Dynasties". Especially in the Northern Song dynasty Kaifeng was arguably one of the most important places in the world. (The Chronology of Chinese Dynasties—the Western Zhou Dynasty (1046 BC–771 BC); the Spring and Autumn Period (770 BC–476 BC); the Warring States Period (475 BC–221 BC); the Qin dynasty (221 BC–207 BC); the Western Han dynasty (206 BC–24 AD); Sui and Tang dynasties (581 AD–907 AD); the Northern Song dynasty (960 AD–1127 AD); the Jin and Yuan dynasties (1115 AD–1368 AD); the Ming dynasty (1368 AD–1644 AD); and the Qing dynasty (1636 AD–1912 AD).), From the Yuan, Ming and Qing Dynasties to 1954 AD, Kaifeng was the provincial capital of Henan province, but this changed in 1954 AD, when the provincial capital moved westward to Zhengzhou and Kaifeng became a regional level city.

Since the third century BC at least seven floods have devastated Kaifeng city (Table 2). From the Southern Song Dynasty to the late Qing dynasty, Yellow River flooding has occurred more than 300 times near Kaifeng city, more than 10 of which besieged the city whilst six actually entered the city [23–25]. In addition, more than 10 of these floods entered the protection earth ramparts and besieged Kaifeng city. During 1448–1492 AD, the Yellow River traversed around Kaifeng city and separated Kaifeng from the north bank of the Yellow River. Kaifeng was subjected to frequent flooding by the Yellow River after the Jin and Yuan dynasties, and this gave rise to complex changes to its urban forms. There were seven major flood disasters, two of which, in 225 BC and 1642 AD, destroyed the whole city. In 1841, the city was inundated for about 8 months, but it was not until 1843 that the walls were reconstructed, and by this time, the urban form of Kaifeng, including the earth rampart, brick city wall and lake, became the critically important flood-adaptability landscape. More significantly, survivors rebuilt a new city on the same site after each flood, resulting in the old city buried in the cultural layers of different periods (i.e., the seven major floods), forming what has become known as the "city overlap city" landscape. Six Kaifeng cities now lie buried under silt with the oldest "fossil" city buried 10 m underground [26]. Nowadays, the flood disasters of the past are almost totally forgotten, but the suspended river landscape, low city wall, the huge lake in the city together with the "city overlap city" landscape significantly contribute to the Yellow River's natural heritage as well as to the cultural heritage of Kaifeng city.

Table 2. Statistics the major floods in ancient Kaifeng city [27].

Dynasty	Time of the Flood	Impact of the Flood
the Warring States Period (475 BC–221 BC)	225 BC	flood into the city
the Yuan dynasty (1271 AD–1368 AD)	1283 AD	flood into the city
the Ming dynasty (1368 AD–1644 AD)	1387 AD	flood into the city
	1397 AD	flood into the earth rampart
	1399 AD	flood into the city
	1404 AD	flood into the earth rampart
	1410 AD	flood into the earth rampart
	1414 AD	flood into the earth rampart
	1422 AD	destroy the earth rampart several times
	1461 AD	flood into the city
	1478 AD	flood into the earth rampart
	1482 AD	flood besieged the city and suggestion for moving the city
	1489 AD	suggestion for moving the city
	1448–61 AD; 1461–92 AD	channel diversion of the Yellow River
	1606 AD	flood into the earth rampart
	1642 AD	flood into the city
the Qing dynasty (1636–1912 AD)	1761 AD	flood into the earth rampart
	1819 AD	flood into the earth rampart
	1841 AD	flood into the city

3. Methods

3.1. Historical Documents

China possesses a remarkable continuous written history providing information on natural and anthropogenic environmental changes. Such information includes details of floods, droughts, agricultural activities, irrigation, river management, extreme climatic events, geography, territory, population, and economic activities which can be used in detailed reconstructions of river evolution [16,17]. The literature used in this research mainly comes from the *Twenty-Four Histories* (such as *Records of the Historian Wei Aristocratic Family*), *Local Chronicles* (such as *Xiangfu Gazetteer, Shunzhi or Guangxu edition*), *Local Literatures* (such as *Bian Wei Wet Record (Bianwei Shijin Lu)* and *Rumeng Record (Ru meng lu)*, and contemporary literatures on Kaifeng Yellow River Research (such as *Annals of Kaifeng Yellow River and Kaifeng suburbs Yellow River*).

3.2. Archaeological Data

Since the 1980s, the Kaifeng Cultural Relics Team and the Songcheng Archaeological Team have carried out a series of archaeological exploration works on the Kaifeng "city overlap city" site, and have made many important discoveries, most notably, the Song-Jin Palace and the Ming Prince Zhou's Mansion (Figure 2) were found below the Longting Lake in Kaifeng city [28]. Some of their results provided clues for the establishment of the chronological framework of each core sedimentary cycles for this study (Table 3).

Table 3. Archaeological explorations near the drilling cores [25,28,29].

Core Location	Archaeological Cultural Sites	Burial Depth (m)
Shizhuan(SZ)	Wu Gate(午) of Ming Prince Zhou's Mansion	4.5
	Government offices of Ming Prince Zhou's Mansion	3.78–6.1
	Wu Gate(五) of Jin Imperial Palace	6.3
	Xuande Gate of North Song Imperial Palace	8.2
	The surface of Song Dynasty	7.5
	The surface of Ming dynasty	4.5
Yizhuan(YZ)	Duanli Gate of inner Ming Prince Zhou's Mansion	3.5–4
	West wall of Ming Prince Zhou's Mansion	5
	The surface of Ming Prince Zhou's Mansion	3–4
	South Gate of North Song Palace	6.5
Jinming(JM)	The surface of Qing dynasty	4
	Guzi Gate of Outer North Song Dynasty	8–11
	West wall of Outer North Song dynasty	11.3
	The surface of the Warring States Period	12.5
	Pottery of Spring and Autumn Period (In Jinming Campus of Henan University)	12–14
	Pottery of Spring and Autumn Period (In Campus of Yellow River Conservancy Technical Institute)	13–14.3
Minglun(ML)	North Song Inner-city Wall	3.6–9
	The surface of Qing dynasty	9.15

Figure 2. Yin'an Dian archaeological Site of the Ming Prince Zhou's Mansion (Photographed in 1985).

3.3. Stratigraphic Methods

3.3.1. Drilling Settings

The sedimentary environments in different areas of Kaifeng may display significant differences due to a range of factors including variations of levees protection, flood intensities and terrain differences, therefore, the selection of representative sites is crucial. The core sites were located cognizant that: (i) the cultural layer should embody all cultural layers from the Warring States to the present time, reflecting the overall spectacle of the "city overlap city" landscape; (ii) the inhomogeneous sedimentary environments indicate that the core sites should be located along the major flood-path trend (i.e., from northwest to southeast). Based on these premises, three sedimentary cores (designated as ML, SZ and YZ) from the urban area together with one core (JM) from the suburbs were acquired through the deployment of a drilling machine (Figure 3).

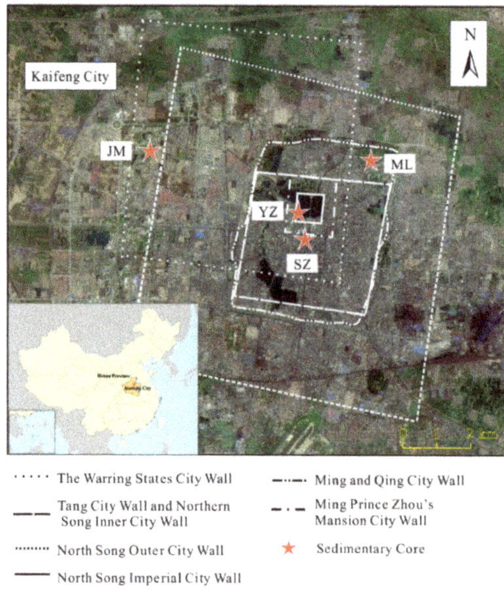

Figure 3. The city walls of Kaifeng city in every dynasty and the core locations.

3.3.2. Sampling and Physical and Chemical Analysis

In April 2012, the four 25 m-long cores (ML, SZ, YZ and JM) were obtained through the use of a corer and large drill in open air. They were then sampled continuously, mostly at 10 cm intervals with a few sandy samples at 20 cm or 30 cm intervals. A total of 861 samples were finally acquired, comprising 213 samples from SZ, 223 from YZ, 204 from ML, and 221 from JM. All samples were subjected to laboratory investigation which included grain size analysis, black carbon (BC) content analysis, and chemical elements (Cu, Zn, Pb, Cd, Al, P, As and Hg) content analysis to classify the various sedimentary cycles (representing flood events). Grain sizes were measured using a laser diffraction particle size analyzer (Mastersizer 3000, Malvern Co. Ltd., Malvern, UK). The black carbon content was measured using a TOC analyzer (Liqui II, Elementar Co. Ltd., Hanau, Germany); whilst the content of other chemical elements were measured using an ICP-MS (Xseries-2, Thermofisher Co. Ltd., Waltham, USA).

From the samples, carbon particles, animal bones, plant seeds and residue (a part of hand-selected uncarbonized plant or a single entity fragment of plant material) as well as clay specimens were diligently selected for dating using the Accelerator Mass Spectrometry of ^{14}C (AMS^{14}C) in the Institute of Archaeology and Culture, Peking University. A silty sample was selected for Optical Stimulated Luminescence (OSL) dating in the Digital Environmental Archaeology Laboratory, Institute of Geography, Henan Academy of Sciences. The 14C date for this research was calibrated using the computer program OxCal v4.2.3 with the IntCal13 atmospheric calibration curve [30]. The dating results are shown in Table 4 below.

Table 4. Results of annual survey of each borehole.

Borehole	Sample Number	Sample Substance	Depth/m	Absolute Age/BP	Calibration Age (2σ, 95.4%)
JM	J72	Plant residue	7.8–7.9	410 ± 20 (^{14}C)	1465 ± 35AD
	J88	Carbon particles	9.7–9.8	360 ± 25 (^{14}C)	1495 ± 40AD
	J120	Clay	13.2–13.3	1145 ± 15 (^{14}C)	920 ± 60AD
	J124	Snail	13.6–13.7	935 ± 20 (^{14}C)	1095 ± 65AD
SZ	S64	Seed	7.1–7.2	660 ± 20 (^{14}C)	1370 ± 20AD
	S76	Bones	8.5–8.6	995 ± 25 (^{14}C)	1015 ± 35AD
	S87	Carbon particle	9.7–9.8	1790 ± 25 (^{14}C)	200 ± 70BC
	S99	Silt	11.05	2404 ± 95 (OSL)	550 ± 95BC
YZ	Y72	Carbon particle	8.0–8.1	740 ± 20 (^{14}C)	1270 ± 20AD
ML	M71	Carbon particle	8.4–8.5	720 ± 25 (^{14}C)	1275 ± 25AD

3.3.3. Sedimentary Cycle Division

The number of sedimentary cycles is a basic indicator which can be used to identify the flood frequency in alluvial strata. The approach usually adopted for the classification of sedimentary cycles based on grain size, however, is sometimes not suitable for urban strata that has been deeply disturbed by human beings. For example, after a Yellow River flood, urban reconstruction work by local residents on the recent sediments and subsequent treasure hunting activities can lead to the disturbance of the normal sequence of sedimentary cycles, which means that classification based on grain size division alone is problematic. In this study, therefore, in addition to the traditional grain size cycle, two new indicators [31,32] were selected; namely, black carbon content (black carbon cycle) and chemical element content (element cycle). During the flood receding stage, residents' production activities and daily life resulted in accumulations of black carbon and other chemical elements (especially anthropogenic elements) near to the ground, the content of which is inevitably different from that occurring during the flood deposition stage. Based on this difference, sedimentary cycles can, therefore, be identified. With the aim of obtaining more accurate sedimentary cycles, this study comprehensively employed the cycle division results of the above three alternative indexes in order to obtain comprehensive cycles for the four cores.

4. Results

4.1. Historical Geography Development of Urban Form

The urban form used in this paper refers to the shape of the city, that is, the shape of the city formed by the enclosure of the city walls. Whether the walls of ancient cities were built or not reflects the development of cities to a certain extent. Therefore, "cities within walls" is one of the main characteristics of ancient cities in China [33]. At the same time, the change of the enclosure area represents the development scale and development level of the city [34,35]. According to the various historical developments of urban form, we can identify four main periods in the history of Kaifeng.

Before 225 BC: "City-Guo" Mode

Although the origin of Kaifeng is undocumented, it is known that a military fort existed near the southern border of the Wei kingdom between 770 BC and 476 BC. In 364 BC, Wei Huiwang moved the capital city from Anyi to Kaifeng.

From the Western Zhou to the Western Han dynasty, the capital city was named as West Cheng and East Guo by Yang Kuan [36]. It comprised a walled inner city (West Cheng) and a walled outer city (East Guo), but the inner-city layout of Kaifeng is unclear. An attempt to deduce the walls of the outer city (East Guo) was made by Wu et al. [37] (Figure 4).

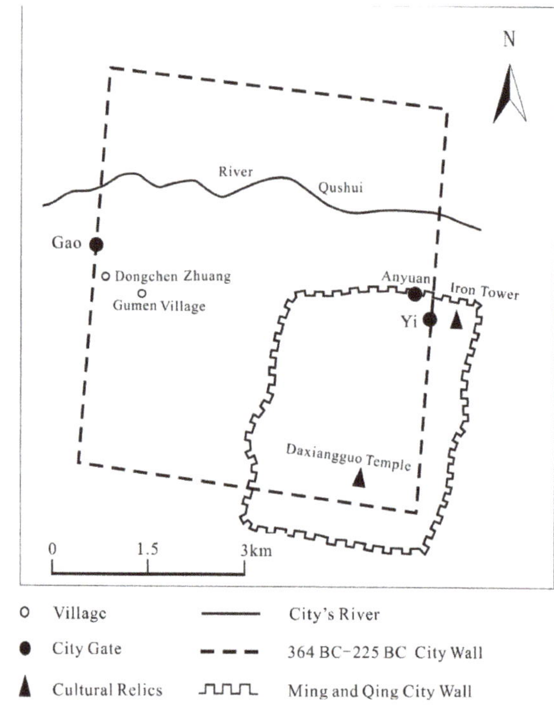

Figure 4. The probable borders of the early city walls of Kaifeng before 225 BC.

The outer city wall of Kaifeng during the Warring States Period was almost square. Only two gates have been confirmed: Yi in the eastern wall and Gao in the western wall. The River Qushui crossed the northern part of the city [37,38].

In 225 BC Kaifeng was completely destroyed by General Wangben. He used the water of Honggou River to flood the entire city. Relics of the city were submerged to a depth of about 12–14 m [39].

225 BC-956 AD: "Government Office–City Wall" Mode

In this period, the city walls underwent two changes. Between 225 BC and 781 AD, the city changed very slowly and there is an absence of evidence of the lines of the city walls. Kaifeng was largely destroyed and drainage systems in the vicinity were disrupted by flood damage in 225 BC. A new city was built on the site, but it developed slowly.

Between 781 AD and 956 AD, the southern line of the city wall was extended to the south of the River Bian. There were seven gates in the wall (Figure 5).

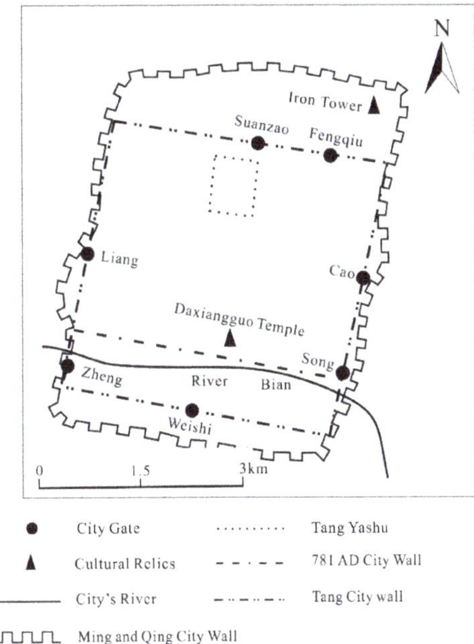

Figure 5. The probable borders of the city walls of Kaifeng in 225 BC–956 AD.

956 AD–1219 AD: "Outer city—Inner city—Imperial city" Mode

In this period, Kaifeng became the richest city in the world, according to Zhou [40]. Reflecting its continuing status as the capital city, more projects of new constructions and city walls were undertaken, and the form of the whole city changed significantly. In 956 AD, the government began to reconstruct the city walls. There were now three city walls: around the imperial city, around the inner city and finally the outer city wall (Figure 6).

It was within the imperial city that the emperor handled state affairs. Its boundaries can be traced back to the Tang Yashu (Figure 5). Emperor Songtaizong enlarged the imperial city wall and it became 5 li (about 2340 m) in length in this period. The imperial city, which was nearly square in shape, was located north west of the inner city. It was occupied in the Jin dynasty. A new palace, Prince Zhou´s Mansion, was built in the Ming dynasty.

The inner city, which was also almost square, was the second-line defense of the city. The eastern and western lines of the inner-city wall were located beneath the city wall of the Ming and Qing Dynasties. In 1219, Emperor Xuanzong rebuilt the new southern and northern walls of the inner city and, consequently, its old southern and northern walls were subsequently destroyed to a large extent. The old southern wall was buried underground to a depth of about 8 m, whilst, the old northern wall was buried to a depth of about 8.5–9.5 m [41]. These walls are now beneath the city wall of the Ming and Qing Dynasties. The whole inner-city wall was built of brick in the Ming dynasty (Figure 7).

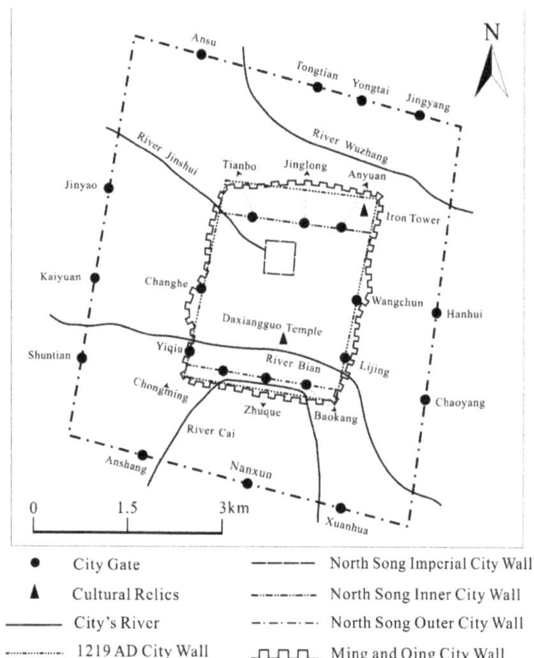

Figure 6. The probable borders of the city walls of Kaifeng, 956 AD–1219 AD.

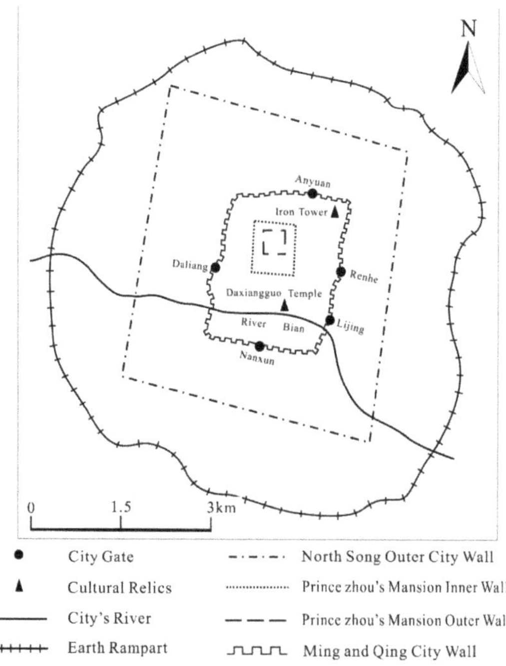

Figure 7. The probable borders of the city walls of Kaifeng in the Ming Dynasty.

The outer city wall was 50 li 165 bu (about 29180 m) in length [42]. During several wars and in the post-war period, it suffered significant destruction, perhaps most notably in 1642 AD, when it was completely destroyed by flooding from the Yellow River.

1219 AD–1907 AD: "Earth Rampart—Original City Form" Mode

During the Jin and Qing Dynasties, flooding by the Yellow River had a considerable impact on the environment of Kaifeng, for instance, the outer city wall gradually disappeared as a result of flooding. An earth rampart of more or less circular form was constructed to protect the city from flooding in 1451 AD. Nevertheless, the city was flooded again in 1642 AD, and this time buried to depth of 8–10 m [39].

In 1719 AD, the Manchu city wall was rebuilt to the north of the abandoned Prince Zhou´s Mansion (Figure 8). Thereafter, the line of the brick wall remained unchanged for some 200 years, until 1907 AD.

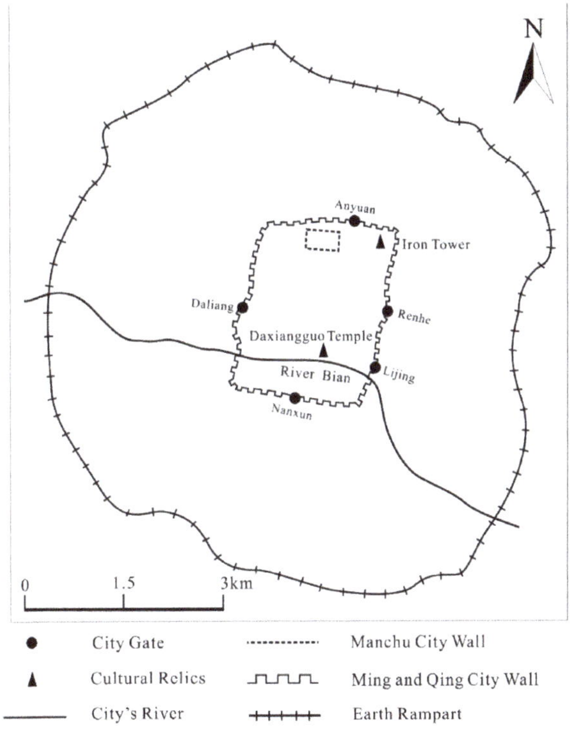

Figure 8. The probable borders of the city walls of Kaifeng in the Qing Dynasty.

Kaifeng's historical geography is, in a number of respects, more complex than that of other cities in the Yellow River floodplain. The early city was flooded in 225 BC. Thereafter, a new county capital, founded on the destroyed area, developed quite slowly, and it was not until 781 AD when evidence suggests that the southern border of the city wall was extended. In the Later Zhou and Northern Song Dynasties, the imperial capital had three city walls comprising: that of the Imperial city, the Inner city and the Outer city. The Outer city wall, however, is on a larger scale than the others in all periods of wall construction. Owing to the flooding of the Yellow River, the following development of the whole city of Kaifeng was limited to within the inner city (brick city) during the Ming and Qing Dynasties. The city wall of today has undergone continuous reconstruction on the basis of previous city walls and formed the special landscape of the sequence of overlapped walls (Figure 9) because of the disastrous Yellow River floods of 1642 AD and 1841 AD.

Figure 9. Cross-section showing the chronological sequence of walls in Kaifeng. (Author´s photograph, 2014.).

4.2. *The Underground City Form of Kaifeng Since the Warring States*

From the analysis results of the grain size, black carbon, and element cycles, the four cores: JM, SZ, YZ and ML were divided-out into 15, 14, 14 and 16 comprehensive cycles respectively (Figure 10). Combined with chronological data, historical documents and archaeological data, the chronological framework of each sedimentary cycle since the Warring States Period can be established.

4.2.1. The Daliang City of the Warring States and the Honggou flood in 225 BC

This was the first time in history that Kaifeng was destroyed. From the results of this study, the ancient ground (cultural layer) of Daliang City is 10–15 m below the present ground level, which is consistent with the relevant archaeological findings [25,39]. The chasm began in 360 BC and was built in about twenty years. The Honggou diverts the water of the Yellow River from Xingyang, where it passes through Xingze and Putianze to Daliang, where it joins the Huaihe River system. At that time, the Daliang City Wall was very tall and was known as the "Seven Gaps City". General Wangben of the Qin State failed to capture the city, and then proceeded to dig the Honggou to the west which resulted in the entrapment of Daliang City by flood water for three months. Finally, a section of the soil-based city wall to the south of the West Wall was immersed and collapsed, and the Wei State was destroyed. One hundred years after Daliang City was destroyed, Sima Qian came to Xunyi (renamed from Daliang) only to be confronted with a shattering scene, the "Ruins of Daliang" [43]. From this we can see that the damage caused by the Honggou flood to the Daliang City was very serious.

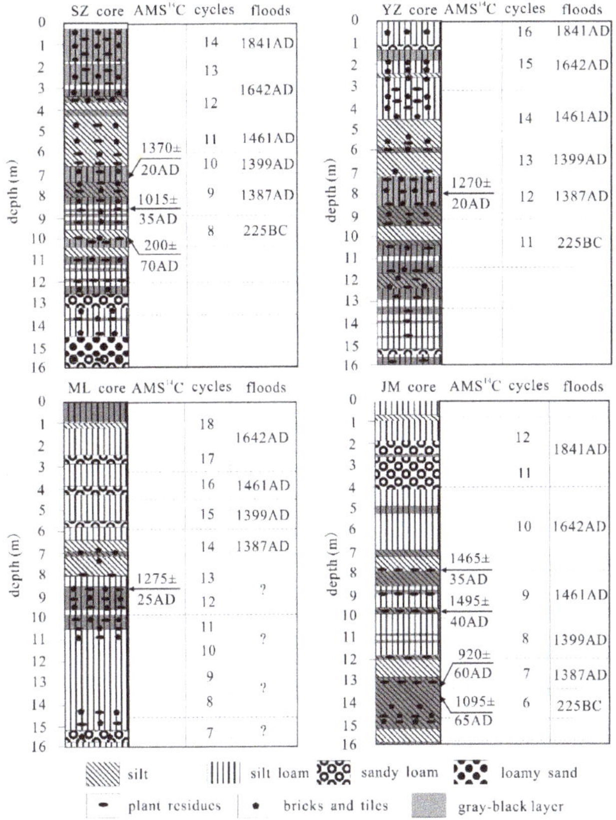

Figure 10. Dating framework of the drilling cores since the Warring States. Gray-black layers were formed under reduced environment or human activities, indicating flood receding periods or cultural layers. In other words, gray-black layers are an important reference indicator for determining sedimentary cycles.

4.2.2. Dongjing City in Northern Song Dynasty and the Yellow River Flood in 1387AD

This Yellow River flood caused the sediment deposited in the drilling area of the SZ and YZ cores to be within 2 m in 1399 AD, and once again, Kaifeng city was buried as it was too late to take mitigating action. From the Qin to Yuan Dynasties, the common ancient ground (cultural layer) in Kaifeng was buried 9–13 m below the present ground level, a finding which is basically consistent with the relevant archaeological discoveries [25,39]. As Kaifeng city in the Jin and Yuan dynasties was rebuilt on the basis of Dongjing City, its magnificence and influence were far less than that of Dongjing City, and the two dynasties lasted for a relatively short time. After 12 years (1399 AD), it was silted up again, so the ancient city flooded by the Yellow River in 1387 AD could still be regarded as Dongjing City.

4.2.3. Kaifeng City in Late Ming Dynasty and the Yellow River Flood in 1642 AD

In the seven floods of Kaifeng, the flooding of the Yellow River in 1642 AD was the most serious. According to statistics, of the 370,000 people who were in Kaifeng at the time, only 30,000 survived the flood [39]. According to the results of this study, the thickness of the flood deposit is 2.3–3.8 m, the depth of the surface culture layer in Kaifeng city is between 3.5~8 m of the late Ming dynasty, and the depth of the suburban area is greater than that in the urban area. It was not until 5 years later (i.e.1647 AD) that the Yellow River breach was completely blocked. It should be pointed out that under the ground of Kaifeng city in the late Ming dynasty, there still existed the surface of the Ming dynasty which was flooded and silted by the Yellow River in 1399 AD and 1461 AD, and the thickness of sediment was 1–2 m. Whether there were ruins of Kaifeng city in these two periods underground needs further investigation. The authors believe that the two floods (i.e., 1399 AD and 1461 AD) were close to the Yellow River flood in 1387 AD, and the sedimentary thickness is not too large. On the basis of the silted ruins and broken walls, the houses were rebuilt, and the city walls and horse paths were heightened in turn. Kaifeng city may not have existed during this period, as only the cultural layer remains.

4.2.4. Kaifeng City in the Qing Dynasty and the Yellow River Flood in 1841 AD

This Yellow River flood overflowed from the South Gate with flood waters proceeding northward from east to west, converging in the north of Longting Pavilion, resulting in serious water accumulations in some parts of the city [44,45]. It was not until February 8, 1842 AD that the closure of the Yellow River's breached embankment was completed. The flood between Kaifeng brick city and the moat and the accumulated water in the city gradually subsided, but the city was besieged by floods for eight months. This Yellow River flood did not cover the whole city, so Kaifeng city was not completely buried underground. For example, the ML core which is located in Gongyuan of Henan Province, showed no signs of deposits from this flood, whilst the deposits in SZ and YZ cores are not too thick, being less than 1 m. The surface of Kaifeng city therefore comprises the upper parts of the sediment layers from both the 1642 AD and 1841 AD Yellow River floods. Nowadays, the surface of the Xiangguo temple and the Yanqing Taoist temple before the whole uplift is higher than that of its base (Figure 11). However, the flood besieged Kaifeng for 8 months, and deposited a large amount of sediment between the outside of the brick city and the moat, as evidenced in the JM core, where the corresponding sedimentary thickness is 3.2 m. This is basically the same as the burial depth of the two Qing dynasty courtyards at the Xinzheng Gate site (Figure 12).

Figure 11. Comparison of the Yanqing Taoist temple before and after the 1841 AD flood which resulted in the lifting of the ground surface by some 3.1 m. The photograph on the right was taken in 2009.

The evidence derived from these stratigraphic, archaeological, and literature investigations, all indicate that there are at least "three and a half ancient cities" located underground in Kaifeng city today; namely, the Daliang City of the Warring States (the representative time is before 225 BC, and the depth is 10–15 m), the Dongjing City in the Northern Song Dynasty (the representative time is 225 BC–1387 AD, and the depth is 9–13 m), the Kaifeng city in late Ming dynasty (the representative time is 1387 AD–1642 AD, and the depth is 3.5–8 m) and the ancient ground, of the Qing dynasty, which existed in certain areas (the representative time is 1662 AD–1841 AD, and the depth is 1–4 m).

Figure 12. The Qing courtyard at the site of Xinzheng Gate in Northern Song outer city wall.

5. The Heritage Value and Utilization of Urban Form

UNESCO seeks to encourage the identification, protection and preservation of cultural and natural heritage around the world considered to be of outstanding value to humanity. This admirable endeavor is embodied in an international treaty called the Convention concerning the Protection of the World Cultural and Natural Heritage, which was adopted by UNESCO in 1972. Since then, not only has the term "world heritage" come into being, but there is also a worldwide common understanding of cultural and natural values. The Convention thus embodies the values of the human subject today. Up to now, 845 cultural heritage sites, 209 natural heritage sites and 38 mixed cultural and natural heritage sites have been selected for the World Heritage List (http://whc.unesco.org/en/list). Among them, China has 36 cultural heritage sites, 13 natural heritage sites and four mixed heritage sites. According to the definition of "World Heritage Convention" and operational guidelines for the Implementation of the World Heritage Convention, the case of this study, Kaifeng city, together with the Yellow River Dams and the lower Yellow River, should, arguably, be included in the World Heritage List as cultural and natural mixed heritage sites. According to the study of Zhou [46,47], the Yellow River Dams meet the standards of cultural heritage iii, iv and v, and the lower Yellow River meets the standards of natural heritage vii and viii. Furthermore, we suggest that Kaifeng city meets the standards of natural mixed heritage. According to the first article of cultural heritage in the World Heritage Convention, Kaifeng city meets the three definitions of cultural relics, group of urban buildings or historic towns and urban center, and sites. At the same time, it is in accordance with the UNESCO Recommendation on the Historic Urban Landscape [48]. Kaifeng city has nine Major Historical and Cultural sites protected at National Level, nine provincial-level key cultural relics' protection units in Henan province, and more than 200 non-removable cultural relics. There are three historical blocks: Shudian Street, Imperial Song Street and Shuanglong lane. The sites of the Northern Song Dynasty ruins were included in the 13th Five-Year special plan for the protection of China's great national ruins. The immovable cultural relics on the ground and the underground piled city sites in Kaifeng together constitute the remarkable characteristics of the centralized distribution of cultural relics, high level of protection and great value impact.

The historic urban landscape is the urban area understood as the result of a historic layering of cultural and natural values and attributes, extending beyond the notion of "historic centre" or "ensemble" to include the broader urban context and its geographical setting. Kaifeng City can meet the four standards of operational guidelines for the Implementation of the World Heritage Convention. The suspended rivers in the lower Yellow River are generally 2~5 m above the ground. But in the Kaifeng reach, the height difference between the riverbed and the urban area is more than 13 m, which is the most typical reach of the lower Yellow River floodplain. The three-dimensional urban flood control system of Kaifeng city, formed in the long-term struggle with the flood and sediment disasters of the Yellow River, is composed of the "Yellow River Dams–City earth rampart moat-brick wall–inner lake" and other adaptive landscapes. It is, also, an outstanding achievement of adapting or transforming the natural environment and has important values regarding landscape and urban planning and design in the lower Yellow River floodplain.

Although Kaifeng city had two catastrophic disasters (i.e., 225 BC and 1642 AD) and there have been several suggestions of moving the city to other areas, the succession and development of different periods remained within the earth rampart, a testament to the struggling spirit of Kaifeng people. According to the analysis of urban archaeology and stratigraphy, the inheritance of the cities buried under Kaifeng city in different periods is more prominent and outstanding, and they need to be protected for the world to share. As a whole historic urban landscape, there are only three sites selected for the world cultural heritage list in China, namely, the Ancient City of Ping Yao, Old Town of Lijiang and Historic Centre of Macao [49]. As long as we fully understand and explore the heritage value of Kaifeng city, it is anticipated that the city will be included in the list of world cultural and natural mixed heritage sites.

6. Conclusions

In this paper, evidence from historical sources as well as current stratigraphic investigations has been presented, demonstrating that Kaifeng, a typical Yellow River city, has both natural and human heritages that are significant in terms of their authenticity and integrity, not just to China, but also to the rest of the world. Such evidence supports the view that Kaifeng meets the requirements as specified by the World Heritage Convention to be included as a World Heritage site. However, Kaifeng city is only one of some 93 disaster cities in the Yellow River floodplain and it is, therefore, recommended that those cities, which includes Kaifeng, with such a unique urban form, as shaped by Yellow River floods and the responses to them, be jointly declared as World Heritage sites together with the Yellow River Dams and the lower Yellow River.

The sedimentary archive at Kaifeng is remarkable for its thickness and exceptional preservation. The environmental and archaeological record here is almost unmatched in China, if not worldwide. Research at Kaifeng, therefore, provides an opportunity for exploring long-term change processes associated with the environment, the relationships between human behavior, technology, and nature, as well as the roles and influences of the Yellow River on history, politics, economics, and society.

Author Contributions: P.W. and D.L. designed the research and wrote the paper. J.M. and C.M. performed the research. L.C., L.G. and J.T. helped to edit the manuscript. All authors read and approved the final manuscript.

Funding: The research was funded by National Social Science Foundation of China (15BZS024), the Youth Backbone Teachers Project of Henan Provincial Universities (2016GGJS-027), the Humanities and Social Science Projects by Ministry of Education of China (12JJD790023) and Program for Innovative Talents of Zhongyuan Post-Doctoral in 2018.

Acknowledgments: The authors are grateful to Professor Liritzis Ioannis and Jeremy Whitehand for helping us to improve the manuscript their comments on a draft of the paper. The authors would also like to thank the anonymous reviewers of the original draft manuscript for their comments, which ultimately helped to improve this paper.

Conflicts of Interest: The authors declare no conflict of interest.

References

1. Cleere, H. *European Convention on the Protection of Archaeological Heritage (1992)*; Springer: New York, NY, USA, 2014.
2. O'keefe, P.J. The European Convention on the protection of the archaeological heritage. *Antiquity* **1993**, *67*, 406–413. [CrossRef]
3. Willems, W.J. Archaeological resource management and academic archaeology in Europe. Some observations. *Quale Futuro Per L'archeologia* **2009**, *10*, 89–99.

4. Willems, W.J. The work of making Malta: The Council of Europe's archaeology and planning committee 1988–1996. *Eur. J. Archaeol.* **2007**, *10*, 57–71. [CrossRef]
5. Niknami, K.A. Iran: Archaeological heritage in crisis: Developing an effective management system for archaeology. *J. Cult. Herit.* **2005**, *6*, 345–350. [CrossRef]
6. Butterworth, A.; Ray, L. *Pompeii: The Living City*; St. Martin's Press: New York, NY, USA, 2005.
7. Laurence, R. *Roman Pompeii: Space and Society*; Routledge: London, UK, 2010.
8. Jing, Z.C.; Rapp, G., Jr.; Gao, T.L. Holocene Landscape Evolution and Its Impact on the Neolithic and Bronze Age Sites in the Shangqiu Area, Northern China. *Geoarchaeol. Int. J.* **1995**, *10*, 481–513. [CrossRef]
9. Jing, Z.C.; Rapp, G.; Gao, T.L. Geoarchaeological aids in the investigation of early Shang civilization on the floodplain of the lower Yellow River, China. *World Archaeol.* **1997**, *29*, 36–50. [CrossRef]
10. Gao, T.L.; Murong, J.; Jing, Z.C.; Niu, S.S. The brief report of the East Zhou Town site reconnaissance in Shangqiu, Henan. *Archaeology* **1998**, *12*, 18–27.
11. Zhao, M.Q. Characteristics and reasons of city overlap city in Xuzhou City. *Chin. Hist. Geogr.* **2000**, *2*, 129–138.
12. Zou, Y.L.; Tan, Q.X.; Shi, N.H. *Yellow River, Changes of River System in China History. Physical Geography of China, Historical Physical Geography*; Science Press: Beijing, China, 1982.
13. Zou, Y.L. (Ed.) *Historical Geography of Huang-Huai-Hai Plain*; Anhui Education Publishing House: Hefei, China, 1993.
14. Shen, Y.; Zhao, S.X.; Zheng, D.L. *The Chronicle of the Yellow River*; Military Committee and Resources Committee of the Republic: Nanjing, China, 1935.
15. Li, S.; Finlayson, B. Flood management on the lower Yellow River: Hydrological and geomorphological perspectives. *Sediment. Geol.* **1993**, *85*, 285–296.
16. Yellow River Conservancy Commission. *The Chronicle of Events of the Yellow River*; Yellow River Water Conservancy Publishing House: Zhengzhou, China, 2001.
17. Chen, Y.Z.; Syvitski, J.P.M.; Gao, S.; Overeem, I.; Kettner, A.J. Socio-economic Impacts on Flooding: A 4000-Year History of the Yellow River, China. *Ambio* **2012**, *41*, 682–698. [CrossRef]
18. Chen, Y.Z.; Overeem, I.; Kettner, A.J.; Gao, S.; Syvitski, J.P.M. Modeling flood dynamics along the superelevated channel belt of the Yellow River over the last 3000 years. *J. Geophys. Res. Earth Surf.* **2015**, *120*, 1321–1351. [CrossRef]
19. Wu, Q.Z. *Study on Flood Control of Chinese Ancient Cities*; China Architecture & Building Press: Beijing, China, 1995.
20. Chen, X. *The Flood Adaptive Landscapes in Ancient Chinese Cities of Shangqiu Area*; Peking University: Beijing, China, 2008.
21. Yu, K.J.; Zhang, L. The flood and watering adaptive landscapes in ancient Chinese cities in the Yellow River Basin. *Urban Plan. Forum* **2007**, *5*, 85–91.
22. Xu, J.Q. The Formation and Evolution of the "Ancient Waters City" in the Yellow River Floodplain. Ph.D. Thesis, Southeast University Academic Degree Dissertation, Nanjing, China, 2009.
23. Cheng, S.Y. The safety flow and flooding of the Yellow River in Kaifeng city around the 12th century. *J. Henan Univ. Soc. Sci.* **2003**, *43*, 32–36.
24. Li, R.T.; Ding, S.Y.; Li, Z.H. The historical evolution of Kaifeng City under the influence of the Yellow River. *Areal Res. Dev.* **2006**, *25*, 1–7.
25. Ma, J.H.; Gu, L.; Wu, P.F.; Liu, D.X.; Chen, Y.F. *Flood Records in Alluvial Stratum by the Yellow River and Its Disaster Inversion in Kaifeng City*; Science Press: Beijing, China, 2016.
26. Qiu, G. An investigation into piled-up cities under the Kaifeng City. *China's Cult. Relics* **2004**, *26*, 70–77.
27. Wu, P.F. Analysis on life cycle of Kaifeng city. *Jianghan Trib.* **2013**, *1*, 121–128.
28. Kaifeng Cultural Relics Team. *The Archaeological Discoveries and Studies of Kaifeng*; Zhongzhou Ancient Books Publishing House: Zhengzhou, China, 1998.
29. Kaifeng Municipal Institute of Cultural Relics and Archaeology. The Excavation of the Office Dwelling of the Prince of Zhou of the Ming Dynasty at Xinjiekou in Kaifeng, Henan. *Cult. Relics* **2017**, *3*, 39–72.
30. Reimer, P.J.; Bard, E.; Bayliss, A.; Beck, J.W.; Blackwell, P.G.; Ramsey, C.B.; Buck, C.E.; Cheng, H.; Edwards, R.L.; Friedrich, M.; et al. IntCal13 and Marine13 radiocarbon age calibration curves 0–50,000 years cal BP. *Radiocarbon* **2013**, *55*, 1869–1887. [CrossRef]

31. Ma, J.H.; Lu, J.; Gu, L. Black carbon as an indicator for dividing sedimentary cycle from the Yellow River flood sediments in Kaifeng. *Acta Geogr. Sin.* **2015**, *70*, 730–738.
32. Ma, J.H.; Chen, Y.F.; Gu, L.; Lou, D.; Wu, P. Sedimentary cycle division from the Yellow River sediments under Kaifeng based on anthropogenic-biogenic elements. *Geogr. Res.* **2017**, *36*, 1147–1158.
33. Chang, S.D. Some observations on the morphology of Chinese walled cities. *Ann. Assoc. Am. Geogr.* **1970**, *60*, 63–91. [CrossRef]
34. Chang, S.D. The Morphology of Walled Capitals. In *The City in Late Imperial China*; Skinner, G.W., Ed.; Stanford University Press: Stanford, CA, USA, 1977; pp. 75–100.
35. Ioannides, Y.M.; Zhang, J.F. Walled cities in late imperial China. *J. Urban Econ.* **2017**, *97*, 71–88. [CrossRef]
36. Yang, K. *An Institutional History of Ancient Chinese Cities*; Shanghai People's Publishing House: Shanghai, China, 2003; pp. 85–88.
37. Wu, P.F.; Xu, J.A.; Ma, J.H. Preliminary study of "Yin Honggou Guan Daliang". *Zhongyuan Cult. Relics* **2016**, *1*, 54–61.
38. Ge, Q.F. A new research of the layout of Daliang city during Warring States and Wei Dynasty. *Zhongyuan Cult. Relics* **2012**, *4*, 27–33.
39. Liu, C.Y. *Uncovering Ancient Cities Built in Different Dynasties under Kaifeng City*; Science Press: Beijing, China, 2009.
40. Zhou, B.Z. *Research of Dongjing in the Northern Song Dynasty*; Henan University Press: Kaifeng, China, 1992.
41. Kaifeng Songcheng Archaeological Team. Exploration and excavation of the Dongjing inner city wall in Song dynasty. *Cult. Relics* **1996**, *5*, 69–75.
42. Kaifeng Songcheng Archaeological Team. Exploration and excavation of the Dongjing outer city wall in Song dynasty. *Cult. Relics* **1992**, *12*, 52–61.
43. Zhou, B.Z.; Xu, B.Y. *Ancient City of Kaifeng*; Henan University Press: Kaifeng, China, 2011.
44. Dodgen, R.A. *Controlling the Dragon: Confucian Engineers and the Yellow River in Late Imperial China*; University of Hawai'i Press: Honolulu, HI, USA, 2001.
45. Wu, P.F.; Lu, J.; Ma, J.H. Spatial reconstruction of the Yellow River flood sieged Kaifeng City in 1841 and the reasons of the flood. *J. Henan Univ. Nat. Sci.* **2014**, *44*, 299–304.
46. Zhou, Z. On the Nomination at the Lower Yellow River and Its Levees for Inscription on the World Heritage List. *Areal Res. Dev.* **2009**, *28*, 142–144.
47. Zhou, Z. Reconsideration of the Nomination of the Lower Yellow River and Its Levees for Inscription on the World Heritage List. *Areal Res. Dev.* **2010**, *29*, 144–148.
48. Araoz, G.F. World-Heritage Historic Urban Landscapes: Defining and Protecting Authenticity. *Apt Bull.* **2008**, *39*, 33–37.
49. Zhao, Y.X. *Zhongguo Shijie Yichan Yingxiang Zhi [Gazetteer of China World Heritage Video]*; China Photography Press: Beijing, China, 2014.

© 2019 by the authors. Licensee MDPI, Basel, Switzerland. This article is an open access article distributed under the terms and conditions of the Creative Commons Attribution (CC BY) license (http://creativecommons.org/licenses/by/4.0/).

 sustainability

Article

Mobile Augmented Reality for Cultural Heritage: Following the Footsteps of Ovid among Different Locations in Europe

Răzvan Gabriel Boboc [1,*], Mihai Duguleană [1], Gheorghe-Daniel Voinea [1], Cristian-Cezar Postelnicu [1], Dorin-Mircea Popovici [2] and Marcello Carrozzino [3]

1 Department of Automotive and Transport Engineering, Transilvania University of Brasov, RO-500036 Brasov, Romania; mihai.duguleana@unitbv.ro (M.D.); daniel.voinea@unitbv.ro (G.D.V.); cristian-cezar.postelnicu@unitbv.ro (C.-C.P.)

2 Department of Mathematics and Computer Science, Ovidius University of Constanta, 8700 Constanta, Romania; dmpopovici@univ-ovidius.ro

3 Scuola Superiore Sant'Anna, 56127 Pisa, Italy; carrozzino@sssup.it

* Correspondence: razvan.boboc@unitbv.ro; Tel.: +40-745-987-710

Received: 31 January 2019; Accepted: 18 February 2019; Published: 22 February 2019

Abstract: There are many examples of cultural events that distinguish people nationally. Celebrating this can bring people closer, as inhabitants of different countries share similar cultural values. This study investigates a sustainable way to enhance these types of events. On the occasion of the 2000-year anniversary of the death of the Roman poet Ovid, we propose a mobile augmented reality (MAR) application that contains historical information related to his life. As Ovid often stated in his last poems, he feared his work would be forgotten after his exile from Rome. This paper focuses on assessing whether this is the case, while also disseminating factual, historic data to people who tested the application. Experiments were conducted in Italy and Romania, in three different cities: Sulmona, Rome, and Constanta. Based on the results collected, four constructs were investigated: comprehensibility, manipulability, enjoyment, and usefulness. The results revealed that the usability of the MAR application, and the interaction metaphor, are appropriate for the general public. The MAR application provided a positive experience, and thus, increased the extent of the occasion.

Keywords: augmented reality; mobile phones; cultural heritage; evaluation

1. Introduction

The development of technology during the last decades has allowed the implementation of new ways of interacting between people and computers. The progress of digital information has also significantly affected the evolution of cultural heritage (CH) dissemination [1], offering new possibilities for developing, e.g., the market of tourist services [2]. Whether it is a mobile application, an online catalog, or a social media exchange, these new paradigms are affecting all aspects of our lives, reshaping the way we communicate, learn, and approach the world around us [3]. One of these technologies that offers new ways of interacting is augmented reality (AR). This technology has a tremendous potential for the promotion and preservation of CH. People are starting to be receptive—they want social and collaborative experiences that combine learning with having fun. CH organizations have to address their needs by creating innovative applications [4], and AR and mixed reality can make this happen. Computers have become more accessible and cheap enough for widespread applications in various fields [5]. CH can take advantage of this development by bringing cultural vestiges to light with the help of systems that can integrate computer-generated digital information with the physical world.

AR technology gives a different perception of reality, as it enriches reality with a computer-generated layer containing visual, audio, and tactile information. In this way, any object can be recreated and used as content for AR applications. Using a 'virtual' representation of a classic museum item has several advantages: it may incorporate dynamic elements, and it allows access to aspects of the artifact that may otherwise be hidden [6]. However, in order to build a complete and complex cultural representation via digital heritage technologies, developers must also understand how users interact with the system or interface [5]. When designing AR applications, it is important to choose the best combination of techniques for presenting the appropriate electronic information to the users. The content must meet their interests and needs in order to be accepted [7]. Thus, the usability evaluation of AR applications is of prime concern. Since AR is a relatively new technology, there are few papers that address the evaluation of AR interfaces, and furthermore, their sustainability in cases such as the one presented in this paper. For this reason, we have developed an AR application with the aim of investigating its usability, taking into account perceptual and ergonomic aspects.

The 2000th anniversary of the death of the Roman poet Publius Ovidius Naso, known as Ovid, was in 2017. This poet was "one of the most brilliant personalities of the Augustan Age", as described in [8]. Ovid was born in 43 BC, in the town of Sulmo (today: Sulmona), located in the middle of Italy, approximately 140 km from Rome. He lived in Rome after his studies, becoming famous and appreciated for his work. After his masterpiece, the "Metamorphoses", he was exiled in 8 AD to the isolated town of Tomis (today: Constanta, Romania), on the shore of the Black Sea at the outer edge of the Roman Empire. He was abruptly removed from the stimulating artistic and social community, which had nourished his extraordinarily successful poetic career, for reasons which were lost in the millennia. He died in Tomis 9 years later, still hoping for a pardon, which never came [9].

In line with the linguistic community, this anniversary highlights a good opportunity to give people a variety of ways to remember this great personality, whose vast influence and popularity continues to this day [10]. This heritage should not be lost, yet keeping it is not an easy task. The methodology of this study is based on bringing to life some creative moments from Ovid's existence. We have chosen three significant cities for the poet: Sulmona, Rome, and Constanta. By means of AR technologies and using mobile devices, we've created an application that was presented to the general public.

In this regard, the application tested had appropriate content for each location: Sulmona, where Ovid was born; Rome, where he has lived most of his life; and Tomis, where he died (Figure 1). Each version of the application recreates some aspects of his life, presenting his 3D animated model in the three stages of his life: as a child, as a famous public person, and as an exiled person in decline, but still artistically creative.

Figure 1. Three cities significant in Ovid's life, where the application was tested by the general public.

AR and virtual reality (VR) are technologies that allow us to bring to life lost historical periods of time, while protecting the real artifacts for generations to come. From this point of view, the main objective of this research was to introduce mobile augmented reality (MAR) as a means to present CH, and to sustainably reuse digital content for general public dissemination.

We proposed the corresponding research objectives: to model and design an attractive application that could be used by tourists and local citizens to learn more about the life of Ovid. To recreate the historical context of that time by using graphical representations enhanced with audio content. Finally, to evaluate the effectiveness of that application.

This article is structured as follows: Section 2 describes related work and sets the context of this paper. Section 3 describes the application and the experiment setup. Section 4 presents the evaluation results. In Section 5, the findings are summarized and, finally, Section 6 reports conclusions and future work.

2. Related Work

Preserving cultural heritage provides the reasoning to investigate the opportunities offered by digital technology. There is a vast body of work within the field of AR in CH contexts, especially for education, learning, and improving tourist experience [11–17].

From a historical perspective, AR was developed from VR, with the specificity that allows the visualization of both real and virtual objects at the same time. The first VR system is considered to be Sensorama machine, developed in the late 1950s [18], while the history of AR started in 1968 with the head-mounted display system developed by Ivan Sutherland. The term "augmented reality" was coined only in 1990 by Tom Caudell and David Mizell [19]. With the development of technology over time, the concept has evolved and the advancements in handheld computers have opened new opportunities for AR [1]. The first handheld mobile phone was presented by Motorola in 1973, and the first personal digital assistant (PDA) appeared on the market in the early 90s. Only a few years later, Bruce H. Thomas developed ARQuake, the first outdoor mobile AR game on PDA [20]. The mobility offered by handheld devices, and the increased progress in terms of software, led to the spread of AR applications as well as the development of new interfaces and new technologies. In 2014, Google announced Project Tango, an Android solution for AR that combines three core technologies: depth perception, motion tracking, and area learning [21].

Since then, stand-alone AR has gradually become more common as computing power on PDAs and smartphones has increased. AR applications implemented on mobile devices were named mobile AR, or handheld mobile AR (HMAR). Such a system is, in its basic form, a mobile device with a display to show an augmentation using the appropriate software [22], with the following characteristics [23]:

- it combines real and virtual objects in a real environment;
- it is interactive, works in real time;
- it registers and aligns real and virtual objects with each other.

There are 2 key issues that need to be considered when developing MAR applications: the accuracy of user tracking and the registration of 3D models with real-world features [24]. The tracking, or the estimation of camera position, can be realized in two ways:

- marker-based tracking, which requires labels with a colored or black and white pattern;
- marker-less tracking, which uses the mobile device's GPS or image recognition systems to identify a location [25].

In the case of a CH application, placing markers in the scene is not practical; thus, marker-less methods are required for extracting features from the environment. These features are used for correct alignment of virtual information with real information, a process known as registration. There are two methods of registration:

- without prior knowledge—the registration is realized 'on-line', without any prior scene knowledge;
- with prior knowledge—predefined patterns are taken into account from the environment in order to identify a match. An overview on the history of tracking for MAR is given in [26].

The surveys on MAR work performed in different application domains [27], or strictly related to the touristic context [28], highlight some determinants and issues that have impact on the utilization of this type of application [15]. Although MAR applications were first designed and tested in indoor environments, the advantage of portability provided by mobile devices facilitated their expansion to the outdoors. This allowed visualization of 3D models of heritage sites to be superimposed over the video stream from the outside environment [29], visualizing virtual points of interest (POIs) overlaid on top of the phone's camera view [30], or textual annotation of outdoor locations based on GPS measurements [31]. Reference [32] proposed an application for outdoor site exploration throughout different periods of time, enriching travel experiences with important historic facts. In other research papers, the authors have used MAR applications "in situ" so that the visitors were able to explore and visualize building details in an intuitive way [33], to educate students visiting an outdoor archaeological site in a game-like approach—"Oracle of Delphi app" [34], or for huge reconstructed models of heritage monuments (e.g., Roman Theatre of Byblos) [35]. A novel paradigm is proposed in [36] based on presenting user artwork, in terms of 3D models, videos, and photos, in virtual kiosks augmented on the camera view of a handheld device.

AR has been introduced in the CH sector over the last decade primarily as a useful technology for assisting visitors inside museums or in heritage sites [37–39]. It allows visitors to explore and appreciate the objects showcased by overcoming barriers in time, space, and language [40]. It also improves the learning experience in CH with the aid of better user interaction methods [24,41]. For instance, in [42] the information is organized in finite triplets (visual representation, context, and corresponding audiovisual content) in order to be presented in a personalized, interactive manner on users' personal mobile devices. Using AR techniques, geomatic resources were also developed, allowing the valorization of the geological heritage [43]. MAR applications can provide real-time information based on the user's preferences and context [44], and can also generate revenue or economic returns [45]. A thorough review of AR applications for CH is provided in [46].

Moreover, AR technologies have proven their importance in the virtual reconstruction of historical monuments, having the potential to provide a new approach to the past by reproducing on-site historical experiences [13]. In the last decade, many objects, monuments, or historical sites were reconstructed into digital formats. The significance of reconstruction is to preserve, protect, and interpret culture and history [4], and to bring to life elements from the past [47,48]. According to [49], every two years the digital heritage is doubling in size and is expected to grow tenfold between 2013 and 2020. While digital heritage is primarily concerned with preservation of CH, virtual heritage (VH) involves the synthesis, conservation, reproduction, representation, digital reprocessing, and display [50] of monuments, artifacts, buildings, and culture, aiming to be more open to global audiences [11]. CH does not end with objects and collections. It also includes information such as traditions, performing arts, social practices, rituals, festive events, activities related to nature and the universe, or skills to produce traditional crafts [51]. According to UNESCO this is the intangible CH, and in order to be celebrated and contemplated more easily, requires the creation of specific conditions to ensure its viability when it is materialized in the form of a media resource [2].

However, the development of applications that targets reactivation of CH in local citizens remains a challenge [52]. In this sense, it is of the utmost importance to take into account the process of evaluation of digital cultural resources. Digital resources are a combination of cultural resources that are created using computer and multimedia technologies. They can be accessed and used in digital form [53].

Given their diversity, evaluation is not a simple process. In this paper we focus only on the evaluation of MAR applications. These should be carefully designed and improved based on user feedback [54]. An important component, which is studied in several papers, is the usability of the application. Relying on five usability principles developed in [55], and on a compilation of heuristic evaluation checklists for mobile interfaces proposed in [56], an MAR application was evaluated in [57] in terms of usability and user expectations. Usability was evaluated in [58] using four different methods. In [52] a methodological framework was proposed, which allowed developers to categorize the type of CH application and to determine which resources should be used. Santos designed a questionnaire for usability evaluation (HARUS), which was composed of two sub-questionnaires, namely the comprehensibility scale and the manipulability scale [59].

Tourist requirements for the development of mobile AR applications were investigated in [60], concluding that tourists are increasingly expecting methods to access information instantly. Other components of experience with AR applications in public were examined in [61], including designing interface features, augmentation features, and so on. Good results after an evaluation of user experience in a mobile AR tourist guide, in terms of ease-of-use and intuitiveness, were obtained in [62]. Users' perceptions and experiences were measured by applying a modified technology acceptance model (TAM) in [63]. Immersion, seen as a form of cognitive and emotional absorption in AR settings, was studied in [64].

In conclusion, there are various studies regarding augmented reality applications in the CH area. Most of them show that the use of new technologies creates innovative and attractive options that increases the users' interest in learning more about CH. However, there are just a few that try to target people from more than one country. Project Tango represents a high-end technology that provides stable and reliable motion tracking. Based on this technology, we've developed an application and assessed its adoption using a questionnaire adapted from [54] and [65], which characterizes the following four subjective parameters: perceived comprehensibility, perceived manipulability, perceived enjoyment, and perceived usefulness. To the best of our knowledge, this is one of the most extensive user studies in AR conducted in outdoor environments.

3. Materials and Methods

3.1. Mobile Augmented Reality (MAR) Setup

The AR application presented in this paper was developed with the aim of bringing to life specific moments from Ovid's life. It was designed for the Lenovo Phab 2 Pro phablet (which incorporates the Google Tango technology). At the time of the research, this was the only Tango-enabled device available on the market. Phab 2 Pro used a Qualcomm Snapdragon 652 processor, 4 GB of RAM, and 64 GB storage. It used specialized Tango-related sensors—a fisheye camera that can handle the motion tracking, and an infrared depth sensor that used the "time-of-flight" principle to corroborate data with the video stream outputted by the camera [66]. Google Tango incorporates three core components:

1. Motion tracking—Tango implements visual–inertial odometry that is used to estimate the trajectory of the device (i.e., its position in the scene relative to the starting point). The inertial sensors of the mobile device are used to provide data regarding the acceleration and rotation of the device.
2. Area learning—This allows the ability to see and remember certain visual features of a physical space (edges, corners, unique features). Thus, the device is able to "recognize" the area. The motion tracking alone can be affected by "drift", which is caused by the integration of inertial data. Area learning improves the accuracy of the trajectory (through a process called "drift correction") and allows a Tango device to perform "localization" (orient and position itself when it recognizes a previously learned area).
3. Depth perception—Tango-enabled devices can understand what is the distance to any object in the real world. A common use is to obtain point clouds of scanned objects in real time.

The AR application that was developed by our team was called OvidAR and used motion tracking and area learning. It required a map in order to start the 3D scene. The workflow for using a Google Tango-enabled device is presented in Figure 2. The area learning application was a demo provided by Tango Project and can only be used to create new area descriptions. The OvidAR application started the 3D scene if there was a previously saved map and if it detected the key visual features that were saved in the area's description file.

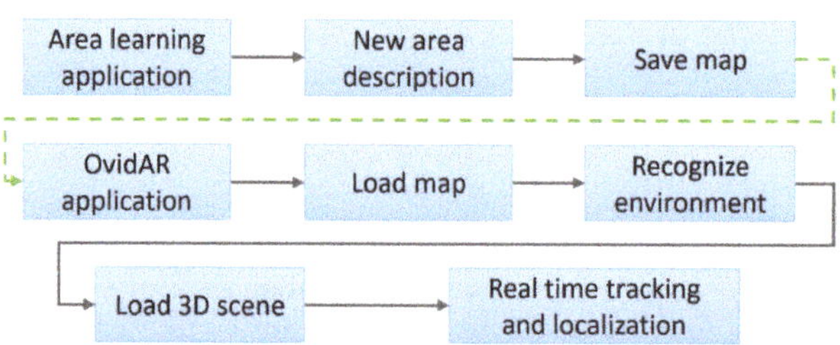

Figure 2. Proposed augmented reality workflow using Project Tango.

Real-time tracking and localization ensures that the user is immersed in the scene. Various animations of the 3D versions of Ovid created a feeling of interaction. Another key element that enhances the experience is the sound, specifically the reverberation zones that distort the audio depending on the location of the audio listener in the reverb zone.

The developed application can run only on Tango-compatible devices. There were some devices that supported Google Tango, like Asus Zenfone AR, but now the project is closed. Google introduced a new AR platform, known as ARCore. Users can still develop AR application using Tango technology, but there are no other releases or provided support.

3.2. 3D Reconstruction and Application Development

A preliminary stage of the work was the design of the 3D objects that served as content for the MAR application. For this purpose, we used historical references for the reconstruction of archaeological artifacts discovered in Dobrogea, a historical region situated in the southeast part of Romania, enhanced with animated elements. The ruins of a Roman house situated in an archeological site in the center of Constanta, known as Tomis in the past, were photographed and measured in order to create a 3D object. Photogrammetry allowed us to obtain a good representation of the walls, which we used as a base for the house. A 3D model of an ancient Roman house was designed using Autodesk 3D Studio Max software, version 2016 (Autodesk, California, USA) [67] (Figure 3). Local experts from the Museum of Archeology from Constanta provided information regarding the design of the house, more specifically, how the roof and windows should look in order to accurately present a building from that period of time.

Figure 3. The 3D Roman house: (**a**) Outdoor view; (**b**) Indoor view.

The virtual 3D character of Ovid included animations and an audio source to play the poem. Visitors used the Lenovo phablet to explore the inside and outside of the Roman house, observe the gestures of Ovid while he recited his own poem, or when he sat at the table writing a poem. If someone passed by or stood next to the virtual character, the device displayed both that person and Ovid on the screen, as in real life.

Ovid's 3D models, as child, adult in Rome, and elder in Tomis, were obtained using MakeHuman, version 1.1.0 [68] and Blender, version 2.77a (Blender Foundation, Amsterdam, Netherlands) [69], which are free, open-source software. They were animated using Adobe Mixamo [70]. The characters were designed based on the historical figure of Ovid. The selected animations corresponded to activities such as walking, sitting, writing, and idle behaviors. These models were integrated in the MAR application using Unity software, version 5.2.5 (Unity Technologies, San Francisco, USA) [71]. In Figure 4 is presented the development process of Ovid3D.

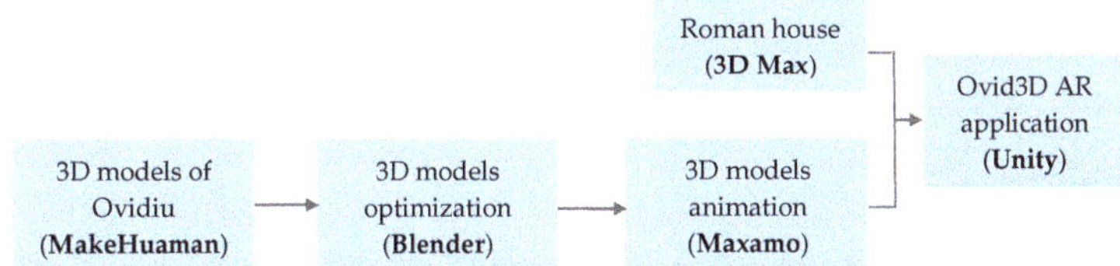

Figure 4. Development stages of Ovid3D.

3.3. Research Method

For the purpose of this study, several locations were selected to perform the experiment. The reason for choosing these specific places was that we wanted to present the application in cities where there was a greater possibility that tourists or residents have heard of the poet Ovid. As mentioned above, the three locations were the following:

- Constanta, Romania—In an archeological site of ancient Tomis, near the cathedral of the Apostles Saint Peter and Saint Paul;
- Rome, Italy—At the south part of Piazza de Colosseo, near Colosseum, the famous Roman amphitheater located in the center of the city;
- Sulmona, Italy—A historic town located in the province of L'Aquila in Abruzzo. The experiment took place in one of the city's main squares, Piazza XX Settembre, which is adjacent to the main street, Corso Ovidio, and very close to Piazza Garibaldi.

Not only was the location different, but also the content of the application was slightly different. For instance, in Constanta the virtual building was placed on top of the ruins of a real ancient Roman house. By doing so, users had a more realistic experience when navigating inside and outside the house (Figure 5a). This version of Ovid was reciting a short part of the poem "Sorrows", in which he expresses his complaints and fears regarding the place where he was exiled. The animated virtual model of Ovid writing at a table was located in the last room of the house, and is presented in Figure 5b.

In Rome, the application was similar, except that Ovid could be seen in Roman clothes, walking through the square (Figure 5c) while reciting a small part of his work The Metamorphoses, Book I—Fable 1. The Roman poet spent his childhood in Sulmona; as such, we decided to use a child version of Ovid. In this scenario, Ovid was walking outside the house, while in the background users could hear a poem recited by a child in an old Italian dialect (Figure 5d).

Therefore, there were three different applications, with different scenarios and audio content heard in the background. Each application build in Unity contained a scene with one or two characters and some objects, as follows:

- For Sulmona, the scenario featured a Roman house, the child version of Ovid walking in circles near the house, and a child voice that recited the poem "Metamorphoses", Fable I in old Italian language. There were also some objects around the house, such as marble columns or barrels. The poem can be heard in a loop, and the animation of the child created with Mixamo repeated itself after 30 s. The time needed to discover the whole MAR scene was roughly 150 s, depending on the pace of the participants.
- For Rome, the application showcased a Roman house with two rooms and different pieces of furniture. The adult version of Ovid was walking in front of the house in a 15-second animation loop, and as audio content, the "Metamorphoses" narrative poem was recited by an adult voice. To discover the entire MAR scene it took around 110 s.
- For Constanta, the scene was comprised of a Roman house and two versions of elder Ovid: one moving around the house and one writing at the table located in the second room of the house. In the background, users could hear a voice reciting Ovid's poem "Sorrows". Around 150 s were needed to discover the whole MAR scene.

A survey was performed following the same procedure in all locations. Participants that agreed to take part in the experiment received a verbal introduction about the purpose of the experiment and a short tutorial regarding the usage of the AR application. They received the Lenovo phablet with the OvidAR application, after which they were allowed to explore the 3D scene as much as they wanted (each participant used the app between 4 and 12 min). After the experiment, participants completed a questionnaire in order to provide feedback on the experience they had using the application.

From the previous literature regarding the evaluation of MAR applications, we adopted four measurement items: perceived comprehensibility, perceived manipulability [59], perceived enjoyment, and perceived usefulness [65].

Figure 5. Photos taken during the experiments—screenshots taken in different places: (**a**) user looking from the inside of the house, in Rome; (**b**) user looking at Ovid writing at his table, in Constanta; (**c**) user exploring the outside of the house, in Rome; (**d**) user exploring the outside of Ovid's home from childhood, in Piazza XX Settembre, Sulmona.

3.4. Participants

Participants included citizens and tourists that showed interest in trying a mobile AR application. A total of 63 subjects, 37 males and 26 females, aged between 17 and 62 (mean = 33.8, SD = 11.1) got the opportunity to try the application. Approximately 1/3 of the respondents were between 20 and 29 years old (34.9%), and another 1/3 were between 30 and 39 years old (33.3%). Participants were asked to rate their usage of mobile devices on a Likert-type scale (from 1 = "never use a mobile device" to 7 = "daily use of mobile devices") and if they have ever previously used an AR application. Each of them used a smartphone or tablet almost every day (mean = 5.9, SD = 1.1) and about half of them (32) had never used or seen an AR application.

Figure 6 summarizes the demographic characteristics of the respondents. Individual characteristics for each of the places where the application was tested are briefly presented in Table 1.

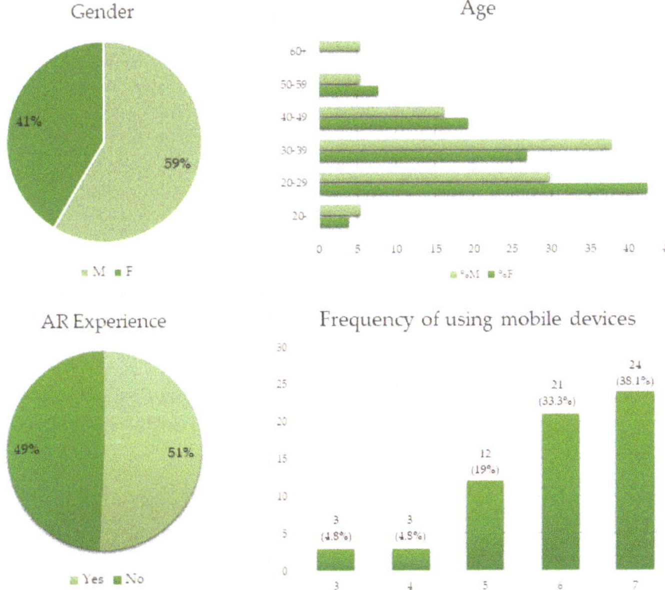

Figure 6. Showing individual variables of all the participants.

Participants that were not familiar with AR applications received a brief introduction of how AR works and how to use the OvidAR application. This was necessary to allow participants to familiarize themselves with the application so that they could evaluate the AR more accurately, even if they had never heard or seen such an application. Then, visitors were asked to use the application for approximately 5 min or more, and to then participate in the survey. The users who had never used AR applications were advised to evaluate OvidAR as objectively as possible, referring strictly to the application and to its content.

The first research stage involved mostly Romanian tourists with very little AR experience, but the application was also tested and evaluated by people who had previous AR knowledge, such as the attendees of the 6th Edition of creatiVE Summer School in Virtual Environments, organized at Ovidius University of Constanta. Participants from Rome were mostly foreign tourists from Germany, USA, Canada, or Romania, but also from Italy. In Sulmona, all participants were local, Italian inhabitants of the area.

Table 1. Individual characteristics of participants for each location.

Variables	No.	Percent (%)	No.	Percent (%)	No.	Percent (%)
	Constanţa		Rome		Sulmona	
Gender						
Male	18	55.0	12	63.0	7	64.0
Female	15	45.0	7	37.0	4	36.0
Age						
Under 20	1	3.03	1	5.26	1	9.09
20–29	11	33.33	7	36.84	4	36.36
30–39	11	33.33	7	36.84	3	27.27
40–49	6	18.18	3	15.78	2	18.18
50–59	3	9.09	0	0.00	1	9.09
60+	1	3.03	1	5.26	0	0.00
AR experience						
Yes	22	67.00	6	32.00	4	36.00
No	11	33.00	13	68.00	7	64.00

3.5. Questionnaire Design

The measurement items of the proposed questionnaire (22 in total) are summarized in Table 2 by each construct: perceived comprehensibility (eight items), perceived manipulability (six items), perceived enjoyment (four items), and perceived usefulness (four items). The participants had absolute freedom to fill the questionnaire to their own discretion.

The questionnaire was divided into three parts:

- The first section intended to collect some individual variables such as age and gender, participants' degree of familiarity with mobile devices and AR applications, and if they had knowledge of who Ovid was;
- The second section (statements 1 to 14) intended to evaluate the usability of the application using the handheld augmented reality usability scale (HARUS) [54]. The HARUS questionnaire had been developed for evaluating the comprehensibility and manipulability of MAR applications. The HARUS questionnaire contained eight statements for the comprehensibility scale and eight statements for the manipulability scale, but we decided to remove two of the manipulability questions ("I found it easy to input information through the application", "I think the operation of this application is simple and uncomplicated") because they were not relevant to the evaluation of the proposed application;
- The last section (statements 15 to 22) aimed to assess the perceived enjoyment and usefulness of the application. The questions have been extracted from the questionnaires proposed in [65]. QU 3 and QU 4 was added in order to evaluate the feature that allowed users to hear audio content.

Table 2. Questionnaire items.

No.	Construct	Abbreviation	Statement
1	Comprehensibility	QC 1	The interaction with this application requires a lot of mental effort
2		QC 2	The amount of information displayed on screen was appropriate
3		QC 3	The information displayed on screen was difficult to read
4		QC 4	The information display was responding fast enough
5		QC 5	The information displayed on screen was confusing
6		QC 6	The display was flickering too much
7		QC 7	The information displayed on screen was consistent
8		QC 8	The position error of the information displayed on the screen was negligible or null
9	Manipulability	QM 1	Interaction with this application requires a lot of body muscle effort
10		QM 2	Using the application was comfortable for my arms and hands
11		QM 3	The device was difficult to hold while operating the application
12		QM 4	My arm or hand became tired after using the application
13		QM 5	I felt that I was losing grip and dropping the device at some point
14		QM 6	The application is easy to control
15	Enjoyment	QE 1	I liked to use this application
16		QE 2	I found using the application unpleasant
17		QE 3	I found using the application exciting
18		QE 4	I found the application boring
19	Usefulness	QU 1	Using the application, I could quickly and easily find historical pictures and information
20		QU 2	Using the application, I have improved my knowledge about the life of Ovid
21		QU 3	Listening to a poem of Ovid and was useful and interesting
22		QU 4	Listening to Ovid in some circumstances was useful and interesting

All the questions used a 7-point Likert scale, ranging from 1—"strongly disagree", to 7—"strongly agree". Depending on the time availability, users were asked to fill in the questionnaire on site, or instead, access a survey link sent to their e-mail to be completed at a later time. Out of the 63 questionnaires collected, 21 were completed online and 2 incomplete questionnaires were eliminated. A total of 61 questionnaires were used for this study (96.82%).

The online survey questionnaire was developed in English, and the paper-based questionnaire was translated into Romanian and Italian for participants who had difficulties understanding English.

4. Results

Responses were put together to obtain a value for each of the four constructs defined by the questionnaire: comprehensibility, manipulability, enjoyment, and usefulness. The questionnaire contained both positive and negative worded items. For the negatively formulated questions, we first reversed the results in order to have the same scale. We chose to alternate positive and negative items. In this way, the user may be more careful when completing the questionnaire, and the answer was, thus, more relevant for the survey. After that, the scores were converted to a range of 0 to 6. All the values were then summed, and the sum was mapped to a range of 0 to 100. This method of data aggregation was done according to the instructions from the original HARUS questionnaire assessment [54].

Figures A1–A3 from Appendix A show the obtained scores for each of the questions for the three experiments, represented as box charts. In each figure, mean, median, interquartile values, whiskers, and outliers were reported.

For validating our measurement model, we evaluated reliability and sampling adequacy. Reliability was assessed by calculating Cronbach's α, and the proportion of variance was assessed by the Kaiser-Meyer-Olkin index. Each of the measures exceeded the recommended threshold (Cronbach's alpha = 0.902, Kaiser-Meyer-Olkin = 0.796).

When participants were asked: "Do you have any knowledge of who Ovid is?", the majority of them answered "Yes" from all three locations. Results showed that Ovid was known by 95% of the participants in Sulmona, 91% of the participants from Rome admitted to have heard of Ovid, while only 73% of the participants in Constanta heard of him. The values obtained for each question for all the users are displayed in Figure 7.

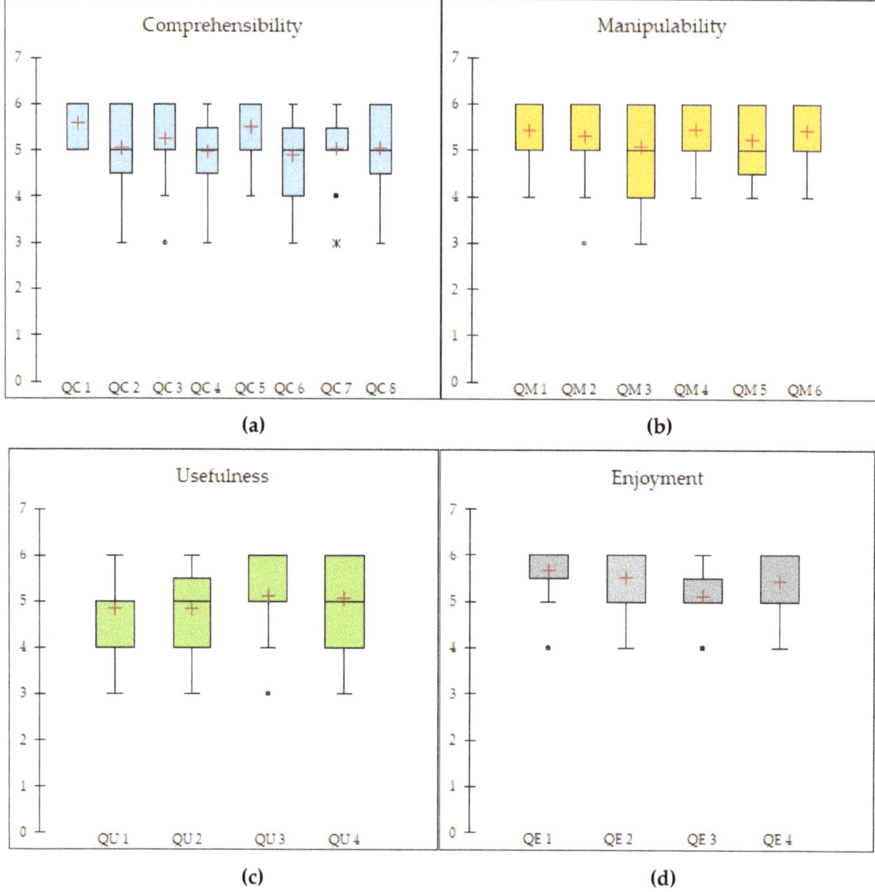

(a) **(b)** **(c)** **(d)**

Figure 7. Boxplots showing the outcomes for the questions, grouped into the four constructs.

The values for each construct—comprehensibility (mean = 91.06; SD = 4.46), manipulability (mean = 93.81; SD = 2.66), enjoyment (mean = 95.88; SD = 4.11) and usefulness (mean = 87.57; SD = 2.61)—are reported in Figure 8, separately for each experiment, and overall. The vertical axis is displayed from 80 to 105 to make the results visible.

At the end of the experiment, some participants showed interest in learning more about AR and how it could change the way people explore and learn new things about heritage and education.

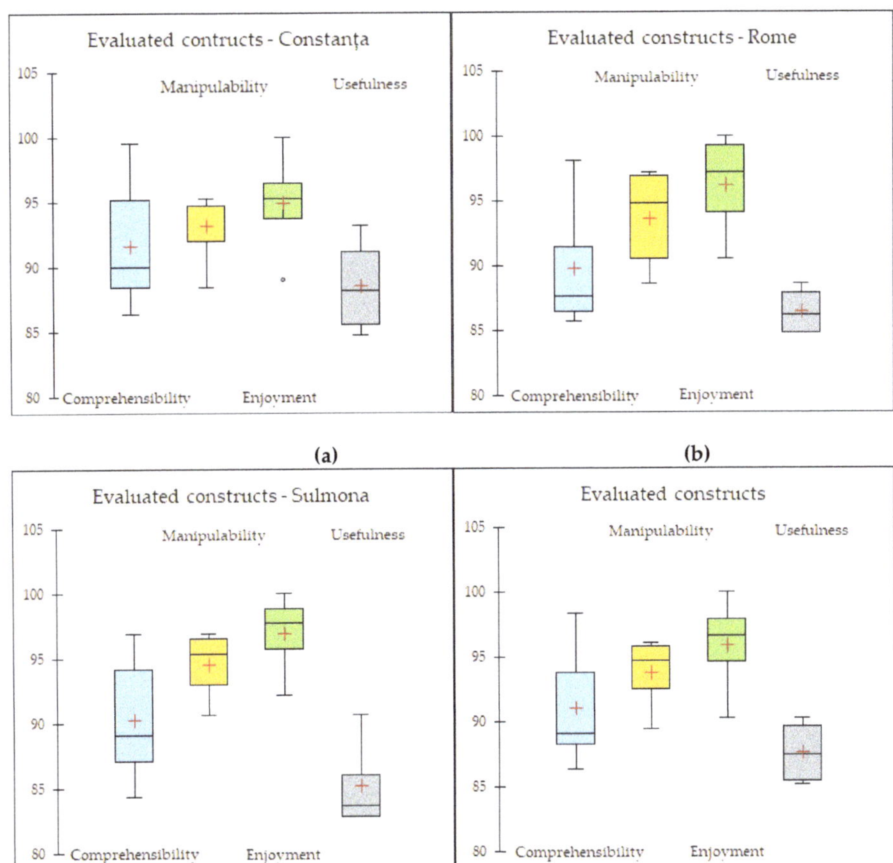

(a) (b) (c) (d)

Figure 8. Boxplot showing the results for the four constructs.

5. Findings and Discussion

The survey was performed to identify users' appreciation for mobile AR applications in the context of urban heritage. The following sections discuss the results from the survey, for each of the four parameters that were evaluated.

5.1. Comprehensibility

The values obtained for each question of comprehensibility indicated a positive evaluation. The responses to QC1 and QC5 had a median of 6.0 and a mean over 5.5. All the others had a median of 5.0 and a mean between 4.91 and 5.27. Item QC6 received the lowest, and item QC1 obtained the highest score. This showed that the application was easy to use and it was intuitive, but there were still small stability tracking issues that determined a slight flickering of the 3D content.

At the same time, the overall values obtained from the three surveys were very similar: Constanta (mean = 91.58), Rome (mean = 89.76), and Sulmona (mean = 90.23).

5.2. *Manipulability*

On perceived manipulability in the survey, QM4 and QM6 items had a median of 6.0 and a mean over 5.4. The others had a median of 5.0 and a mean between 5.08 and 5.44. While QM4 obtained the highest score, QM3 had the lowest median. The device was small, and the users did not have problems related to their arm comfort. Only small difficulties in holding the device while operating the application were reported.

Similar to the comprehensibility construct, very close values were obtained for all the experiments for the manipulability construct: Constanta (mean = 93.16), Rome (mean = 93.65), and Sulmona (mean = 94.53).

5.3. *Enjoyment*

For perceiving enjoyment, only QE3 had a median of 5.0, while the other three items had a median of 6.0. The mean ranged from 5.13 to 5.68. For this construct we obtained the highest values. QE1 achieved the highest score from all the questions that were proposed in the survey. The users liked to use the application. Good results were also obtained for QE3, which indicated that the objective set at the beginning was met. Only eight participants rated 5 in the Likert-scale for this question, and the others rated it 6 and 7.

The values obtained for this construct were: Constanta (mean = 94.87), Rome (mean = 96.19), and Sulmona (mean = 96.88).

5.4. *Usefulness*

On perceived usefulness, all four items had a median of 5.0 and a mean between 4.84 and 5.12. Item QU1 had the lowest average score. Some users suggested that the application might be improved with text to better reflect its purpose and to have more historical information about Ovid. Better scores were obtained for the items related to audio content, i.e., when Ovid's poem was heard in the background.

Perceived usefulness achieved the lowest score in the three cases: Constanta (mean = 88.55), Rome (mean = 86.43), and Sulmona (mean = 85.16).

Usefulness was the area of the survey where the users had the most variable results. The standard deviation ranged from 0.81 to 0.90. However, the highest value for standard deviation was obtained for item QM3 (SD = 0.94).

Briefly, we can conclude that participants enjoyed using the application and were mostly satisfied with its content. Although some historical data were missing, the application was easy to use, it was useful, and it confirmed the fact that the development of new interfaces for promoting CH is a sustainable action well-received by the public. Relevant photos and screenshots from the experiments are presented in Figure 9.

In Figure 9a2 the modelled house of Ovid is presented, placed on the top of the ruins of the real ancient house from Constanta. Figure 9c represents another modelled house, and the child version of Ovid walking.

The technology allowed registration of the model in different places accessible to tourists. It took into account their orientation, and adapted to their viewpoint. A wide variety of tourists can use the device (Figure 9a1,a3,b).

(a1) (a2) (a3)

(b) (c)

Figure 9. Mobile Augmented Reality (MAR) application used in (**a**) Constanta, (**b**) Rome, and (**c**) Sulmona.

6. Conclusions

Despite the fact that MAR applications show great potential in the CH area, the number of works that provide an analysis of user acceptance is relatively low. This work aimed to add value in this respect, and proposed a user study to assess participants' perception regarding the use of an MAR application with historical information related to CH. A prototype AR application was implemented as a subject for the study, and a questionnaire was used for collecting the data. Three sessions were conducted in three different locations, with participants from several countries. The questionnaire items, other than demographic information, intended to analyze the following four constructs: comprehensibility, manipulability, enjoyment, and usefulness.

The results show that the usability of the MAR application and the interaction was considered appropriate. Good results were obtained for all four constructs that were analyzed.

It should be noted that MAR technology has proved to increase the accessibility of tourists to intangible heritage. Tango technology allows users to create and save maps of the environment that are linked to the starting point of a scene, thus making it easy for developers to create AR applications.

The application meets the proposed objectives. Most of the users agreed that OvidAR had attractive content and met the usability requirements, providing a positive experience during the interaction. AR is a technology that will be more accessible in the upcoming years, and people will be more receptive in using it. In the field of CH, institutions and companies started to use this technology in order to provide more realistic experiences for tourists and museum visitors. In this regard, we think that AR technology represents a proper media for the sustainability of CH conservation and/or documentation.

Nevertheless, this study is not exhaustive. Our aim was not to obtain data that was statistically reliable, but to present a method that can be used sustainably to disseminate CH events at an international level. Future developments can include: improving tracking stability by fine tuning the Tango-related parameters, improving the 3D models to make them more realistic, or conducting further usability evaluations that involve more constructs and participants. That being said, we conclude that MAR could be one of the answers to various problems faced by the CH area in the 21st century, such as the lack of funding, the poor impact on cultural growth, and the weak cultural cohesion of neighboring countries.

Author Contributions: Conceptualization, R.G.B., C.-C.P. and D.-M.P.; methodology, G.D.V. and C.-C.P.; software, R.G.B. and G.D.V.; validation, D.-M.P., and M.C.; data curation, G.D.V.; writing—original draft preparation, R.G.B.; visualization, M.C.; supervision, M.D. and M.C.; project administration, M.D.

Funding: This paper is supported by European Union's Horizon 2020 research and innovation program under grant agreement No 692103, project eHERITAGE (Expanding the Research and Innovation Capacity in Cultural Heritage Virtual Reality Applications).

Acknowledgments: We also acknowledge the participants of the 6th Edition of the Summer School on Virtual Environments organized at Ovidius University of Constanta, Romania.

Conflicts of Interest: The authors declare no conflict of interest.

Appendix A

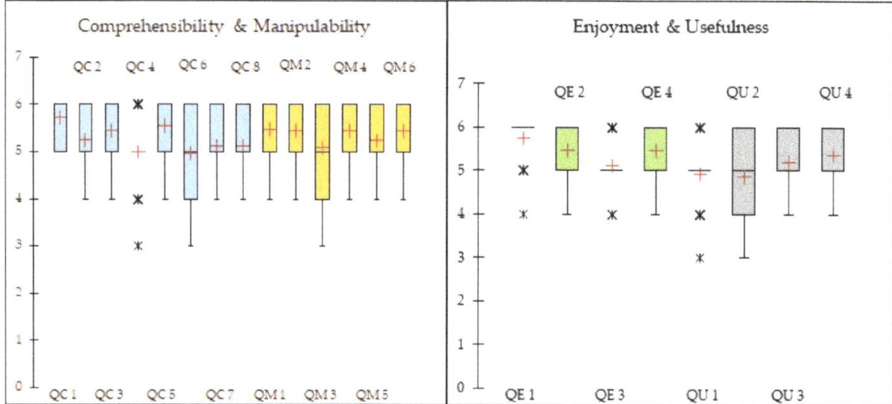

Figure A1. Boxplots showing the outcomes for the questions for the first experiment (Constanta).

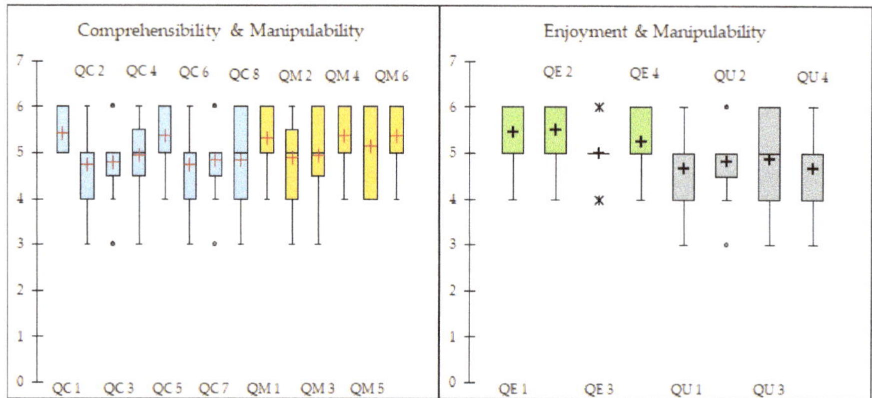

Figure A2. Boxplots showing the outcomes for the questions, for the second experiment (Rome).

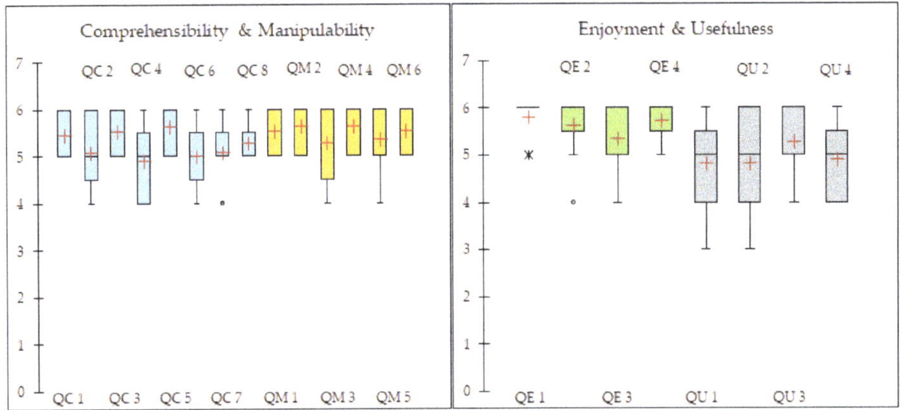

Figure A3. Boxplots showing the outcomes for the questions, for the third experiment (Sulmona).

References

1. Palombini, A. Storytelling and telling history. Towards a grammar of narratives for Cultural Heritage dissemination in the Digital Era. *J. Cult. Herit.* **2017**, *24*, 134–139. [CrossRef]
2. Madirov, E.; Absalyamova, S. The influence of Information Technologies on the Availability of Cultural Heritage. *Procedia-Soc. Behav. Sci.* **2015**, *188*, 255–258. [CrossRef]
3. Economou, M. *Use and Impact of Digital in Cultural Heritage: Insights from the Scottish Network of Digital Cultural Resources Evaluation, MW17: Museums and the Web 2017, Cleveland, OH, USA, 19–22 April 2017*; Museums and the Web 2017: Silver Spring, MD, USA, 2017. Available online: https://mw17.mwconf.org/paper/evaluating-impact-and-use-of-digital-cultural-resources-lessons-from-the-scotdigich-network/ (accessed on 17 October 2018).
4. Creed, C.; Sivell, J.; Sear, J. Multi-Touch Tables for Exploring Heritage Content in Public Spaces. In *Visual Heritage in the Digital Age*; Ch'ng, E., Gaffney, V., Chapman, H., Eds.; Springer London: London, UK, 2013; pp. 67–90.
5. Thwaites, H. Digital Heritage: What Happens When We Digitize Everything? In *Visual Heritage in the Digital Age*; Ch'ng, E., Gaffney, V., Chapman, H., Eds.; Springer London: London, UK, 2013; pp. 327–348.
6. Malpas, J. Cultural heritage in the age of new media. In *New Heritage: New Media and Cultural Heritage*; Kalay, Y.E., Kvan, T., Affleck, J., Eds.; Routledge: London, UK, 2008; pp. 13–26.
7. Tahoon, D.M.A. Case Study: Pyramids and Giza Plateau cultural heritage site, Egypt. In Proceedings of the 1st BUE Annual Conference & Exhibition (BUE ACE1), Cairo, Egypt, 7–9 November 2016. Available online: https://pdfs.semanticscholar.org/5365/cce55e220ea8ee1034d5b70c170e890fabaf.pdf?_ga=2.66583277.2118423092.1550735871-1930073487.1544169653 (accessed on 24 October 2018).
8. Boldrer, F. Recalling Ovid 2000 years later between lights and shadows: Memory and rework of a Latin poet in the Italian literature of the 20th century. In Proceedings of the Classical antiquity & memory in 19th–20th–21st century International conference, Bonn, Germany, 28–30 September 2017. Available online: http://memory.strikingly.com/blog/recalling-ovid-2-000-years-later-between-lights-and-shadows (accessed on 17 October 2018).
9. Wise, C.N. Banished to the Black Sea: Ovid's poetic transformations in Tristia 1.1. Master's Thesis, Georgetown University, Washington, DC, USA, 16 October 2014.
10. Currie, H.M. Ovid's Personality. *Class. J.* **1964**, *59*, 145–155.
11. Noh, Z.; Sunar, M.S.; Pan, Z. A Review on Augmented Reality for Virtual Heritage System. In *Learning by Playing. Game-based Education System Design and Development: 4th International Conference on E-Learning and Games, Edutainment 2009, Banff, Canada, 9–11 August 2009*; Springer: Berlin, Germany, 2009.
12. Wu, H.-K.; Lee, S.W.-Y.; Chang, H.-Y.; Liang, J.-C. Current status, opportunities and challenges of augmented reality in education. *Comput. Educ.* **2013**, *62*, 41–49. [CrossRef]

13. Chung, N.; Han, H.; Joun, Y. Tourists' intention to visit a destination: The role of augmented reality (AR) application for a heritage site. *Comput. Hum. Behav.* **2015**, *50*, 588–599. [CrossRef]
14. Garrido, R.M.; Jimenez, D.V.; Baldiris, S.; Fabregat, R. "Social Heritage" Augmented Reality Application to Heritage Education. In Proceedings of the Second International Conference on Augmented and Virtual Reality, VR 2015, Lecce, Italy, 31 August–3 September 2015. Available online: https://www.springer.com/cda/content/document/cda_downloaddocument/9783319228877-c2.pdf?SGWID=0-0-45-1524302-p177647007 (accessed on 17 September 2018).
15. Kečkeš, A.L.; Tomičić, I. Augmented Reality in Tourism—Research and Applications Overview. *Interdiscip. Descr. Complex Syst.* **2017**, *15*, 157–167. [CrossRef]
16. tom Dieck, M.C.; Jung, T.H. Value of augmented reality at cultural heritage sites: A stakeholder approach. *J. Destin. Mark. Manag.* **2017**, *6*, 110–117. [CrossRef]
17. Velázquez, F.D.C.; Méndez, G.M. Augmented Reality and Mobile Devices: A Binominal Methodological Resource for Inclusive Education (SDG 4). An Example in Secondary Education. *Sustainability* **2018**, *10*, 3446. [CrossRef]
18. Martins, J.; Gonçalves, R.; Branco, F.; Barbosa, L.; Melo, M.; Bessa, M. A multisensory virtual experience model for thematic tourism: A Port wine tourism application proposal. *J. Destin. Mark. Manag.* **2017**, *6*, 103–109. [CrossRef]
19. Berryman, D.R. Augmented reality: A review. *Med Ref. Serv. Q.* **2012**, *31*, 212–218. [CrossRef]
20. Clemens, A.; Lukas, G.; Raphael, G.; Tobias, L.; Alessandro, M.; Dieter, S.; Daniel, W. The History of Mobile Augmented Reality. Inst. for Computer Graphics and Vision, Graz University of Technology, Austria. 11 November 2015. Available online: https://arxiv.org/pdf/1505.01319.pdf (accessed on 17 September 2018).
21. Tomaštík, J.; Saloň, Š.; Tunák, D.; Chudý, F.; Kardoš, M. Tango in forests—An initial experience of the use of the new Google technology in connection with forest inventory tasks. *Comput. Electron. Agric.* **2017**, *141*, 109–117. [CrossRef]
22. Gjøsæter, T. Affordances in Mobile Augmented Reality Applications. *Int. J. Interact. Mob. Technol.* **2014**, *8*, 45–55. [CrossRef]
23. Chatzopoulos, D.; Bermejo, C.; Huang, Z.; Hui, P. Mobile Augmented Reality Survey: From Where We Are to Where We Go. *Ieee Access* **2017**, *5*, 6917–6950. [CrossRef]
24. Bostanci, E.; Kanwal, N.; Clark, A.F. Augmented reality applications for cultural heritage using Kinect. *Hum.-Cent. Comput. Inf. Sci.* **2015**, *5*, 20. [CrossRef]
25. Mota, J.M.; Ruiz-Rube, I.; Dodero, J.M.; Arnedillo-Sánchez, I. Augmented reality mobile app development for all. *Comput. Electr. Eng.* **2017**, *65*, 250–260. [CrossRef]
26. Wagner, D.; Schmalstieg, D. History and Future of Tracking for Mobile Phone Augmented Reality. In Proceedings of the 2009 International Symposium on Ubiquitous Virtual Reality, Gwangju, South Korea, 8–11 July 2009. Available online: https://ieeexplore.ieee.org/iel5/5232241/5232242/05232244.pdf (accessed on 14 August 2018).
27. Mekni, M.; Lemieux, A. Augmented reality: Applications, challenges and future trends. In Proceedings of the 13th International Conference on Applied Computer and Applied Computational Science (ACACOS '14), Kuala Lumpur, Malaysia, 23–25 April 2014; pp. 205–214. Available online: https://www.semanticscholar.org/paper/Augmented-Reality-%3A-Applications-%2C-Challenges-and-Mekni-Lemieux/a77e78e46a0cdfdc191e732e23da3deac57fd366 (accessed on 12 September 2018).
28. Yovcheva, Z.; Buhalis, D.; Gatzidis, C. Overview of Smartphone Augmented Reality Applications for Tourism. *e-Rev. Tour. Res. (Ertr)* **2012**, *10*, 63–68.
29. Han, J.-G.; Park, K.-W.; Ban, K.-J.; Kim, E.-K. Cultural Heritage Sites Visualization System based on Outdoor Augmented Reality. *Aasri Procedia* **2013**, *4*, 64–71. [CrossRef]
30. Dünser, A.; Billinghurst, M.; Wen, J.; Lehtinen, V.; Nurminen, A. Exploring the use of handheld AR for outdoor navigation. *Comput. Graph.* **2012**, *36*, 1084–1095. [CrossRef]
31. Larabi, S. Augmented Reality for Mobile Devices: Textual Annotation of Outdoor Locations. In *Augmented Reality and Virtual Reality: Empowering Human, Place and Business*; Jung, T., tom Dieck, M.C., Eds.; Springer International Publishing: Manchester, UK, 2018; pp. 353–362. Available online: https://link.springer.com/chapter/10.1007/978-3-319-64027-3_24 (accessed on 13 September 2018).

32. Duguleana, M.; Brodi, R.; Girbacia, F.; Postelnicu, C.; Machidon, O.; Carrozzino, M. Time-Travelling with Mobile Augmented Reality: A Case Study on the Piazza DeiMiracoli. In Digital Heritage. Progress in Cultural Heritage: Documentation, Preservation, and Protection: 6th International Conference, EuroMed 2016, Nicosia, Cyprus. 2016. Available online: https://link.springer.com/chapter/10.1007/978-3-319-48496-9_73 (accessed on 25 September 2018).
33. Puyuelo, M.; Higón, J.L.; Merino, L.; Contero, M. Experiencing Augmented Reality as an Accessibility Resource in the UNESCO Heritage Site Called "La Lonja", Valencia. *Procedia Comput. Sci.* **2013**, *25*, 171–178. [CrossRef]
34. Ekonomou, T.; Vosinakis, S. Mobile Augmented Reality games as an engaging tool for cultural heritage dissemination: A case study. *Sci. Cult.* **2018**, *4*, 97–107.
35. Younes, G.; Kahil, R.; Jallad, M.; Asmar, D.; Elhajj, I.; Turkiyyah, G.; Al-Harithy, H. Virtual and augmented reality for rich interaction with cultural heritage sites: A case study from the Roman Theater at Byblos. *Digit. Appl. Archaeol. Cult. Herit.* **2017**, *5*, 1–9. [CrossRef]
36. Sekhavat, Y.A. KioskAR: An Augmented Reality Game as a New Business Model to Present Artworks. *Int. J. Comput. Games Technol.* **2016**, *2016*, 7690754. Available online: https://www.hindawi.com/journals/ijcgt/2016/7690754/ (accessed on 30 July 2018). [CrossRef]
37. Hammady, R.; Ma, M.; Temple, N. Augmented Reality and Gamification in Heritage Museums. In *Serious Games: Second Joint International Conference, JCSG 2016, Brisbane, QLD, Australia, 26–27 September 2016*; Springer International Publishing: Manchester, UK, 2016; pp. 181–187.
38. Madsen, J.B.; Madsen, C.B. Handheld Visual Representation of a Castle Chapel Ruin. *J. Comput. Cult. Herit* **2015**, *9*, 1–18. [CrossRef]
39. Pedersen, I.; Gale, N.; Mirza-Babaei, P.; Reid, S. More than Meets the Eye: The Benefits of Augmented Reality and Holographic Displays for Digital Cultural Heritage. *J. Comput. Cult. Herit.* **2017**, *10*, 1–15. [CrossRef]
40. Chung, N.; Lee, H.; Kim, J.-Y.; Koo, C. The Role of Augmented Reality for Experience-Influenced Environments: The Case of Cultural Heritage Tourism in Korea. *J. Travel Res.* **2018**, *57*, 1–17. [CrossRef]
41. Damala, A.; Hornecker, E.; van der Vaart, M.; van Dijk, D.; Ruthven, I. The Loupe: Tangible Augmented Reality for learning to look at ancient greek art. *Mediterr. Archaeol. Archaeom.* **2016**, *16*, 73–85.
42. Deliyiannis, I.; Papaioannou, G. Augmented Reality for archaeological environments on mobile devices: A novel open framework. *Mediterr. Archaeol. Archaeom.* **2014**, *14*, 1–10.
43. Martínez-Graña, A.; González-Delgado, J.Á.; Ramos, C.; Gonzalo, J.C. Augmented Reality and Valorizing the Mesozoic Geological Heritage (Burgos, Spain). *Sustainability* **2018**, *10*, 4616.
44. Chen, W. Historical Oslo on a Handheld Device—A Mobile Augmented Reality Application. *Procedia Comput. Sci.* **2014**, *35*, 979–985. [CrossRef]
45. Cranmer, E.E.; tom Dieck, M.C.; Jung, T. How can Tourist Attractions Profit from Augmented Reality? In *Augmented Reality and Virtual Reality: Empowering Human, Place and Business*; Jung, T., tom Dieck, M.C., Eds.; Springer International Publishing: Manchester, UK, 2018; pp. 21–32.
46. Bekele, M.K.; Pierdicca, R.; Frontoni, E.; Malinverni, E.S.; Gain, J. A Survey of AugmentedVirtual, and Mixed Reality for Cultural Heritage. *J. Comput. Cult. Herit.* **2018**, *11*, 7:1–7:36. [CrossRef]
47. Kysela, J.; Štorková, P. Using Augmented Reality as a Medium for Teaching History and Tourism. *Procedia-Soc. Behav. Sci.* **2015**, *174*, 926–931. [CrossRef]
48. Boboc, R.G.; Girbacia, F.; Duguleana, M.; Tavčar, A. A handheld Augmented Reality to revive a demolished Reformed Church from Brasov. In *Proceedings of the VRIC 2017—19th ACM Virtual Reality International Conference, Laval, France, 22–24 March 2017*; ACM: New York, NY, USA, 2017. Available online: https://dl.acm.org/citation.cfm?id=3110311 (accessed on 15 October 2018).
49. Tan, K.L.; Lim, C.K. Digital heritage gamification: An augmented-virtual walkthrough to learn and explore historical places. In Proceedings of the 2nd International Conference on Applied Science and Technology 2017 (ICAST'17), Langkawi, Kedah, Malaysia, 3–5 April 2017. Available online: https://aip.scitation.org/doi/pdf/10.1063/1.5005472?class=pdf (accessed on 16 October 2018).
50. Roussou, M. Virtual Heritage: From the Research Lab to the Broad Public. In *Proceedings of the VAST 2000 Euroconference, Arezzo, Italy, 24–25 November 2000*; Oxford, A., Ed.; Archaeopress: Oxford, UK, 2000; pp. 93–100.
51. ICH. Available online: https://ich.unesco.org/en/what-is-intangible-heritage-00003 (accessed on 5 December 2017).

52. Hincapie, M.; Diaz, C.; Zapata, M.; Mesias, C. Methodological Framework for the Design and Development of Applications for Reactivation of Cultural Heritage: Case Study Cisneros Marketplace at Medellin, Colombia. *J. Comput. Cult. Herit* **2016**, *9*, 1–24. [CrossRef]
53. Gao, Y.; Wu, X.; Zhang, Y.; Shen, Y.; Yang, X. The value evaluation system and applications of digital cultural resources. In Proceedings of the 6th International Conference on Information Management, Innovation Management and Industrial Engineering, Xi'an, China, 23–24 November 2013. Available online: https://download.atlantis-press.com/article/20591.pdf (accessed on 18 September 2018).
54. Santos, M.E.C.; Taketomi, T.; Sandor, C.; Polvi, J.; Yamamoto, G.; Kato, H. A usability scale for handheld augmented reality. In *Proceedings of the 20th ACM Symposium on Virtual Reality Software and Technology, Edinburgh, Scotland, 11–13 November 2014*; ACM: New York, NY, USA, 2017; pp. 167–176.
55. Ko, S.M.; Chang, W.S.; Ji, Y.G. Usability Principles for Augmented Reality Applications in a Smartphone Environment. *Int. J. Hum. Comput. Interact.* **2013**, *29*, 501–515. [CrossRef]
56. Yáñez Gómez, R.; Cascado Caballero, D.; Sevillano, J.-L. Heuristic Evaluation on Mobile Interfaces: A New Checklist. *Sci. World J.* **2014**, *19*, 434326.
57. Tsai, T.-H.; Chang, H.-T.; Yu, M.-C.; Chen, H.-T.; Kuo, C.-Y.; Wu, W.-H. Design of a Mobile Augmented Reality Application: An Example of Demonstrated Usability. In *Universal Access in Human-Computer Interaction. Interaction Techniques and Environments: 10th International Conference, Toronto, ON, Canada, 17–22 July, 2016*; Springer International Publishing: Basel, Switzerland, 2017.
58. Schaeffer, S.E. Usability Evaluation for Augmented Reality. Department of Computer Science, University of Helsinki, 2014. Available online: https://core.ac.uk/download/pdf/33725105.pdf (accessed on 20 February 2019).
59. Santos, M.E.C.; Polvi, J.; Taketomi, T.; Yamamoto, G.; Sandor, C.; Kato, H. Toward Standard Usability Questionnaires for Handheld Augmented Reality. *Ieee Comput. Graph. Appl.* **2015**, *35*, 66–75. [CrossRef] [PubMed]
60. Han, D.-I.; Jung, T. Identifying Tourist Requirements for Mobile AR Tourism Applications in Urban Heritage Tourism. In *Augmented Reality and Virtual Reality: Empowering Human, Place and Business*; Jung, T., tom Dieck, M.C., Eds.; Springer International Publishing: Manchester, UK, 2018; pp. 3–20.
61. Javornik, A. Directions for Studying User Experience with Augmented Reality in Public. In *Augmented Reality and Virtual Reality: Empowering Human, Place and Business*; Jung, T., tom Dieck, M.C., Eds.; Springer International Publishing: Manchester, UK, 2018; pp. 199–210.
62. Střelák, D.; Škola, F.; Liarokapis, F. Examining User Experiences in a Mobile Augmented Reality Tourist Guide. In Proceedings of the 9th ACM International Conference on Pervasive Technologies Related to Assistive Environments, Corfu Island, Greece, 29 June–1 July 2016; pp. 1–8. Available online: https://www.fi.muni.cz/~{}liarokap/publications/PETRA2016.pdf (accessed on 29 September 2018).
63. Rese, A.; Baier, D.; Geyer-Schulz, A.; Schreiber, S. How augmented reality apps are accepted by consumers: A comparative analysis using scales and opinions. *Technol. Forecast. Soc. Chang.* **2017**, *124*, 306–319. [CrossRef]
64. Georgiou, Y.; Kyza, E.A. The development and validation of the ARI questionnaire: An instrument for measuring immersion in location-based augmented reality settings. *Int. J. Hum. Comput. Stud.* **2017**, *98*, 24–37. [CrossRef]
65. Haugstvedt, A.C.; Krogstie, J. Mobile augmented reality for cultural heritage: A technology acceptance study. In *2012 IEEE International Symposium on Mixed and Augmented Reality (ISMAR), Georgia Tech, Atlanta, USA, 5–8 November 2012*; IEEE Computer Society: Washington, DC, USA, 2017.
66. Tango. Available online: https://developers.google.com/tango/ (accessed on 27 December 2017).
67. 3DS MAX. Available online: https://www.autodesk.com/products/3ds-max/overview# (accessed on 4 January 2018).
68. Makehuman. Available online: http://www.makehumancommunity.org/ (accessed on 2 January 2018).
69. Blender. Available online: https://www.blender.org/ (accessed on 2 January 2018).
70. Mixamo. Available online: https://www.mixamo.com/#/ (accessed on 2 January 2018).
71. Unity. Available online: https://unity3d.com/ (accessed on 3 January 2018).

© 2019 by the authors. Licensee MDPI, Basel, Switzerland. This article is an open access article distributed under the terms and conditions of the Creative Commons Attribution (CC BY) license (http://creativecommons.org/licenses/by/4.0/).

Article

Archaeometry's Role in Cultural Heritage Sustainability and Development

Ioannis Liritzis [1,2,*] and Elena Korka [3]

1 Key Research Institute of Yellow River Civilization and Sustainable Development & College of Environment and Planning, Henan University, Kaifeng 475001, China
2 Laboratory of Archaeometry, Department of Mediterranean Studies, University of the Aegean, 85100 Rhodes, Greece
3 Honorary G. D. of Antiquities and Cultural Heritage, Ministry of Culture and Sports, 11525 Athens, Greece; ekorka@otene.gr
* Correspondence: liritzis@rhodes.aegean.gr or ioliritzis@gmail.com; Tel.: +30-6932-275757

Received: 27 January 2019; Accepted: 18 March 2019; Published: 3 April 2019

Abstract: The interdisciplinary field of archaeometry covers a wide range of subject categories and disciplines in relation to science and humanities. It is a well-established academic field of study and accredited part of higher education. Since its inception, the nomenclature designation of archaeometry signifies the appropriate methodology applied to archaeological materials and questions emerging from this field, regarding monuments, artifacts, and the reconstruction and management of landscape bearing cultural assets. The measurements of tangible culture denote significant information, such as chronology, authenticity, technology, characterization, provenance, discovering buried antiquities, ancient-day life activities, and three-dimensional (3D) reconstructions and modelling; furthermore, proxy data collected from environmental dynamic non-liner perturbations, which link local ecosystems with dwellings, are gathered by academia to study the past. The traditional rooting signifies the cultural legacies of people, which define the human desire and the confidence of memory and future trends. Beyond the mere study of the past, archaeometry's role increasingly proves affinity to prosperity, if properly managed. The major archaeometrical contributions in cultural heritage and archaeology in general are reviewed herein, and we present the policies that could develop archaeometrical data into a sustainable stage of local, regional, and national economic development. The United Nations Educational, Scientific, and Cultural Organization (UNESCO) conventions for the documentation and protection of cultural heritage via new technologies and archaeometry are reviewed and connected to development strategies and sustainable development goals.

Keywords: archaeological sciences; buried antiquities; prospection; documentation; digitalization; management; UNESCO; cultural tourism; economic values; new technologies; information

---

1. Introduction

1.1. Cooperation between Scientific and Humanities Disciplines

The relationship that links the interdisciplinary field of archaeometry with the concept of sustainability, as well as the way these two fields cooperate, alerted authorities and institutions of the public and private sector.

Archaeometry, or archaeological sciences, is the application of natural sciences to solving archaeological or cultural heritage problems. This interdisciplinary subject combines many disciplines such as physics, chemistry, geology, astronomy, mathematics, geophysics, geology, biology, informatics, etc. The main objective of the archaeological sciences is the comprehension of past societies through the study of natural sciences and new technologies of material culture [1,2]. By using the latest technological achievements in a combined and flexible manner, archaeometry is capable of detecting and extracting hidden information from heritage objects, which may comprise evidence that can determine the object's origin, construction period or technique, etc. The strategy consists of the use of technological achievements through areas of science, and the creation of innovative models, protocols, and web platforms, as well as, given direct or indirect options to legacy, the enhancement of cultural values and the creation of ecosystems and cultural heritage services. Hence, the cooperation between scientific and historic disciplines is now prevalent, only beginning in the 1960s, due to what seemed to be a lack of mutual understanding. This allowed bridging the gap between the fields of physics and chemistry on the one side and the needs and problems of archaeologists or historians on the other.

Archaeology today is an extremely diverse discipline, encompassing interest groups, which focus on distinct periods, regions, theoretical perspectives, and methodological techniques. Although this diversity is a positive development, bringing to light problematic issues due to miscommunication from different components of the discipline [3,4], or the development of epistemological arguments to coin progress perplex the archaeological–archaeometrical connection, and each new term is triggered by archaeological theorists and archaeological scientists [5,6].

At any rate, archaeometry applies a wide range of scientific techniques for the study of both heritage objects and historic/prehistoric events. Such techniques are based on basic principles and phenomena of physics (e.g., radioactivity, electricity and magnetism, atomic theory, electromagnetic radiation), chemistry (e.g., diffusion, reactions, melting, affinities), geology (e.g., geomorphology, sedimentology, petrology), geophysics (e.g., paleoclimate, geomagnetic field, atmosphere), astronomy (e.g., solar system, celestial mechanics), and mathematics (e.g., algorithms, statistics). The important contribution of archaeometry to cultural heritage and archaeology for most of the years of recent developments remained known either to a few open-minded archaeologists or to a narrow group of academia.

Eventually, understanding of the usefulness of cultural tourism and the valorization of cultural heritage assets, supported by United Nations Educational, Scientific, and Cultural Organization (UNESCO) and European Union (EU) projects placed archaeometry as one of the top priorities for sustainability on national and regional levels [7–9]. However, when overcoming the past theoretical approaches using archaeological dialogues, one should consider the perpetually accredited scientifically holistic approach involving theory and practice, covering processual and post-processual approaches for the ever-developing archaeology (PASHA). This approach was designed to account for new diversity and virtual reconstruction in the field, integrating micro and macro perspectives from human life stories (bio-archaeology, ancient a-DNA, and isotopes) to their larger social, cultural, and environmental framework (travels, interaction, networks, major genetic shifts, paleoclimates, and paleoenvironments). This, in turn, led to the introduction of new scientific branches emerging from science, technology, engineering, and mathematics (STEM), as well as the addition of arts (STEAM) [10,11] and, more recently, the addition of culture (STEMAC) [12]. The evolution of scientific techniques opens new doors for cultural and archaeological evolution [13–15], the details of which are outlined here in Sections 2 and 3.

Today, humanities and archaeology, in particular, need to engage in discussing the implications of the expanding frontier of knowledge, ranging from archaeo-genetics to the diet and mobility of individuals, incorporating demography and sustainability in the long-term. We can now revisit museums and archaeological sites and select materials in order to reconstruct the whole life stories of individuals, in addition to their diet, mobility, and close family stories, as well as their larger genetic family stories from prehistory until the present. Sedimentological and geoarchaeological data can be used to reconstruct the ancient environment. Thus, a new door is now open to previously hidden knowledge, which will once again reduce the amount of qualified guessing, thereby refining and redefining theory and interpretation. Moreover, these new techniques bring archaeology much closer to public awareness.

All these new methodologies allow the development of a new strategy for connecting our more complete knowledge to a sustainable level for present society, with economic, political, and social implications. The present paper focuses on archaeometry's role in cultural heritage sustainability and development. Hence, the triptych of archaeometry, added value, and sustainability is approached for the first time. At this point in time, one can take note of Oscar Wilde verse of the 19th century, "nowadays, people know the price of everything and the value of nothing" (from *The Portrait of Dorian Gray*).

Sustainable development is a balance between society, the environment, and the economy, which, in turn, interact and/or grow with specific groups of people, companies, and stakeholders, for whom archaeometry can play a significant yet neglected role (Figure 1). Thus, the objectives and working policy of the present paper address sustainable development through the lens of archaeometry.

Figure 1. The holistic approach of sustainable development to the environment, the economy, and society, where archaeometry is encompassed and potentially may emerge from within the proper handling of each named agent.

The novel development of cultural tourism due to archaeometric contribution implies by-product growth in the field of cultural heritage such as public names, bilingualism, protection and care of local culture and traditions [15]. The realization of current methods presumes disseminating archaeometric knowledge by:

- Virtual Reality in museums, documentaries with 3D goggles, etc.
- Information Communication Technology (ICT), computer and mobile applications (educational games, interactive tools, etc.).
- Summer schools, volunteer programs, workshops, community-led programs, participatory archaeology.

Therefore, heritage sites, especially when properly documented with scientific methods, are important cultural attractions in the World, with adults visiting a site, participation of children due to family and school visits, inbound tourism, and especially international tourists, thus extending economic benefit beyond the discovered archaeological sites and cultural heritage monuments [16,17]. Having said that above, the present article is a first-time written concept, fully documented, and detailed in cases, to cover this wide interdisciplinary field. As such, and, due to the diversified nature of the journal, the reader (s), owe to be fully aware and the text must be digestive, to serve its purpose.

1.2. Sustainability of Cultural Heritage through New Technologies: A Critical Evaluation

Under the trend of technological progress, the methods used for interpretation, documentation and promotion of cultural heritage have evolved significantly over the last decades [18]. Modern technology and the digital revolution have fundamentally altered the way in which cultural images and applications are produced, distributed and accessed. The evolving field of using information and communication technology (ICT) for the benefit of cultural heritage has produced in a short period of time a wide range of applications for scholarly research, public information and marketing as well as management of cultural sites. It has also provided new tools for cultural heritage preservation, as well as access, interaction, and knowledge-sharing. In terms of documentation and interpretation, in particular, the contribution of sciences such as physics, chemistry, biology, earth sciences, mathematics, statistics, and computing are regularly used to enlighten specific questions in regard to the archaeological data. This multidisciplinary field engulfs the discipline of Archaeometry, which is a combination of archaeology by ultimate aim (αρχαον) and natural science by approach (μέτρον), thus introducing a new scope to the solution of questions [19]. Archaeometry techniques can document and communicate ancient artifacts, places and practices faster, in greater detail and with better perception amongst a broader public than ever before.

The archaeological work, which is by nature costly in time and money, is assisted by efficient methods which, in most cases, require a small amount of equipment and less labor expenses (For more information about non-intrusive archaeological survey techniques see Imogen Burrelli "What are non-intrusive archaeological survey techniques, and how they are used to archaeological advantages during excavations" in The Post Hole, Issue 36, March/April 2014, p. 8–13.). This is in accordance with the provisions of the European Convention on the Protection of the Archaeological Heritage (Valletta-Convention, Article 3), which states that non-destructive investigation methods should be applied wherever possible (the European Convention for the Protection of the Archaeological Heritage (revised) replaced and updated the original London Convention of 1969. It reflected the change in the nature of threats to the various types of archaeological heritage It established a pool of new basic legal standards for Europe, to be met by national policies for the protection of archaeological assets as sources of scientific and documentary evidence, in line with the principles of integrated conservation.).

Within UNESCO (The United Nations Educational, Scientific and Cultural Organization (UNESCO) is a specialized agency of the United Nations (UN) whose main objective is to contribute to peace and security by promoting international collaboration through educational, scientific, and cultural reforms in order to increase universal respect for law and human rights along with fundamental freedom proclaimed in the United Nations Charter http://www.unesco.org.) the specialized cultural agency of the United Nations, systematic efforts are made to ensure that all countries benefit from scientific and technological progress and innovation. This approach is in consistence with the United Nations 2030 Agenda for Sustainable Development, with its 17 Sustainable Development Goals. Within this framework the Organization promotes ICT that address specific challenges for the preservation of cultural properties, such as the impact of climate change and natural hazards. The 2005 Convention on the Protection and Promotion of the Diversity of Cultural Expressions recognizes that the protection, promotion and maintenance of cultural diversity are an essential requirement for sustainable development for the benefit of present and future generations and therefore promotes the use of new technologies and encourages partnerships to enhance information sharing and cultural understanding. In addition to the various investigations that have been carried out to date, the Parties to the 2005 Convention are gradually including digital applications in their quadrennial periodic reports, as they move forward in drafting operational guidelines exclusively focused on this topic.

The International Council on Monuments and Sites (ICOMOS) [20] on the other hand, which is the competent professional network that works for the conservation and protection of cultural heritage sites around the world, as well as the official technical consultant to UNESCO, is also promoting the use of modern technologies such as digital image processing, digital orthophoto production, terrestrial laser scanning, and 3D model processing in order to fully document the existing architectural heritage. Therefore, in case a monument is destroyed or damaged, and in order for the cultural managers to reconstruct, at least digitally, it is necessary for a full scanning project to have taken place beforehand. Although the subject of an actual reconstruction is still debatable among scholars the perspective of being able to retrieve, with the use of modern technology, data concerning a ruined or extant monument with the use of modern technology is undoubtedly worthwhile. The introduction of digital technology in the field of preservation and documentation of monuments has engaged extensive conversation on principles of reconstruction of cultural heritage and has contributed in somewhat shifting the overall negative position towards it. According to the Burra Charter (Australia), reconstruction can only be accomplished when a site is incomplete through damage or alteration, and enough evidence exists to reproduce the earlier state of the fabric of the monument. Here is where ICT come into the picture. The information system generated by digital photogrammetry and geographical information systems (GIS) leads to the efficient use of data, in a cost-efficient way, which is very valuable for saving cultural properties. The loss of many monuments due to armed conflicts raging across Syria, Iraq, and other countries in the Middle East has spurred the international scientific community and many cultural organizations to apply all available technologies, such as 3D scans, drones, lid methodology, and satellites. Even individuals were encouraged to post photographs of monuments and sites before destruction, thus providing valuable tools to prevent the possible permanent loss of data for cultural heritage.

The application of ICT definitely has an impact on several aspects of heritage preservation and enhancement management policies. In earlier years, both UNESCO and ICOMOS were generally opposed to reconstructions, following the provisions of the Venice Charter, with very few exceptions (for example the reconstruction of the historic center of Warsaw in 1980s and the Mostar Bridge). The theoretical framework for adopting new technologies and archaeometry in the field of culture protection and preservation has been formally introduced within ICOMOS with the approval of the Interpretation and Presentation of the Cultural Heritage Sites Charter, also known as Ename, which was the first international text ratified by ICOMOS to recognize the importance of using virtual reconstructions in the field of archaeological heritage. Among others, in Article 4.2 it recommends that *"Visual reconstructions, whether by artists, architects, or computer modelers, should be based upon detailed and systematic analysis of environmental, archaeological, architectural, and historical data, including analysis of written, oral and iconographic sources, and photography. The information sources on which such visual renderings are based should be clearly documented and alternative reconstructions based on the same evidence, when available, should be provided for comparison"*. Through this text archaeometry became officially the prime collaborator of cultural sciences.

Furthermore, a special Committee has been established, within ICOMOS, the International Scientific Committee on Interpretation and Presentation of Cultural Sites, with the task to study the evolving technologies and techniques of data interpretation and presentation, to evaluate their potential to enrich contemporary scientific discourse and to focus on the experiential dimension of visits to cultural heritage sites, particularly by means of various media and methods of public communication.

In 2009, the London Charter on the Computer-based Visualization of Cultural Heritage was signed and approved and until today it constitutes the most important document of the international community in the field of cultural heritage and new technologies.

The Charter of Krakow (2000, Article 5) on the Principles for the Conservation and Restoration of Built Heritage, included for the first time, as Article 5, a specific recommendation for the use of new technologies in the field of archaeological heritage: *"In the protection and public presentation of archaeological sites, the use of modern technologies, databanks, information systems and virtual presentation techniques should be promoted"*. This addition, unprecedented in other previous charters, marks an important turning point in the use of information technology as a tool in the regular work of conservation and presentation of archaeological heritage. Consequently, it should be considered as an important milestone in the history of virtual archaeology, leading through these new tools to the sustainability of cultural resources.

Several issues regarding the implementation of modern technologies to assist archaeological research have been thoroughly examined and evaluated by experts and cultural organizations, in an effort, to draw an effective framework with respect to the cultural heritage ethics. Under this perspective, for example, the European Archaeological Council (EAC), realizing the new possibilities that digital technologies are opening up for the promotion and presentation of archaeological research and investigation, EAC has organized in 2016, on the occasion of its annual meeting, a conference titled "Digital Futures: Archaeology in Europe" with the scope to explore developments in digital technologies and to consider how they may become embedded in general archaeological policy and practice over the next few years. The economic growth from new technologies implemented via the archaeometry applications in cultural heritage is at high stake if properly managed and becomes a new way to sustainable Development. Additionally, culture is *"a drive and a catalyst of sustainable development" (UNESCO).*

2. How Archaeometry Works for Sustainability

Archaeometry inheres two roles and needs reformations towards sustainability and development: the indirect (related to intangible culture) and the direct (tangible) [21]. That is a revisit of existing monuments and artifacts with new methods and techniques from physical sciences (Figure 2).

Figure 2. Some monuments and artifacts studied by archaeological sciences methods.

This requires proper implementation of cultural politics accompanied by strategies, to fulfill the provisions of Article 5 of the London Charter: "*Strategies should be planned and implemented to ensure the long-term sustainability of cultural heritage-related computer-based visualization outcomes and documentation, in order to avoid loss of this growing part of human intellectual, social, economic and cultural heritage*".

Several questionnaires to the public regarding attractive presentation of museum exhibits and cultural monuments, point to the urgent need for new technologies to show up archaeological and archaeometrical data. For example, comparing the user response before and after virtual reality (VR) demo, there is a clear sign of increased level of interest and cultural heritage awareness with increase in likelihood that the user would tell others about the sites after viewing the VR demo [22,23].

The means of reforming are:

(A) Applied informatics, geoinformatics (computer-human interaction, IT, computer sciences, geographical information systems (GIS) [24].

(B) Documentation of museum objects [25,26].

(C) Management of archaeological sites and museum tasks concerning cultural heritage (improve in quality, accessibility, reclaiming of digital content) [27–29].

(D) Multimedia technologies, digital reconstructions, simulation of artifacts and monuments, and 3D VR. Combining never-before-published aerial and remote sensing photography and high-definition, 3D computer reconstructions with clear, informative descriptions of the various sites, innovative descriptions of the various terrestrial or underwater sites. These ways use the latest scientific and archaeological tools to visualize the wonder of the past [30–35].

(E) Discovering new buried monuments, making cultural parks, and developing cultural tourism [36,37].

(F) Exhibition, dissemination, pedagogy fostering of archaeometrical results, made popular to the society via digital means [38,39].

The development targets derived thereof are on:

- Cultural tourism;
- Social coherence;
- Employment; and
- Development of innovative applications (digital).

The "tourism" has a multidimensional face and "cultural tourism" is a vital component, as each country has a cultural rooting which fits as cornerstone in the global cultural unity (Figure 3). Using techniques borrowed from the entertainment industry, more and more archaeologists are boosting their imaginations and insights with virtual worlds.

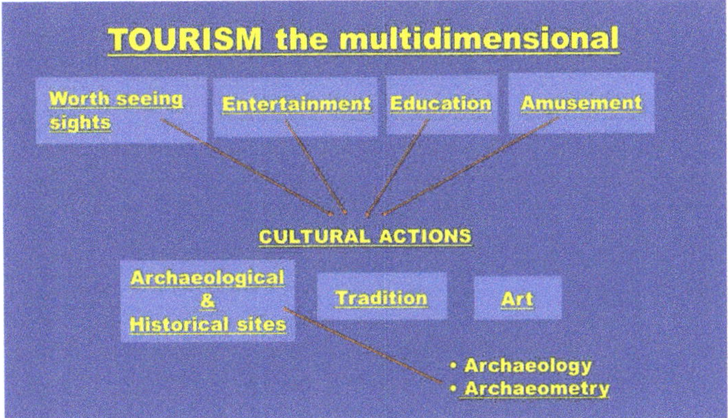

Figure 3. The multi-faceted tourism.

With combined new technologies, in museums, archaeological sites and parks, and reanimating their seemingly "dead" nature, it adds getting closer to the interest of layman, professional and trade market (visitors, scientists, students, stakeholders).

Some selective major archaeometrical case studies with respective economic impact are described below.

3. Some Archaeometric Case Studies with Economic Impact

3.1. Characterization and Provenance

The analysis of ancient archaeological finds is of interest, and it provides clues on the technology of production, the used ingredients, the raw materials, the local production or traded goods. Additionally, the possibility of the reconstruction roots develops scenarios of reconstruction of implements used by ancient societies. Amongst thousands of relevant studies one example is chosen here. For example, the study of prehistoric amphorae provenance based upon their chemical analysis via neutron activation analysis provided clues regarding their mobility in the prehistoric Aegean (Figure 4). Modern techniques allow non-destructive analysis of artifacts, an advantage in the sampling procedure and issues regarding preservation and legalities of evasiveness of precious works of art.

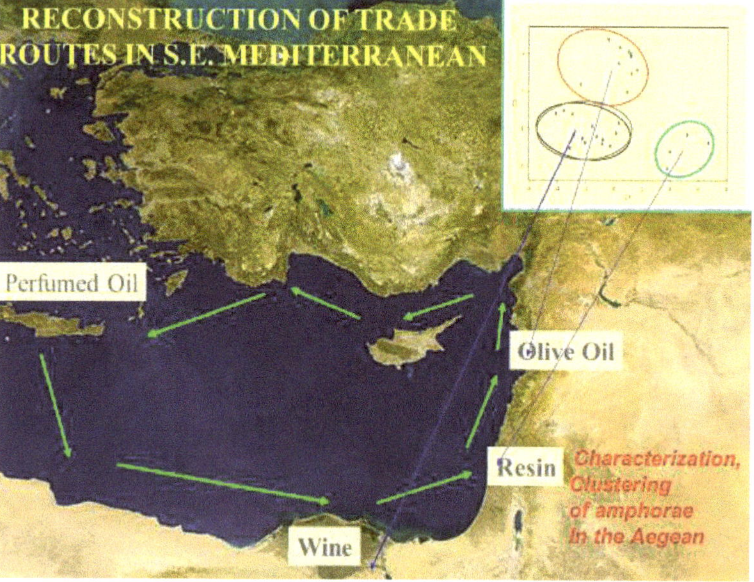

Figure 4. An example of trade in the prehistoric Southeast Mediterranean. Late Bronze Age prehistoric amphorae chemical analysis via neutron activation provides clues regarding their mobility in the prehistoric Aegean. Statistical clustering with principal component analysis (PCA) identified the origin of traded contents (oil, resin, wine, perfumed oils). In the inset, a plot of PCA components C1 vs. C2 and the resulting groups from respective regions (e.g., 39).

The portability of instruments has solved scientific issues allowing analysis in situ or of museum exhibits and not transferring the artifacts to the laboratory. Moreover, non-destructive methods of physico-chemical analysis contribute in the conservation work of monuments and artifacts and in their preservation and restoration of organic and inorganic materials [18,40–44].

In addition to the conservation procedures applied for exposition in museums more information on the investigation of contents in these artifacts helps our knowledge of the past and increases the interest and attention to visitors and strengthens cultural tourism.

3.2. Locating Buried Monuments

Geophysical prospecting aims to detect and map concealed antiquities and has been employed in the archaeological practice since the end of the Second World War.

Geophysical prospection has been also employed in non-conventional manners to tackle specific archaeological problems. Such problems might be, for example, the cases of locating tombs under tumuli embankments, assessing the moisture content in walls, the depth of fractures in sculptures, exploring the space behind walls, mapping the waterways along which the water drains out or in monuments, investigating in urban environment, etc. Relatively recently, the potential of this kind of operations has been the subject of numerous papers. However, many of these operations require a great amount of expertise and innovation [37,45–48].

Further, surface magnetic susceptibility may have been mapped, or electromagnetic data may have been collected. Additionally, aerial and satellite images might be available for the particular site, plus lidar images and accurate digital elevation models (DTM) [49,50].

Each one of the aforementioned methods is sensitive to different physical properties. Presumably, the combination of the information that each method provides could yield much better constrains in identifying, mapping and finally documenting the archaeological targets.

Geophysical prospecting at archaeological sites contributes to the sustainable development of specific provinces and areas. In particular, wide areas can be easily explored by (archaeo-) geophysical prospection, while under favourable conditions this approach may produce ground or underwater views of the buried ancient vestiges. Hence the following digs focus on pointed targets; as a result of saving capital, time, and effort. Accordingly, areas can be evaluated and included in local or regional development planning.

An archaeological dig may be shown in a holistic manner making use of cyber-archaeology, exhibiting the unearthed parts and displaying the parts that still reside underground in plates, leaflets, movies, etc.

The non-evasive archaeo-geophysical surveys are most significant in unearthing buried antiquities and offer a cultural attraction and development. It contributes a great deal to sustainability. Hidden antiquities are revealed. Many projects have been implemented where geophysical investigations have helped to exhibit new archaeological sites, either underground or under the sea, and restrict illicit excavations and trafficking of antiquities, and protecting at-risk antiquities from either public works or environmental risks.

There exist a number of examples from all over the world, especially from countries with a rich cultural heritage. The numerous archaeo-geophysical prospection (electrical, magnetic, georadar, remote sensing satellite imaging, and seismic sounding) first detect the buried target, which was followed by archaeological excavation, a study of the finds, and conservation and restoration tasks, to finally manage the opening the archaeological site and associated museum to the public, acquiring a sustainable character, such as in Italy with a significant project in South Etruria by the British School in Rome and Italian authorities (http://www.bsr.ac.uk/research/archaeology/completed-projects/tiber-valley-project/south-etruria-survey) [51], and in Egypt, China, Turkey and other parts of the world [52].

Reports of a number of examples exist from the Aegean region (Greece) that geophysical investigations have driven the archaeological investigations to a surgical kind of excavation and, at the same time, contributed to the promotion of the sites and the general prominence of them. Just to mention some of them, next to the Neolithic settlement of Dimini, integrated geophysical approaches (magnetic, soil resistance, electromagnetic techniques) were responsible for mapping the residues of two large monumental megaron-type compartments that were identified with the foundations of the Mycenaean palace that was subsequently excavated fully and was identified with the settlement of Iolkos in Thessaly, Greece, the home of Homeric King and hero Achilles. Today, it is accessible to visitors and it is one of the main archaeological attractions of the region [53]. Similar kinds of investigations with a subsequent steady increase of visitors have been carried out in the area of the Sanctuary of Poseidon at Kalaureia on Poros island. The subsequent excavations by the Swedish Institute at Athens revealed a number of structural remains that were associated with the daily activities of the site [54,55]. On Therasia island, despite its remote location and the hardness of the volcanic environments, geophysical measurements at the monastery of Koimisi and Orycheia produced substantial evidence related to the habitation of the island during the Early and Middle Cycladic periods (and most probably abandoned before the large volcanic eruption), as it was confirmed from the excavations that followed [56]. Even more examples can be presented for such cases where the geophysical survey has been able to contribute not only to the promotion of the sites, but also to their sustainability and their openness to the wider public since they have been excavated and offered to the visitors (the ancient Greek city of Sikyon [57]; the site of Bedenaki-Walls of Venetian Herakleio [58]; and the Minoan site of Sissi in Eastern Crete [59].

The cemetery of the Roman Era of Europos, Northern Greece, comprises another example. It is situated at the foothill of a topographic table. In fact, the resistance prospecting at this particular area yielded the distribution of resistances displayed in Figure 5a [37]. Its interpretation is straightforward since pronounced high resistivity anomalies are surrounded by a rather uniform low resistivity environment. The "twin probe" electrode arrangement was used having the roving electrodes 1 m apart, one form the other, and 1 and 2 m, in line and cross line spacing, respectively. Therefore, it was a low-resolution survey aiming only to detect and map the position and the areal extent of large monumental tombs. Presumably, each one of the well-defined in space high resistance anomalies has been caused by such a concealed structure. On the other hand, the relatively sizeable blurred anomaly at the west side of the image was attributed to a hidden gravel deposition. In fact, after the excavations, the aforementioned interpretation proved true in all its predictions.

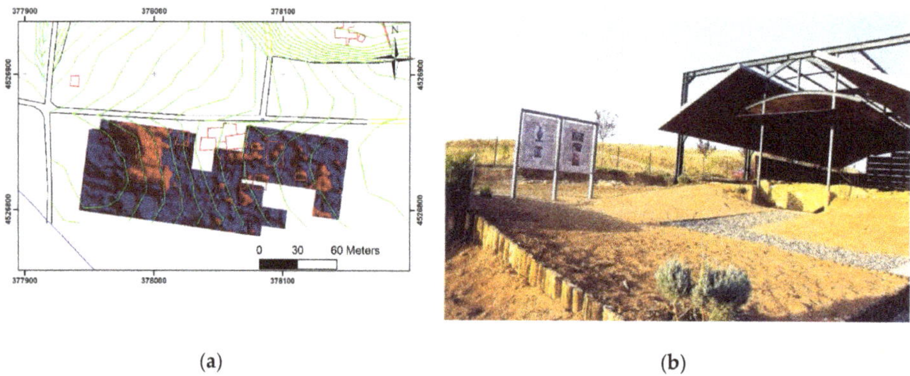

(a) (b)

Figure 5. (**a**) Resistivity prospecting of the Roman era Europos site exhibited the distribution of iso-resistances and pseudo-colors displayed, where high resistivity anomalies are surrounded by a rather uniform low resistivity environment [37]; (**b**) The site as an archaeological park. The plate shown introduces the visitors to the whole adventure to unearth and promote the monuments, from the geophysical image to the final stages of designing the shelters and organize the accessibility and promotion of the site where revealed tombs and architectural remains are now a tourist attraction to visitors (© G. Tsokas pers. comm., 2019).

The geophysical investigations guided the excavations (headed by archaeologist Dr. Thomais Savvopoulou) to the revelation of many monumental tombs, unfortunately all of them almost completely looted. However, the area was recreated to an excellent archaeological park (Figure 5b) contributing to the development of the particular province of the Greek State.

Why, after all, is geophysical prospection useful? It reveals hidden antiquities from a tomb to a city and others, by discovering buried ancient relics, it results to the creation of archaeological parks, museum exposition, other cultural activities, and in the updating of historical information with digital technology, all of which contribute to the local sustainable market. The ability to provide reliable images of the subterranean antiquities renders geophysical prospecting as an invaluable tool for the archaeological research, cultural heritage planning. Consequently, by assisting the archaeological research and promotion of monuments, which in addition, are studied by other archaeometrical methods, all is a decisive factor in the cultural tourism and economic growth.

3.3. *Cyber-Archaeology and Bio-Archaeological Issues*

The novel technologies and instrumentation development have achieved a goal: use of "Big Data", the new quantitative modelling and the results from aDNA, strontium and other isotopes and related scientific methods, have produced a new concept in contemporary archaeology [18,60,61].

The new technologies indeed alter our lives and the way in which we perceive it beyond the imaginable. This further ulterior over is the point in space-time in which coalition of science, technology, and art openly combined for the 3rd Cultural Revolution, and for environmentally sustainable abundance. However, this time, the "beyond" not only explores the dynamic 3D screen, it moves on from the bits to the atoms and incorporates 3D-printing and digital cloud-distribution which combined to relevant scanning or photographic technologies create a virtual environment as a real world. We are entering the central source for current and emerging trends in cultural heritage informatics with new disciplines, sub-disciplines and terminology to emerge. Virtual, cyber-archaeology, and cultural heritage to cyber-archaeometry are matters that have been recently tackled [18,62].

The virtual archaeology case studies, over the world, as a result of advanced technology emerging from computer sciences, however, stress the naturalistic methodology, and challenges digital reconstructions and serious games. There may also be provocation and harassment and the emergence of fundamental hermeneutical questions which serve as the basis of a synoptic and synthetic philosophy that combines art and science corresponding to classical techne, logos, and ethos.

The need for objective methodologies point out the need of hyper-taxonomies for interpreting the past, leading computing archaeology to an objective "scientific" interpretation [63–65].

The VR, 3D modelling and metadata, serious gaming, digital 3D reconstructions of bioarchaeological and cyber-archaeology data, all build a virtual collection and dataset and enrich the cultural heritage repertoire via virtual museums and make such archaeological heritage accessible to the public [64,66]. Hence, "*the websites of museums seem far more reluctant to display the dead online ... It might also relate to the fear of de-contextualising human remains*" [67].

Increasingly, institutions are also understanding that much of their public engagement will now take place digitally, with platforms acting as a "*medium through which information is published or exchanged*" [68] though bio-archaeological data are contrasted with a critical review of current ethical and technical guidelines, indicating potentially ethically compromising practices, particularly the lack of contextualising metadata for some models [69].

3.4. *The Antikythera Mechanism*

The Antikythera mechanism is the world's first analogical computer, used by ancient Greeks to chart the movement of the sun, moon and planets, predict lunar and solar eclipses and even signal the next Olympic Games and bearing inscriptions [70] (Figure 6).

(a) (b)

Figure 6. Antikythera mechanism, one of the pieces displayed in Athens Archaeological Museum, and a reconstruction of back dials ((**a**): National Archaeological Museum, Athens, photographer: Kostas Xenikakis, copyright Hellenic Ministry of Culture and Sports/Archaeological Receipts Fund; (**b**): Model constructed by John Gleave, according to Derek de Solla Price [71,72]).

The 2000-year-old astronomical calculator is a small size metal device. The detailed X-ray tomography imaging of the interior revealed at least 30 meshing bronze gears, was used for the determination of time, and included a user's guide of operation; new data proved its use as an astronomical device, and a 3D reconstruction has been made. Despite its poor condition, crammed insides, obscured by corrosion, it showed traces of technology that appear utterly modern: gears with neat triangular teeth (just like the inside of a clock) and a ring divided into degrees. It is a unique discovery from antiquity, and beyond doubt the only sophisticated, unparalleled object appears again for more than a thousand years. In particular, it calculated the prediction of eclipses, recorded the zodiac and the Egyptian and Greek Calendars, the solar-lunar calendar, the 19-year Cycle of Meton, the Saros cycle or exeligmos, Greek inscriptions, markings of astronomical symbols on discs, and the determination of major cultural events (Olympia, Pythia etc.), and cycles of 19, 76, 18+, and 54+ years [73–76].

The management of the revealed data and use of modern archaeometrical techniques had a respectful impact of increased visits in the National Archaeological Museum, Athens.

Moreover, a review of the published news in the media have shown a social interest and awareness, contributed to the cultural tourism, added to the international visibility-cultural diplomacy, become a paradigm of true interdisciplinarity, with the 3D reconstruction to provide an attraction to visitors, especially students (Table 1).

Table 1. Visits in the national archaeological Museum of Athens, 6 April 2012–28 April 2013. For 2012–2014 the total visitors were 800,000.

2011 (before the Exposition)	**Visitors**	**32,723**
	Students	**38,474**
2012 (Year of exposition)	Visitors:	153,581
	Students	47,958
2013 first 3 months (January to March)	Visitors	56,896
	Students	29,141

3.5. The Ice Man "Otzi" in the Alps

The ice man 'Otzi' (this name refers to the discovery site in the Ötztal Valley in the Alps) is a glacier mummy from the Early Bronze age (3300–3100 B.C.) Central European Alps and has been preserved to the present day. He was discovered accidentally by hikers in 1991, together with his clothing and equipment, on the Schnalstal/Val Senales Valley glacier and has been the subject of intensive research. Ötzi and his artefacts have been exhibited at the South Tyrol Museum of Archaeology in Bolzano, Italy since 1998. Following a thorough archaeometrical research the mummy is stored in a specially devised cold cell—a glass vitrine with controlled temperature (−6 °C) and humidity (98%) at glacier-like conditions. Ötzi's numerous pieces of equipment and clothing have been painstakingly restored. The magnificent work of conservation, exposition, and reconstruction of this mummy has been initiated by the holistic archaeometric contribution. Techniques applied and materials measured include: (a) radiocarbon (^{14}C) dating of clothing, wooden bow, and bone [77,78]; (b) X-ray flourescence (XRF) of hair discovering traces of copper and arsenic, implying his involvement in early pyrotechnology of smelting copper [79]; (c) X -ray radiography of his whole body discovering the fatal flint arrow in his left back shoulder, and other injuries [80,81]; (d) bio-archaeology used to decoding the Iceman's genetic make-up through aDNA, as well as carbon, oxygen, nitrogen, and strontium isotopic analysis in his teeth and bones proved his southern of Alps (Italy) origin, while he had eaten three meals during the last day or so, including a final meal about two hours before he was killed. Methods of thermal ionization (TIMS), inductively-coupled plasma (ICP-MS) and gas mass spectrometry included isotope ratios of $^{18}O/^{16}O$ ($\delta^{18}O$), $^{87}Sr/^{86}Sr$, and $^{206}Pb/^{204}Pb$, in order to reveal the Iceman's origin and migration behaviour. Analysed samples include tooth enamel, bones, and contents of his intestine, which all represent different ontogenetic (developmental) stages [80,82–85]; (e) palynology for pollen contained within the foods consumed by Ötzi, along with other palaeopathology evidence, colleagues were able to reconstruct his hectic itinerary in the hours before he died and defined the late spring/early summer death incident as well as the archaeobotanical environment [86]' and (f) virtual reconstruction of the Ice Man Otzi and his equipment, showing how he was equipped for the harsh temperatures of the high mountains and his clothing, in a reconstruction [87].

The impact to cultural tourism through increased number of visitors in south Tyrolo, and especially the Otzi museum, was perceptible (Figure 7). The museum attracts a variety of groups: school groups, the local population, visitors from further afield and from abroad. Ötzi and the South Tyrol Museum of Archaeology represent the main cultural destination in adverts promoting tourism in South Tyrol abroad, attracting a great many visitors and are, therefore, a not insignificant economic factor. A large number of them have not come to the regional capital for a holiday, but are day-trippers to Bozen-Bolzano, enjoying a perfect mix of sightseeing, shopping and local specialties, and of course to visit Ötzi in the South Tyrol Museum of Archaeology. This mixture of activities has an economic ripple effect: South Tyrol scores as a holiday destination with its attractive museums, and the many people who come to Bozen-Bolzano have a positive influence on the creation of added value and on the tourist facilities and services on offer. Since its opening on 28 March 1998, the museum has been visited by 5 million people; in 2017 by 286,972 visitors [88].

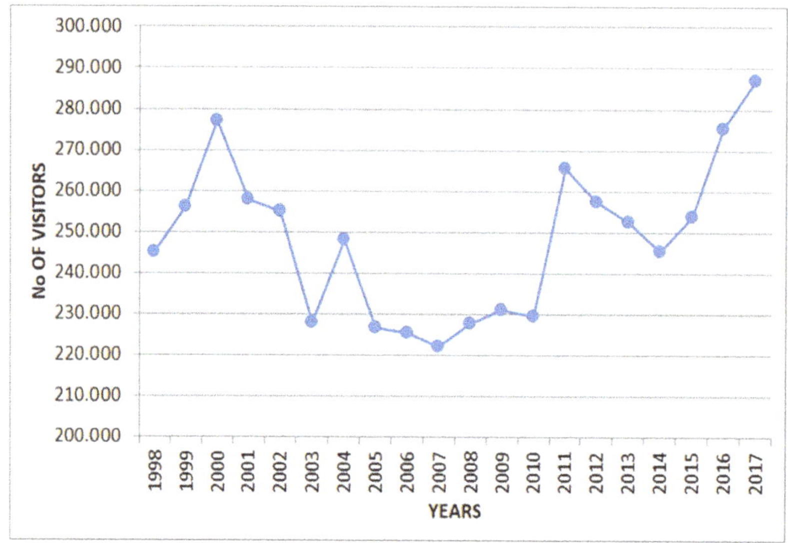

Figure 7. Trend of visitors per year from 1998 (opening) to 2017 ([89]. The peaks at 2000 and 2011 denote significant news information in the previous or running year from published data derived from more archaeometrical analysis of Otzi and properly disseminated through multimedia. Since 1998 (opening) it attracts 250,000 visitors/year. From a financial point of view, the museum has revenues from tickets sales, merchandising, sponsors, and publishing (see also [90]).

3.6. Archaeoastronomy in Stonehenge, England

Cultural heritage related to the sky (archeoastronomy) is a vital component of cultural heritage, in general, and recent UNESCO's Astronomy and World Heritage Initiative, supported by the International Working Group on Astronomy and World Heritage and by the International Astronomical Union through its Commission C4 on World Heritage and Astronomy emphasizes its added value [91].

The prehistoric monument of Stonehenge has long been studied for its possible connections with ancient astronomy (Figure 8). The site is aligned in the direction of the sunrise of the summer solstice and the sunset of the winter solstice. Archaeoastronomers have made a range of further claims about the site's connection to astronomy, its meaning, and its use.

(a)

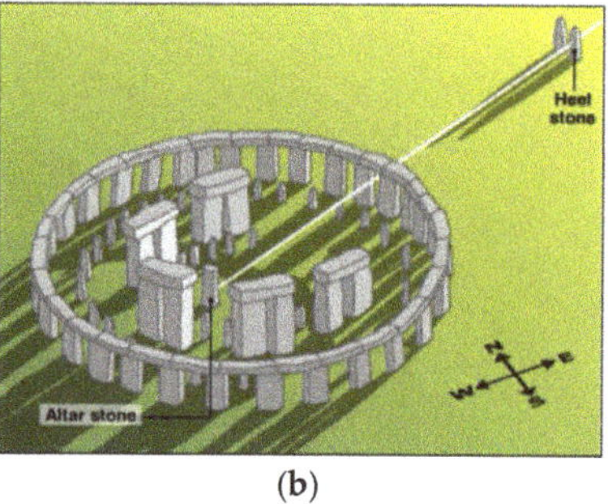

(b)

Figure 8. (a) Stonehenge from Google Earth and (b) 3D rendering and reconstruction showing the summer solstice sunrising light beam [92].

Early attention regarding the astronomical significance of the monument include: Stukeley [93], famous astronomer Edmund Halley, Sir Norman Lockyer (founding Editor of the journal Nature), and others [94–98].

These 3D studies and reconstructions and laser scans, video tours [99] together with ^{14}C dating both have boosted and enhanced its archaeological and astronomical significance [100–102].

Stonehenge's use is still not known but it is accepted as part of a ritual landscape. Whatever religious, mystical, or spiritual elements were central to Stonehenge, its design includes a celestial observatory function, which might have allowed prediction of eclipse, solstice, equinox, and other celestial events important to a contemporary religion.

Stonehenge has become an increasingly popular destination during the summer solstice, with 20,000 people visiting in 2005, scholars have developed growing evidence that indicates prehistoric people visited the site only during the winter solstice (Figure 9). An increased number of visitors is observed, especially after the opening in December 2013 of the new visitor centre which houses permanent and temporary exhibitions, enhanced with more archaeometrical results as they become available annually for the monument.

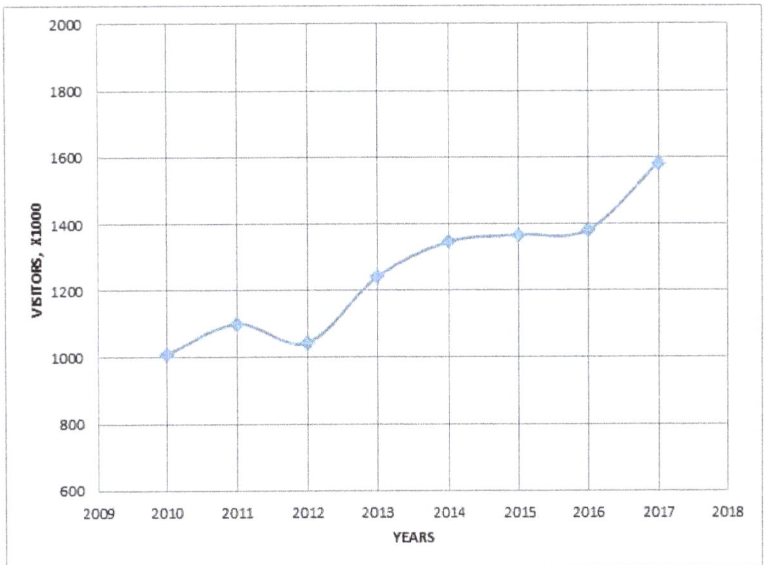

Figure 9. Visitors in the Stonehenge during 2010–2017 [103].

The only megalithic monuments in the British Isles to contain a clear, compelling solar alignment are Newgrange and Maeshowe, which both famously face the winter solstice sunrise.

The most recent evidence supporting the theory of winter visits includes bones and teeth from pigs which were slaughtered at nearby Durrington Walls. Their age at death indicating that they were slaughtered either in December or January every year

Archaeological research and archaeometric measurements consider it was constructed from 3000 BC to 2000 BC but re-used at later times to at least the Late Bronze Age.

One of the most famous landmarks in the United Kingdom, Stonehenge is a legally protected Scheduled Ancient Monument since 1882 when legislation to protect historic monuments was first successfully introduced in Britain. The site and its surroundings were added to UNESCO's list of World Heritage Sites in 1986.

Other monuments built in accordance with astronomical alignments are potential tourist attractions if properly managed with new technologies.

Several potential or ongoing archaeometrical projects properly managed could with certainty become an added value and lead to an enhanced cultural tourism. Prior to completion of an archaeological-archaeometrical project, that is, the archaeological site and exhibition museum open to the public, but during the implementation of the program, at the level of summer schools and field work by groups of students and research institutes, the necessary expenses on a daily basis for the research investigators contributes as the economic stimulation to the local society. Hence, three factors are reinforced; *social cohesion*, *economic benefit*, and *scientific development*.

3.7. A Recent Combined Project in Kaifeng, Henan (China)

Geophysical prospection, photogrammetric techniques, (LIDAR, GIS, GPS, etc.) were applied to locate and reveal the ancient capital of China at Kaifeng. It was the geophysical detection of sub-terrain relics that drove the archaeological service to start a large excavation enterprise. Six ancient cities buried deep underground and spanning several dynasties over 2000 years have been unearthed in Kaifeng, in the east of China's Henan province. Archaeologists and archaeometrists worked collaboratively at the Xinzheng Gate site, a gate on the wall of the ruins of Bianliang (Kaifeng's name during the Northern Song Dynasty, 960 to 1127 AD) [104,105] (Figure 10).

(a) (b)

Figure 10. (**a**) Stratigraphy of the Kaifeng six city-upon-city, Xinzheng Gate site, (**b**) view of the ongoing excavation. Objects found are in the adjacent museum (photo © by IL).

The 2000-square-meter site boasts cultural relics from the Warring States Period (475–221 BC) to the Qing Dynasty (1644–1911 AD), with each city piled one on top of another.

It includes city gates and walls from the Song Dynasty (960–1279 AD), roads from the Song to the Qing Dynasties, civilian homes and courtyards from the Qing Dynasty, plus courtyard walls from Yuan Dynasty (1271–1368 AD). An increased number of visits to the site and associated museum in Kaifeng has been noted as a result of government's efforts to capitalize on tourism has kicking into high gear.

3.8. Conservation of Monuments, Castles

The application of archaeometric techniques and conservation works to monuments and castles has proven the added value to the economy of a nation. Undertaken projects of such scale mainly require decision-makers' attention in the authorities of government [106].

Projects of this kind follow a series of protocols and necessary steps regarding documentation, evaluation, archaeometry, conservation, restoration, revitalization, and management [107] as outlined below (in bold those steps related to present paper's aims):

STEP 1: Natural environment (location, climate, geomorphology, geopolitics, GIS, flora and fauna)

STEP 2: Socioeconomic status (ownership, legal frame of the state and local community, economical activities of the region, resources, main occupation educational level)

STEP 3: Architectural, structural engineering, and technological data (designs of topography and architecture of the structures included. Survey of pathology and construction techniques. Documentation of old materials. Old regional techniques of construction).

STEP 4: Historical data (ethnology, recursion of the history, historic events, references, everyday life inside and outside the castles)

STEP 5: Evaluation of the selected data (definition of the basic axis or eminent characteristics which should be alleged and of the priorities of activities)

STEP 6: Archaeometry-conservation work (spectroscopic analysis, dating, mechanical properties, 3D documentation. Interventions for solidification conservation, cleaning, consolidation, pointing reconstruction, reinforcement of foundation sheltering, conservation of work of arts included in the ensemble)

STEP 7: Excavation for further archaeological documentation. Geophysical prospection (find data for "dark" historical periods)

STEP 8: Plans for the functional rearrangement of the site (access, protection of natural and architectural expression, lightening, readability of the history, outbuildings).

STEP 9: Plans for connection with other monuments of architecture (pedestrian, archaeological walk, place of sightseeing and relaxation)

STEP 10: Establishment and operation of monitoring system (Long-time behaviour of repaired parts, record of visitors. Regulations)

STEP 11: Digital reconstruction, digital scenarios, Virtual museum.

STEP 12: Use for educational purposes (Plans for regular and occasional activities in cooperation with local and state authorities)

STEP 13: Enhancement of cultural activities (festival, infrastructure for theatre performance, meetings, events).

STEP 14: Plans for development of local economy (ecotourism, local bus, guides for visitors, activation of the municipality and periphery in participation of joint development projects)

The development and gives a statistical result that helps unemployment and economy in general. Conservation and restoration projects in which the wide archaeometrical techniques and methods were applied has provided economic benefits to some countries.

In 2004 Italy's income of 14 billion Euros came from cultural tourism from conservation and restoration works (towns with history, monuments/works of art, restoration/conservation/preservation).

In 2002 15% of France's annual income, equal to about 15 billion Euros, came from cultural tourism of castles and monasteries, yet about 855 million Euros per year was used for archaeometry-conservation-restoration of cultural heritage.

Another characteristic and recent example is the "Integrated Diagnostic Research Project and Strategic Planning for Materials, Interventions Conservation and Rehabilitation of the Holy Aedicule of the Church of the Holy Sepulchre in Jerusalem". New technologies, including LIDAR, sonar, laser scanning, thermal imaging, georadar (GPR), and luminescence dating and characterization, were used to preserve this important site, as well as, the geometric documentation with materials analysis, the identification of building phases and the diagnosis of decay and pathology through the use of analytical and non-destructive techniques were included. The resulted outcome led experts in the completion of a challenging and seemingly impossible restoration of a universal monument, with the result of securing increased number of visitors with the new findings and the safety measures taken [103,108].

In other parts of the World integrated projects of castles, monasteries of interdisciplinary and multi-scientific nature also were made aiming to increased number of tourists and thus income [109–114].

4. Discussion

According to the above, it becomes, by now, easy to perceive that the new technological evolution has dramatically reshaped the value chain, in the way that it impacts the sustainability of the systems of governance for culture. Cultural heritage has a historical, social, and anthropological value and it is considered as an enabler of sustainable development (United Nations' Sustainable Development Goals (SDGs) 11 and 8).

Cultural heritage, on the other hand, contributes significantly to a more balanced and sustainable urban or regional development and economy, providing many opportunities for social interaction and economic growth of host communities. It is, indeed, widely acknowledged that the various aspects and methods of new technology represent a remarkable and valuable tool in terms of economy. Even though the traditional methods of research, assessment, and collection of archaeological data are still being used, sciences and archaeometry come to the aid of professionals in order to assist conservation, archaeology, restoration, and so many others, but also to save time and money. Non-invasive geophysical surveys and remote sensing activities, for example, are nowadays often implemented before the dig in order to provide substantial indications as to the existence of antiquities, a piece of information which would otherwise take long periods of trial trenches to obtain and by sequence large amounts of money and human resources. The documentation of the finds is also easily assured with the use of 3D representations, laser scanning, and several other imaging techniques of high technology. Collections data can be far more detailed and stored in far less space than ever before. Databases can hold complete descriptions along with photographs and links to the raw elements of analysis and interpretation. In addition, the artefact can be compared to previous data to track any changes, such as deterioration or fading. The three-dimensional building models are also increasingly necessary for urban planning within areas including buildings and structures of cultural heritage. In a sense it is perhaps the only though method preserving for the condition of a monument in view of natural and man provoked hazards.

Archaeometry's impact in development applies also to at-risk cultural heritage protection. On the other hand, laboratory tests of high accuracy are also employed against all sorts of fake provenance documentation.

Equally important is that ICT applications making also use of archaeometric results amplify the access of people to cultural information and services, since the public can enjoy them more easily, quickly and cheaply. In terms of economy, ICT provide a great opportunity for cultural industries, especially of tourism and computing, to adopt a sustainable and cultural-social centric approach and to develop into attractive and profitable stakeholders.

The use of archaeometry helps make the cultural heritage industry more widely and effectively extroverted, through the documentation and dissemination of cultural objects, movable and immovable monuments, tangible or intangible heritage, and provides democracy and globalization of the knowledge, participation, and accessibility of local communities and the wider universal community to the local cultural heritage, highlighting regional, territorial sustainable development and cultural tourism.

Archaeometric studies consist of the links between the scientific information derived from cultural sites and objects, museums as cultural transmission areas-either real or virtual—and the global community as the final recipient of cultural data. Digitalization provides new opportunities of data processing and analysis and boost cultural development, including the ability of individuals to have, at any time, no limit of access to cultural heritage.

3D digitization technologies create new promotion potentials of archaeometry and give the European and world public the opportunity to take a close look of the specialized work carried out by scientific groups throughout the world, to participate in the fascinating course of science, archaeometry innovative applications, and the experience of cultural documentation.

Virtual world platforms are used for artefacts' and museums' modelling and exposure, while the use of panoramic images is spreading to present captured aspects of a real museum. Many museums, real or fictional, have been created in order to replicate the experience of a physical visit to the museum and make the user feel immersed in the digital space [115].

Sustainable management of the cultural and natural environment aims to the development in an unceasing way. The ultimate goal is the society, science and technology development in such a way as to ensure a durable permanence of future generations (trans-generational) and to reserve in all sectors, mainly environment, but also society and economy, for them [116]. Sustainable development is directly linked to culture and aims at its sustainability. Sustainability as a term links past, present, and future, because people are studying the past and discover those elements that consist of heritage, as perceived at the present time, for the people of the future. As UNESCO states: "*Education for Sustainable Development allows every human being to acquire the knowledge, skills, attitudes and values necessary to shape a sustainable future*".

With digital archaeological and territorial information interconnected an increase in cultural heritage knowledge worldwide is achieved. Archaeometry data analysis transmission, combined with stakeholders and public involvement, allows a wide participation at the area of cultural management and treatment of monuments and enables individuals' awareness to central decisions about heritage sustainability. Meanwhile, organizations are more careful and responsible about their impact on monuments and cultural environment [18].

The natural and cultural heritage of a region and its management is of particular importance for tourism development, it brings economic benefits but also risks due to the exploitation of sensitive resources such as the natural, cultural, and historical environments. Sustainable management of cultural objects and the soft, controlled development of cultural destinations are required to reduce the negative impact of tourism activity and the consequent economic development. The inclusion of the protection and promotion of cultural goods in the national planning and the local-regional strategy for the sustainable development of the regions is inextricably linked with the development of the economy. The interaction is intense and flows in both directions.

Archaeometrical results coupled with sustainability, and vice versa, as a cause-and-effect relationship benefits several fields; from advancing science, enhancing the value of reminiscent relics, induce a pedagogical effect, provide social cohesion, help economy and employment and reinforce cultural diplomacy. The factors of communication of the archaeometric results to the public, i.e., the manner, media of results communication, the target group that the scientific data is addressing to, the accessibility, influence this relationship. Overall, interdisciplinarity in archaeometry is connected with sustainability because it produces a novel and valuable and unpreceded information on the cultural heritage material. This dual relationship is quantified from statistics of visitors in respective museums or open archaeological sites.

Unhappily the common consensus is not a respected connection between archaeometry and sustainability, as the measurable results are confined to a narrow academia. However, if proper use is made with the results in all fields of archaeometry (provenance, characterization, location, dating, technology, intangible heritage from material culture analysis and interpretation of their arrangement, archaeoastronomy, VR, conservation-restoration, digital documentation, dissemination, etc.) then a direct connection between archaeometry and sustainability will emerge.

Along the lines of archaeometry's role in cultural regional economic sustainability and development strategies, the development of geomythology to decipher ancient myths, emerging from archaeometry applied to intangible heritage is of unprecedented essence. The recent upcoming research discipline of the disaster archaeology and attempts to explain beyond a mythical/legendary account by natural sciences, lead unavoidably to the establishment of a new academic discipline of archaeological mythology, much like biblical archaeology with Israel's patrology conveyed orally and finally recorded in the holy Bible [117–119]. Therefore, archaeometry on a perpetually accredited scientifically holistic approach (PASHA) continually and consecutively developed, coupled with the mythologies in the world, the Greek mythology being of paramount importance, adds another dimension in revealing remote past, merging tangible and intangible heritage, and forming a new academic field with social, scientific, and pedagogical dimensions and new learning outcomes.

The systematic way to model the coupling of archaeometry and sustainability needs a combined grounded theory (GT) for quantitative data [120,121], but with traditional comparative ethnography, archaeology, and, if available, ancient literature approach. Archaeometric data are quantitative which are analyzed and treated in such a manner to produce and/or reproduce accurate and precise end product (a date, a content, a reconstruction).

The economic valuation of cultural heritage remains a scientific challenge. Archaeometry's impact targets at the economic effect, either as an external benefit or as a source of tourist attraction. Thus, the application of the contingent valuation method, which has the objective aim to examine the effect of opportunity cost in indirect economic valuation of some special public goods through the supply of voluntary labour may be worth linking to archaeometrical fieldwork [122].

The archaeometry updated by its new technologies may well apply a resilient strategy to cultural heritage, in the domain of fundamental and applied research, education and training, and economy. At any rate the implementation of all these should take into consideration the pertaining uncertainties from natural disasters, terrorist activities, and climatic changes [123].

5. Conclusions

Archaeometry is a scientifically-established international discipline that investigates scientific issues of cultural heritage; it is a multidisciplinary science that develops research and solves archaeological problems. With the help of this interdisciplinary subject new unexplored fields, political, cultural, and social landscapes are discovered, and scientific gaps are covered because science, although divided into subgroups, is unified and indivisible.

Archaeometry results consist of data (such as graphs, statistical information, etc.) which simplify and facilitate the possibility of comparing cultural samples and retrieving maximal information from their micro scale, thus conducting safe conclusions, which can be used globally by researchers, scientists, and government officials, resulting in the dissemination of information and the globalization of science, the scientific and administrative dialogue, the promotion of administrative functions, and the convenience of citizens to fair and proper administrative treatment.

Cultural heritage is, without doubt, a particularly complex field. The protection and preservation of archaeological sites, ancient monuments, vernacular architecture, industrial installations, cultural and historical landscapes, and many other forms of cultural property is and should be the object of many disciplines. The integration of cultural management services with archaeometry and information culture technologies has already shown the successful impact in many of the processes of documenting and monitoring, interpreting and communicating the data, enhancing many aspects of the research, building capacity and achieving public involvement in the integration of the past into our lives. The management of big data from archaeometrical applications and the cyber-archaeology in the field, the museum, and the office, upraises the interdisciplinary direction to the PASHA, which has the potential for addressing the new challenges the heritage sector faces and securing its long-term sustainability and preservation, giving a hopeful prosperous future to local, regional, and national economic development, from the cultural heritage re-treatment to economic benefit.

At any rate, combined multi-scientific archaeometric projects, use "new technologies" ("new" in the sense of updated progressive development in science and technology) and retrieve information in the micro- and macroscopic level. The obtained data of applications to current material culture or in the revealing and documentation of new buried antiquities, with proper interpretation and integration in the wider context of tangible and intangible cultural heritage, are inextricably linked to sustainability.

Author Contributions: Writing: I.L. and E.K.; Providing case and idea: I.L.; Providing revised advice: I.L.

Acknowledgments: Ioannis Liritzis is thankful for the project fund support from Key Research Institute of Yellow River Civilization and Sustainable Development & Collaborative Innovation Center on Yellow River Civilization of Henan Province, Henan University, Kaifeng 475001, China. Colleagues G.Tsokas and A.Sarris for useful first-hand information on geophysical results, to Niki Platia MSc, for initial useful discussions and N.Laskaris and A.Vafiadou for help in plots.

Conflicts of Interest: The authors declare no conflict of interest.

References

1. Jones, A. Archaeometry and materiality: Materials-based analysis in theory and practice. *Archaeometry* **2004**, *46*, 327–338. [CrossRef]
2. Gilberto, A. *Scientific Methods and Cultural Heritage: An Introduction in the Application of Materials Science to Archaeometry and Conservation Science*; Oxford University Press: New York, NY, USA, 2010.
3. Schiffer, M.B. Some relationships between behavioural and evolutionary archaeologies. *Am. Antiquity* **1996**, *61*, 643–662. [CrossRef]
4. Brown, K.A.; Pluciennik, M. Archaeology and human genetics: Lessons for both. *Antiquity* **2001**, *75*, 101–106. [CrossRef]
5. Edmonds, M. Science, technology and society. *Scott. Archaeol. Rev.* **1990**, *7*, 23–31.
6. Dunnell, R.C. Why archaeologists don't care about archaeometry. *Archeomaterials* **1993**, *7*, 161–165.
7. Liritzis, I. The contribution of archaeometry to the development (cultural, economic) of the Mediterranean islands. *J. Mediterr. Stud.* **2001**, *10*, 225–2308.
8. Liritzis, I. Does Archaeometry contribute in the development of cultural educational tourism & sustainable development? In Proceedings of the 17th Scientific Conference of Greek Regionalists Delphi "The Future of Develomental and Urban Planning of Greece", Athens, Greece, 2012; pp. 545–559. (In Greek)
9. Liritzis, I. Searching for buried relics in the Castle of Salona (Greece) with geophysical methods: Examples and proposal. In *Proceedings of the 1st Conference on "Castle of Salona (Amfissa): Historical Prospectives, Possibilities & Cultural Development", 2013*; Papakonstantinou, M., Ed.; Municipality of Delphi in Collaboration with 24th Ephoria of Byzantine Antiquities & 8th Ephoria of Classical-Prehistoric Antiquities: Lamia, Greece, 2013. (In Greek)
10. Psycharis, S. STEAM in education: A literature review on the role of computational thinking, engineering epistemology and computational science. computational steam pedagogy (CSP). *Sci. Cult.* **2018**, *4*, 51–72. [CrossRef]
11. Panou, E.; Violetis, A. Teaching astronomy using monuments of cultural heritage: The educational example of "Horologion of Andronikos Kyrrhestes". *Sci. Cult.* **2018**, *4*, 77–83. [CrossRef]
12. Liritzis, I. STEMAC (Science, Technology, Engineering, Mathematics for Arts & Culture): The emergence of a new pedagogical discipline. *Sci. Cult.* **2018**, *4*, 73–76. [CrossRef]
13. Anderson, C.; Törnberg, A.; Törnberg, P. An Evolutionary Developmental Approach to Cultural Evolution. *Curr. Anthropol.* **2014**, *55*, 154–174. [CrossRef]
14. Liritzis, I. Twelve thousand years of non-linear cultural evolution: The physics of chaos in Archaeology. *Synesis J. Sci. Technol. Ethics Policy* **2013**, *4*, G19–G31.
15. Kintigh, K.; Altschul, J.; Beaudry, M.; Drennan, R.; Kinzig, A.; Kohler, T.; Limp, W.F.; Maschner, H.; Michener, W.; Pauketat, T.; et al. Grand Challenges for Archaeology. *Am. Antiquity* **2014**, *79*, 5–24. [CrossRef]
16. Thomas, B.; Thomas, S.; Powell, L. The development of key characteristics of welsh island cultural identity and sustainable tourism in Wales. *Sci. Cult.* **2017**, *3*, 23–39. [CrossRef]
17. El-Menshawy, S. Traveling within pharaonic Egypt for discovering the past. *Sci. Cult.* **2017**, *3*, 15–21. [CrossRef]
18. Liritzis, I.; Al-Otaibi, F.M.; Volonakis, P.; Drivaliari, A. Digital technologies and trends in cultural heritage. *Mediterr. Archaeol. Archaeom.* **2015**, *15*, 313–332. [CrossRef]
19. Reindel, M.; Wagner, G.A. *New Technologies for Archaeology, Multidisciplinary Investigations in Palpa and Nasca, Peru*; Springer Science & Business Media: Berlin/Heidelberg, Germany, 2009; p. 2.
20. Available online: www.icomos.org (accessed on 18 March 2019).
21. Edwards, H.; Vandenabeele, P. *Analytical Archaeometry, Selected Topics*; RSC Publishing: Cambridge, UK, 2012.

22. Li, F. VR Cultural Heritage Sites: A Case Study on Creating Immersive Visual & Auditory Experience. BSc Senior Thesis, Rose-Hulman Institute of Technology, Department of Computer Science and Software Engineering, Terre Haute, IN, USA, 2016.
23. Kanellos, I.; Antin, S.; Dimou, O.; Kanellos, M.-A. Educational enhancing of virtual expositions. towards visitor-centered storytelling digital museology. *Mediterr. Archaeol. Archaeom.* **2014**, *14*, 117–123.
24. Sarris, A. (Ed.) *Best Practices of GeoInformatic Technologies for the Mapping of Archaeolandscapes. Archaeopress Archaeology*; Archaeopress Publishing Ltd.: Oxford, UK, 2015; ISBN 9781784911621.
25. Liritzis, I.; Zacharias, N.; Papageorgiou, I.; Tsaroucha, A.; Palamara, E. Characterisation and analyses of museum objects using pXRF: An application from the Delphi Museum, Greece. *Stud. Antiqua Archaeol.* **2018**, *24*, 31–50.
26. Levy, T.E.; Sideris, T.; Howland, M.; Liss, B.; Tsokas, G.; Stambolidis, A.; Fikos, E.; Vargemezis, G.; Tsourlos, P.; Georgopoulos, A.; et al. At-Risk World Heritage, Cyber, and Marine Archaeology: The Kastrouli–Antikyra Bay and Sea Project, Phokis, Greece. In *Cyber-Archaeology and Grand Narratives, One World Archaeology*; Levy, T.E., Jones, I.W.N., Eds.; Springer International Publishing AG: Basel, Switzerland, 2018; pp. 143–230.
27. Sarris, A.; Déderix, S. GIS for Archaeology and Cultural Heritage Management in Greece. Quo Vadis? In Proceedings of the 3rd Symposium Archaeological Research and New Technologies ARCH_RNT, Kalamata, Greece, 3–6 October 2012; pp. 7–20.
28. Portalés, C.; Alonso-Monasterio, P.; Viñals, M.J. 3D virtual reconstruction and visualisation of the archaeological site Castellet de Bernabé (Llíria, Spain). *Virtual Archaeol. Rev.* **2017**, *8*, 75–82. [CrossRef]
29. Balla, A.; Pavlogeorgatos, G.; Tsiafakis, D.; Pavlidis, D. Recent advances in archaeological predictive modeling for archeological research and cultural heritage management. *Mediterr. Archaeol. Archaeom.* **2014**, *14*, 143–153.
30. Hatzopoulos, J.N.; Stefanakis, D.; Georgopoulos, A.; Tapinaki, S.; Volonakis, S.; Volonakis, P.; Liritzis, I. Use of various surveying technologies to 3D digital mapping and modelling of cultural heritage structures for maintenance and restoration purposes: The Tholos in Delphi, Greece. *Mediterr. Archaeol. Archaeom.* **2017**, *17*, 311–336. [CrossRef]
31. Zara, J. Virtual Reality and Cultural Heritage on the Web. In Proceedings of the 7th International Conference on Computer Graphics and Artificial Intelligence (3IA 2004), Limoges, France, 12–13 May 2004; pp. 101–112.
32. Georgopoulos, A. 3D virtual reconstruction of archaeological monuments. *Mediterr. Archaeol. Archaeom.* **2014**, *14*, 155–164.
33. Barsanti, S.G.; Caruso, G.; Micoli, L.L.; Rodriguez, M.C.; Guidi, G. 3D visualization of cultural heritage artefacts with virtual reality devices. *Int. Arch. Photogramm. Remote Sens. Spat. Inf. Sci.* **2015**, *XL-5/W7*, 165–172. [CrossRef]
34. Bruno, F.; Lagudi, A.; Barbieri, L. Virtual Reality Technologies for the Exploitation of Underwater Cultural Heritage. In *Latest Developments in Reality-Based 3D Surveying and Modelling*; Remondino, F., Georgopoulos, A., González-Aguilera, D., Agrafiotis, P., Eds.; MDPI: Basel, Switzerland, 2018; pp. 220–236.
35. Osburg, T.; Lohrmann, C. *Sustainability in a Digital World. New Opportunities Through New Technologies*; Springer: Cham, Switzerland, 2017.
36. Sarris, A.; Papadopoulos, N.; Agapiou, A.; Salvi, M.C.; Hadjimitsis, D.; Parkinson, W.; Yerkes, R.; Gyucha, A.; Duffy, P. Integration of geophysical surveys, ground hyperspectral measurements, aerial and satellite imagery for archaeological prospection of prehistoric sites: The case study of Vésztő-Mágor Tell, Hungary. *J. Archaeol. Sci.* **2013**, *40*, 1454–1470. [CrossRef]
37. Tsokas, G.N.; Giannopoulos, A.; Tsourlos, P.; Vargemezis, G.; Tealby, J.M.; Sarris, A.; Papazachos, C.B.; Savopoulou, T. A large-scale geophysical survey in the archaeological site of Europos (nothern Greece). *J. Appl. Geophys.* **1994**, *32*, 85–98. [CrossRef]
38. Pavlidis, G.; Levy, T.; Liritzis, I. Pedagogy and Engagement in At-Risk World Heritage Initiatives. In *Heritage and Archaeology in the Digital Age: Acquisition, Curation, and Dissemination of Spatial Cultural Heritage Data*; Vincent, M.L., Bendicho, V.M., Ioannides, M., Levy, T.E., Eds.; Springer International Publishing: Basel, Switzerland, 2018; pp. 167–183.
39. Tsiafaki, D.; Michailidou, N. Benefits and problems through the application of 3d technologies in archaeology: Recording, visualisation, representation and reconstruction. *Sci. Cult.* **2015**, *1*, 37–45. [CrossRef]
40. Liritzis, I.; Polychroniadou, E. Optical and analytical techniques applied to the Amfissa Cathedral mural paintings made by the Greek artist Spyros Papaloukas (1892–1957). *Revue d' Archaeometrie (Archaeosciences)* **2007**, *31*, 97–112.

41. Pratt, C.E. Critical Commodities: Tracing Greek Trade in Oil and Wine from the Late Bronze Age to the Archaic Period. Ph.D. Thesis, California Digital Library, University of California, Los Angeles, CA, USA, 2014.
42. Elhagrassy, A.F.; Hakeem, A. Comparative study of biological cleaning and laser techniques for conservation of weathered stone in Failaka island, Kuwait. *Sci. Cult.* **2018**, *4*, 43–50. [CrossRef]
43. Abdel Rahim, N.S. Analytical study and conservation of archaeological terra sigllata ware from roman period, Tripoli, Libya. *Sci. Cult.* **2016**, *2*, 19–27. [CrossRef]
44. Kousouni, C.-K.; Panagopoulou, A. Non-destructive physicochemical analysis and conservation of metallic book covers of ecclesiastical books from Saint Mavra and Timotheos Church in Zakynthos (Greece). *Sci. Cult.* **2018**, *4*, 85–95. [CrossRef]
45. Scollar, I.; Tabbach, A.; Hesse, A.; Herzog, I. *Archaeological Prospecting and Remote Sensing*; Cambridge University Press: Cambridge, UK, 1990.
46. Athanasiou, E.; Tsourlos, P.I.; Vargemezis, G.; Papazachos, C.B.; Tsokas, G.N. Nondestructive DC resistivity surveying using flat-base electrodes. *Near Surf. Geophys.* **2007**, *5*, 243–272. [CrossRef]
47. Angelis, D.; Tsourlos, P.; Tsokas, G.; Vargemezis, G.; Zacharopoulou, G.; Power, C. Combined application of GPR and ERT for the assessment of a wall structure at the Heptapyrgion fortress (Thessaloniki, Greece). *J. Appl. Geophys.* **2018**, *152*, 208–220. [CrossRef]
48. Drahor, M.G.; Tsokas, G.N.; Piro, S.; Trinks, I. Integrated Geopysical Investigations for Archaeology Foreword. *Near Surf. Geophys.* **2015**, *13*, 519–521. [CrossRef]
49. Lasaponara, R.; Masini, N. Detection of archaeological crop marks by using satellite QuickBird multispectral 477 imagery. *J. Archaeol. Sci.* **2007**, *34*, 214–221. [CrossRef]
50. Karamitrou, A.; Petrou, M.; Tsokas, G.N. A pixel-based semi-stochastic algorithm for the registration of 479 geophysical images. *Archaeol. Prospect.* **2017**, *24*, 413–424. [CrossRef]
51. Geophysics Techniques. Available online: http://www.bsr.ac.uk/research/archaeology/geophysics-2/geophysics-techniques (accessed on 18 March 2019).
52. El-Qady, G.; Metwaly, M. (Eds.) *Archaeogeophysics: State of the Art and Case Studies*; Springer Nature: Cham, Switzerland, 2019.
53. Sarris, A.; Kalayci, T.; Papadopoulos, N. Geophysical Prospection on the Volcanic Environs of Therasia Island, Cyclades, Greece. *ISAP NEWS* **2013**, *37*, 2–4.
54. Sarris, A.; Topouzi, S.; Adrimi-Sismani, V. Mycenaean Dimini: Integration of Geophysical Surveying and GIS. In Proceedings of the CAA2002 International Conference: Computer Applications & Quantitative Methods in Archaeology, The Digital Heritage of Archaeology, Herakleion, Crete, Greece, 2–6 April 2002; pp. 1–6.
55. Papadopoulos, N.G.; Sarris, A.; Kokkinou, E.; Wells, B.; Penttinen, A.; Savini, E.; Tsokas, G.N.; Tsourlos, P. Contribution of multiplexed electrical resistance and magnetic techniques to the archaeological investigations at Poros, Greece. *J. Archaeol. Prospect.* **2006**, *13*, 75–90. [CrossRef]
56. Pentinnen, A.; Wells, B.; Mylona, D.; Pakkanen, P.; Pakkanen, J.; Karivieri, A.; Hoorton, A.; Savini, E.; Theodoropoulou, T. Report on the Excavations in the Years 2007 and 2008 Southeast of the Temple of Poseidon at Kalaureia. *Opuscula* **2009**, *2*, 89–141. [CrossRef]
57. Sarris, A.; Papadopoulos, N.; Trigkas, V.; Kokkinou, E.; Alexakis, D.; Kakoulaki, G.; De Marco, E.; Seferou, E.; Shen, G.; Karaoulis, M.; et al. Recovering the Urban Network of Ancient Sikyon through Multi-component geophysical approaches. In *Layers of Perception, Proceedings of the 35th International Conference on Computer Applications and Quantitative Methods in Archaeology (CAA), Berlin, Germany, 2–6 April 2007*; Posluschn, A., Lambers, K., Herzog, I., Eds.; Kolloquien zur Vor und Frühgeschichte: Bonn, Germany, 2007; Volume 10, pp. 11–16.
58. Sarris, A.; Papadopoulos, N. Geophysical Surveying in Urban Centers of Greece. In Proceedings of the 16th International Congress "Cultural Heritage and New Technologies", Vienna, Austria, 14–16 November 2011; pp. 96–114, ISBN 978-3-200-02740-4.
59. Sarris, A.; Manataki, M.; Dederix, S. Geophysical Investigations at Sissi 2015–2016. In *AEGIS 13: Excavations at Sissi IV. Preliminary Report on the 2015–2016 Campaigns*; Driessen, J., Anastasiadou, M., Caloi, I., Claeys, T., Déderix, S., Devolder, M., Jusseret, S., Langohr, C., Letesson, Q., Mathioudaki, I., et al., Eds.; Presses universitaires de Louvain: Louvain-la-Neuve, Belgium, 2018; pp. 43–58.

60. Vanek, D.; Brzobohata, H.; Silerova, M.; Horak, Z.; Nyvltova Fisakova, V.G.; Vasinova Galiova, M.; Zednikova Mala, P.; Urbanova, V.; Dobisikova, M.; Beran, M.; et al. Complex Analysis of 700-Year-Old Skeletal Remains found in an Unusual Grave—Case Report. *Anthropol* **2015**, *2*. [CrossRef]
61. Cooper, A.; Green, C. Embracing the Complexities of 'Big Data' in Archaeology: The Case of the English Landscape and Identities Project. *J. Archaeol. Method Theory* **2016**, *23*, 271–304.
62. Moscati, P. (Ed.) *Virtual Museums and Archaeology. The Contribution of the Italian National Research Council*; Archeologia E Calcolatori, Supplemento 1; Publ. All'Insegna del Giglio: Roma, Italy, 2007.
63. Barcelo, J.A.; Moitinho de Almeida, V. Functional analysis from visual and non-visual data an arti-ficial intelligence approach. *Mediterr. Archaeol. Archaeom.* **2012**, *12*, 273–321.
64. Forte, M. *Cyber-Archaeology*; Archaeopress: Oxford, UK, 2010.
65. Levy, T.E.; Petrovic, V.; Wypych, T.; Gidding, A.; Knabb, K.; Hernandez, D.; Smith, N.; Schlulz, J.; Savage, S.; Kuester, F.; et al. On-Site Digital Archaeology 3.0 and Cyber-Archaeology: Into the Future of the Past—New Developments, Delivery and the Creation of a Data Avalanche. In *Cyber Archaeology*; Forte, M., Ed.; BAR International Series 2177; Archaeopress: Oxford, UK, 2010; pp. 135–153.
66. Siebke, I.; Campana, L.; Ramstein, M.; Furtwängler, A.; Hafner, A.; Lösch, A. The application of different 3D-scan-systems and photogrammetry at an excavation—A Neolithic dolmen from Switzerland. *Digit. Appl. Archaeol. Cult. Herit.* **2018**, *10*, e00078. [CrossRef]
67. Williams, H.; Atkin, A. Virtually Dead: Digital Public Mortuary Archaeology. *Internet Archaeol.* **2015**, *40*. [CrossRef]
68. Proctor, N. Digital: Museum as Platform, Curator as Champion, in the Age of Social Media. *Curator. Mus. J.* **2010**, *53*, 35–43. [CrossRef]
69. Ulguim, P. Models and Metadata: The Ethics of Sharing Bioarchaeological 3D Models Online. *Archaeologies* **2018**, *14*, 189–228. [CrossRef]
70. The Inscriptions of the Antikythera Mechanism. *Almagest (J. Netw. Hist. Sci. South. Eur.)* **2016**, *7*, 6–310, (volume dedicated to Antikythera mechanism). [CrossRef]
71. Available online: https://www.bibliotecapleyades.net/ciencia/esp_ciencia_antikythera02.htm (accessed on 18 March 2019).
72. Available online: https://www.bibliotecapleyades.net/ciencia/esp_ciencia_antikythera07.htm (accessed on 18 March 2019).
73. Seiradakis, J.H.; Edmunds, M.G. Our current knowledge of the Antikythera Mechanism. *Nat. Astron.* **2018**, *2*, 35–42. [CrossRef]
74. Freeth, T. The Antikythera mechanism I: Challenging the classic research. *Mediterr. Archaeol. Archaeom.* **2002**, *2*, 21–35.
75. Available online: http://www.antikythera-mechanism.gr/project/overview (accessed on 18 March 2019).
76. Freeth, T. The Antikythera mechanism. II: Is Posidoniu's Orrery? *Mediterr. Archaeol. Archaeom.* **2002**, *2*, 45–58.
77. Bonani, G.; Ivy, S.; Hajdas, I.; Niklaus, T.R.; Suter, M. AMS 14C age determinations of tissue, bone and grass samples from the Ötztal Ice Man. *Radiocarbon* **1994**, *36*, 247–250. [CrossRef]
78. Kutschera, W.; Rom, W. Ötzi, the prehistoric Iceman. *Nucl. Instrum. Methods Phys. Res. Sect. B Beam Interact. Mater. Atoms* **2000**, *164–165*, 12–22. [CrossRef]
79. Artioli, G.; Angelini, I.; Kaufmann, G.; Canovaro, C.; Dal Sasso, G.; Villa, I.M. Correction: Long-distance connections in the Copper Age: New evidence from the Alpine Iceman's copper axe. *PLoS ONE* **2017**, *12*, e0189561. [CrossRef]
80. Macko, S.A.; Lubec, G.; Teschler-Nicola, M.; Andrusevich, V.; Engel, M.H. The Ice Man's diet as reflected by the stable nitrogen and carbon isotopic composition of his hair. *FASEB J.* **1999**, *13*, 559–562. [CrossRef] [PubMed]
81. Price, D.T.; Burton, J.H. *An Introduction to Archaeological Chemistry*; Springer: New York, NY, USA, 2012.
82. O'Sullivan, N.J.; Teasdale, M.D.; Mattiangeli, V.; Maixner, F.; Pinhasi, R.; Bradley, D.G.; Zink, A. A whole mitochondria analysis of the Tyrolean Iceman's leather provides insights into the animal sources of Copper Age clothing. *Sci. Rep.* **2016**, *6*, 31279. [CrossRef]
83. Kutschera, W.; Muller, W. "Isotope language" of the Alpine Iceman investigated with AMS and MS. *Nucl. Instrum. Methods Phys. Res. B* **2003**, *204*, 705–719. [CrossRef]
84. Müller, W.; Fricke, H.; Halliday, A.N.; McCulloch, M.T.; Wartho, J.A. Origin and Migration of the Alpine Iceman. *Science* **2003**, *302*, 862–866. [CrossRef]

85. Rollo, F.; Ubaldi, M.; Ermini, L.; Marota, I. Ötzi's last meals: DNA analysis of the intestinal content of the Neolithic glacier mummy from the Alps. *Proc. Natl. Acad. Sci. USA* **2002**, *99*, 12594–12599. [CrossRef] [PubMed]
86. Bortemschlager, S.; Oeggl, K. (Eds.) *The Iceman and His Natural Environment Palaeobotanical Results*; Springer: Wien, Austria, 2000.
87. Available online: https://www.livescience.com/13150-otzi-reconstruction-iceman-mummy-copper-age.html (accessed on 18 March 2019).
88. Available online: http://www.iceman.it/en/5-mio-visitors/ (accessed on 18 March 2019).
89. Available online: https://www3.astronomicalheritage.net/ (accessed on 18 March 2019).
90. Brida, J.G.; Meleddu, M.; Pulina, M. *Museum Visitors Can Be Regarded as Specific Cultural Tourists? A Length of Stay Analysis*; Working Papers; CREN O S: Cagliari, Italy, 2011; Volume 14.
91. Available online: http://baron-de-synclair.blogspot.com/2012/06/cette-carte-est-elle-representative.html (accessed on 18 March 2019).
92. Available online: https://www.english-heritage.org.uk/visit/places/stonehenge/history-and-stories/stonehenge360/ (accessed on 18 March 2019).
93. Stukeley, W. *Stonehenge, A Temple Restor'd to the British Druids*; W. Innys & R Manby: London, UK, 1740; p. 81.
94. Hawkins, G.S. *Stonehenge Decoded*; Souvenir Press: London, UK, 1966.
95. Thom, A.; Stevenson, T.A.; Strang, T.A. Stonehenge. *J. Hist. Astron.* **1974**, *5*, 71–90. [CrossRef]
96. Hill, R. *Stonehenge*; Profile Books: London, UK, 2008.
97. Ruggles, C.; Hoskin, M. *Astronomy Before History. I: The Cambridge Concise History of Astronomy*; Michael, H., Ed.; Cambridge University Press: Cambridge, MA, USA, 1999; p. 6.
98. Thom, A.; Stevenson, T.A.; Strang, T.A. Stonehenge as a Possible Lunar Observatory. *J. Hist. Astron.* **1975**, *6*, 19–30. [CrossRef]
99. Available online: https://www.statista.com/statistics/586843/stonehenge-visitor-numbers-united-kingdom-uk/ (accessed on 18 March 2019).
100. Unver, E.; Taylor, A. Virtual Stonehenge Reconstruction. In *Progress in Cultural Heritage Preservation*; Ioannides, M., et al., Eds.; Springer-Verlag: Berlin/Heidelberg, Germany, 2012; pp. 449–460.
101. MacDonald, J. New media applications and their potential for the advancement of public perceptions of archaeoastronomy and for the testing of archaeoastronomical hypotheses. *Mediterr. Archaeol. Archaeom.* **2006**, *6*, 181–184.
102. Bayliss, A.; Bronk Ramsey, C.; McCormac, F.G. Dating Stonehenge. *Proc. Br. Acad.* **1997**, *92*, 39–59.
103. Available online: https://www.researchgate.net/publication/323128412_FAITHFUL_REHABILITATION_OF_THE_HOLY_TOMB_OF_CHRIST/references (accessed on 18 March 2019).
104. Masini, N.; Capozzoli, L.; Chen, P.; Chen, F.; Romano, G.; Lu, P.; Tang, P.; Sileo, M.; Ge, Q.; Lasaponara, R. Towards an Operational Use of Geophysics for Archaeology in Henan (China): Methodological Approach and Results in Kaifeng. *Remote Sens.* **2017**, *9*, 809. [CrossRef]
105. Ge, Q. *Report of Archaeological Stratigraphy*; Kaifeng Institute of Archaeology CASS: Kaifeng, China, 2016.
106. Neville, A.; Bridgland, J. (Eds.) *Of the Past, for the Future: Integrating Archaeology and Conservation, Proceedings of the Conservation Theme at the 5th World Archaeological Congress, Washington, DC, USA, 22–26 June 2003*; Getty Conservation Institute Symposium Proceedings Series; Getty Conservation Institute: Los Angeles, CA, USA, 2006; Available online: http://hdl.handle.net/10020/gci_pubs/of_past_for_future (accessed on 18 March 2019).
107. Patsavos, N.; Mergos, G. *Cultural Heritage and Sustainable Development. Economic Benefits, Social Opportunities, Policy Challenges*; Technological University of Crete, and InHerit Erasmus+: Crete, Greece, 2017.
108. Moropoulou, A.; Zacharias, N.; Delegou, E.T.; Apostolopoulou, M.; Palamara, E.; Kolaiti, A. OSL mortar dating to elucidate the construction history of the Tomb Chamber of the Holy Aedicule of the Holy Sepulchre in Jerusalem. *J. Archaeol. Sci. Rep.* **2018**, *19*, 80–91. Available online: https://www.int-arch-photogramm-remote-sens-spatial-inf-sci.net/XLII-5-W1/487/2017/isprs-archives-XLII-5-W1-487-2017.pdf (accessed on 18 March 2019). [CrossRef]
109. Elyamani, A. Re-use proposals and structural analysis of historical palaces in Egypt: The case of Baron Empain Palace in Cairo. *Sci. Cult.* **2018**, *4*, 53–73. [CrossRef]
110. Salama, K.K.; Ali, M.F.; Moussa, A.M. Deterioration factors facing mural paintings in el Sakakeny Palace (problems and solutions). *Sci. Cult.* **2016**, *2*, 5–9. [CrossRef]

111. Amin, E.A. Technical investigation and conservation of a tapestry textile from the Egyptian textile Museum, Cairo. *Sci. Cult.* **2018**, *4*, 35–46. [CrossRef]
112. Elyamani, A.; Roca, P. One century of studies for the preservation of one of the largest cathedrals worldwide: A Review. *Sci. Cult.* **2018**, *4*, 1–24. [CrossRef]
113. Măruţoiu, C.; Bratu, I.; Troşan, L.; Neamtu, C.; Măruţoiu, V.C.; Pop, D.; Tănăselia, C.; Garabagiu, S. Scientific Investigation of the Imperial Gates Belonging to the 17th Century Wooden Church of Sălişca, Cluj County, Romania. *Sustainability* **2018**, *10*, 1503. [CrossRef]
114. Elyamani, A.; Roca, P. A review on the study of historical structures using integrated investigation activities for seismic safety assessment. Part II: Model updating and seismic analysis. *Sci. Cult.* **2018**, *4*, 29–51. [CrossRef]
115. Vosinakis, S.; Tsakonas, Y. Visitor experience in google art project and in second life-based virtual museums: A comparative study. *Mediterr. Archaeol. Archaeom.* **2016**, *16*, 19–27. [CrossRef]
116. Krysiak, F.; Krysiak, D. Sustainability with Uncertain Future Preferences, Environmental & Resource Economics. *Eur. Assoc. Environ. Resour. Econ.* **2006**, *33*, 511–531.
117. Vitaliano, D.B. Geomythology: Geological Origins of Myths and Legends. In *Myth and Geology*; Piccardi, L., Masse, W.B., Eds.; GSL, Special Publications: London, UK, 2007; Volume 273, pp. 1–7.
118. Levy, T.E.; Schnider, T.; Propp, W. (Eds.) *Israel's Exodus in Transdisciplinary Perspective*; Springer: Berlin, Germany, 2015.
119. Liritzis, I.; Raftopoulou, M. *Argolid: Connection of Prehistoric Legends with Geoenvironmental and Archaeological Evidence*; TUBA-AR, II; Turkish Academy of Sciences: Ankara, Turkey, 1999; pp. 87–99.
120. Glaser, B.G.; Strauss, A. *Discovery of Grounded Theory. Strategies for Qualitative Research*; Aldine Transaction: London, UK, 1967.
121. Corbin, J.; Strauss, A. *Basics of Qualitative Research: Techniques and Procedures for Developing Grounded Theory*, 3rd ed.; Sage Publications, Inc.: Thousand Oaks, CA, USA, 2008.
122. Kopsidas, O. The Opportunity Cost as a Critical Determinant Factor of the Cultural Heritage Monuments' Valuation: A Modified Contingent Valuation Method. *J. Environ. Sci. Eng. B* **2018**, *7*, 55–64. [CrossRef]
123. Macalister, F. Preparing for the future: Mitigating disasters and building resilience in the cultural heritage sector. *J. Inst. Conserv.* **2015**, *38*, 115–129. [CrossRef]

© 2019 by the authors. Licensee MDPI, Basel, Switzerland. This article is an open access article distributed under the terms and conditions of the Creative Commons Attribution (CC BY) license (http://creativecommons.org/licenses/by/4.0/).

Article

Archaeoastronomy: A Sustainable Way to Grasp the Skylore of Past Societies

Antonio César González-García [1,*] and Juan Antonio Belmonte [2]

1 Instituto de Ciencias del Patrimonio, Incipit, Consejo Superior de Investigaciones Científicas, Avda. de Vigo s/n, 15705 Santiago de Compostela, Spain

2 Instituto de Astrofísica de Canarias & Universidad de La Laguna, Calle Vía Láctea s/n, 38200 La Laguna, Spain; jba@iac.es

* Correspondence: a.cesar.gonzalez-garcia@incipit.csic.es

Received: 8 March 2019; Accepted: 10 April 2019; Published: 14 April 2019

Abstract: If astronomy can be understood as the contemplation of the sky for any given purpose, we must realize that possibly all societies throughout time and in all regions have watched the sky. The why, who, how and when of such investigation is the pursuit of cultural astronomy. When the research is done with the archaeological remains of a given society, the part of cultural astronomy that deals with them is archaeoastronomy. This interdisciplinary field employs non-invasive techniques that mix methodologies of the natural sciences with the epistemology of humanities. Those techniques are reviewed here, providing an excellent example of sustainable research. In particular, we include novel research on the Bohí Valley Romanesque churches. The results provided go beyond the data. This is because they add new value to existing heritage or discovers new heritage due to the possible relationship to the spatial and temporal organization of past societies. For the case of the Bohí churches the results point to a number of peculiarities of these churches in a valley in the Pyrenees. This links these aspects to the ritual, practical and power sphere of past societies. A wonderful example of such links is the high mountain sanctuaries in Gran Canaria, where archaeoastronomy helps promoting a World Heritage candidacy.

Keywords: cultural astronomy; archaeoastronomy; field techniques; Bohí Romanesque churches; Canary Islands; landscape archaeology

---

1. Introduction

Far from being a particular version of the history of astronomy, or a specialized archaeometry, cultural astronomy is something else. Human knowledge can be specialized (i.e., medical, astronomical, etc.) but when such knowledge is used it becomes social and human, as society is the context where all that is human is produced and used. This is why we must ask for the sense of such uses. When we do this with respect to specialized knowledge, there appear research questions, data and methodologies that go beyond the expertise of the specialist on such knowledge. A social and human version of this knowledge is thus needed to answer those questions. In the case of the sky, such an approach is cultural astronomy.

There is one part of the humanities that focuses on the environment influence on human societies. However, there is only one discipline that deals with the relationship between those societies throughout history and the sky below they dwell, and this is what has been called astronomy in culture or rather cultural astronomy.

If a naïve and simple definition of astronomy could be sky-watching, then cultural astronomy is realizing that any society, present or past, has actually looked at the sky. This is why as astronomers we would like to know what they were watching. However, as cultural astronomers we want to understand how they generated, processed and used such knowledge. Hence, the focus is not in the celestial objects they identified but how those were understood and incorporated by that society from their cultural point of view.

Cultural astronomy thus tries to answer questions that go beyond astrophysical interest. It tries to understand how astronomical knowledge is generated in ancient, traditional and ancestral societies. How is such knowledge transmitted? What are the processes of social production, transference and diffusion of such astronomical knowledge? Was it invented by all societies independently? Or, on the contrary, were some basic principles (like the identification of solstices or equinoxes) generated in some cultures and later transmitted to others? Were 'primitive astronomers' some kind of 'artisans', i.e., a specific social group? Were they privileged? Or, was the astronomical lore simply produced in a general sense, without any concrete author? Other pertinent questions, such as what was the influence of the concept of the sky in the power relations or the structuration of society may also arise.

In a definition given by Stanislaw Iwaniszewski [1], cultural astronomy is the study of the relations between people's perception of the sky and the organization of different aspects of the social life. Such study, according to Edwin Krupp [2] includes several different topics: Calendars, practical observation, celestial cults and myths, symbolic representation of astronomical events, naming celestial objects, astronomical concepts or laws, the astronomical orientation of tombs, temples, sanctuaries and towns, traditional cosmology and the ceremonial use of astronomical traditions.

Cultural astronomy thus includes several other sub-disciplines, such as ethnoastronomy and astronomy in traditional and subsidiary societies (like pastoral groups, etc.), or archaeoastronomy. In this context, archaeoastronomy can be understood as the study of the orientation of buildings and their possible relationship to astronomical events. In short then, archaeoastronomy is the study of cultural astronomy through the material record, while ethnoastronomy would be the study of cultural astronomy via the ethnography of present day or past societies.

Cultural astronomy is a relatively new discipline that complements and deepens, from different points of view, the understanding of several other disciplinary fields. As examples to be considered we may count on: Landscape archaeology, understood in a broad sense, including the construction of spatial materiality in relation to the observation of astronomical phenomena [3]; the history of religions, from the orientation of sacred sites to the ritual conditioning of calendars; historic anthropology, understood as a holistic focus on human actions in the past, where cultural astronomy contributes by studying the different images of the cosmos in different societies. Finally, it could also be seen as a history of science as it studies the knowledge of the universe in different cultures and historical periods.

As with any other historical or archaeological data, archaeoastronomy pretends to shed light into the people in the past, and to do so we must do it by approaching the way of thinking of the society we deal with [4]. This means that the data and hypotheses advanced by cultural astronomy have to be supported on and by the archaeological, ethnographic or historical record [5].

In summary, cultural astronomy offers an increase in the understanding of cultural and archaeological heritage in two ways: It discovers 'new heritage' and provides new value to that already known or helps identifying heritage at risk (see Figure 1).

All in all, it should be noted that cultural astronomy is the way to get in contact with a basic heritage element: The one constituted by the materials, knowledge and values related to the observation of the sky. Cultural astronomy offers a way to gain access to a number of phenomena that otherwise would pass unnoticed. UNESCO has recognized such dimension in the category of Astronomical World Heritage Sites [6,7]. In the subsequent sections of the present paper we will focus on the sustainability of archaeoastronomical research. First we will present how this research is currently done, next we will introduce the results for a particular case study and, finally, we will focus on a qualitative example on how this research helps promoting a candidacy for a World Heritage Site.

Figure 1. Archaeoastronomical fieldwork helps identify heritage at risk. This was the case with the Dahmiyeh dolmen field, one of the Bronze Age megalithic necropolises in Jordan studied by the authors [8]. These megalithic monuments of the local Bronze Age are built with a valuable travertine stone and were located next to the site where this material was obtained. A stone quarry now menaces this dolmen field and several dolmens are already being dismantled without compassion. © A.C. González-García.

2. Sustainability of Archaeoastronomical Research

According to Nature's website [9], sustainability is the ability to endure, for example, by exploiting resources in a way that does not deplete their availability for the future. We will use this definition as a working tool for the next paragraphs when presenting how archaeoastronomy does research and if this is done in a sustainable way. In other words, in such a way that may allow further data acquisition from the same site in the future. A final consideration on another aspect of the sustainability of archaeoastronomical research will be done at the conclusion section.

As indicated above, archaeoastronomy is done through the study of the archaeological record from the perspective of cultural astronomy. It is common lore that the conventional archaeological methodology is non-sustainable as it relies on the excavation and thus on the 'destruction' of the archaeological site. However, it is indeed true that nowadays there are a number of non-invasive techniques helping archaeology, such as different kinds of geophysical surveys, and some of them will be acknowledged and mentioned later on.

From this perspective, and having in mind the sustainability of research in archaeoastronomy, there are different levels open for consideration. For example, do we need excavation for doing research in archaeoastronomy? Are there any alternatives?

All these issues deal with the methodology employed to carryout data acquisition. In general, archaeoastronomical data are obtained through the visual inspection of the site of interest. Once a line of sight is defined for whatever reason (notably a symmetry line of a given building or the alignment of a number of artificial elements) our measuring device is situated either on top of that line of sight or closely parallel to it. Then two readings are usually recorded, the angle of this line with respect to true north (the azimuth) and the altitude of the horizon along that line of sight. Those two measurements together with the location of the observation point read through a GPS will facilitate the translation of these readings into astronomical measurements (see Figure 2; for a complete and detailed description of the measurement procedure see [10]).

Figure 2. The data acquisition process in archaeoastronomy does not involve any kind of disturbance to the archaeological or heritage sites. The picture shows a common procedure for the measurement of a built structure, in this case a Romanesque church in Galicia, Spain. The technique involves deriving a parallel line to the wall to be measured with two rods. © A.C. González-García.

One other kind of data acquisition is done through the landscape inspection and measurement for the verification of potential landscape/heavenly relations. This is done by carefully inspecting the horizon observed from a particular site. There, singular directions, perhaps marked by other sites located at the horizon or by natural singular features such as notches or mountain peaks, could be linked to directions where interesting astronomical phenomena can be observed (see e.g., [11–13]).

In order to verify if the astronomical data is of any use at any archaeological site, one must be very careful. The pertinence of such data ought to be reckoned, making questions such as: Were astronomy or astral deities known to be important for the society we are dealing with? (see e.g., [14–17]). If we do not have any such information, as for prehistoric societies, either we can rely on solid interpretations of the iconography or we can try to proof the intentionality by performing statistical studies that may indicate the existence of orientation patterns that could only be explained by astronomical phenomena (see e.g., [18,19]).

As we can see, there is no need for an excavation or any other invasive technique within this procedure, so in principle the methodology per se is quite non-interventionist, nor harmful and thus sustainable. Another question is that to be able to define the measured line or the observation point we may need previously excavated structures, or not, if new techniques such as geo-radar are applied.

One key issue that we advocate archaeologists should incorporate to their excavation/prospection routine, as a standard procedure, is that of a horizon reconstruction or recording, which is seldom altered during the excavation process. This procedure can be as easy as taking a number of photographs, normally more than eight, to cover the 360° visible from the prospection or excavation site. This would help the researcher to reconstruct the horizon in the laboratory, for instance using customary image treatment software, and then incorporating such landscape into planetarium software, such as Stellarium [20]. This will be useful to verify if there are interesting and/or important astronomical connections from that site. As we have argued, this does not necessarily take much time and the amount of data that can be provided is substantial (see Figure 3).

Figure 3. With the help of a panoramic picture properly referenced we can reconstruct the sky viewed in ancient times. The picture shows a panoramic reconstruction from the newly discovered cultic platform in Petra (Jordan) that has been incorporated into Stellarium to replicate the sky appearance at the epoch of use. © A.C. González-García.

Another way to do this landscape reconstruction implies having a good geo-referenced positioning of the site and employing geographic information systems (GIS) to reconstruct the horizon. This is useful not only for those cases when a 360° photograph of the site is not available, but also for those where vegetation or buildings hamper the visibility of the sites. These can incorporate for instance data from LiDar flights. Indeed, the resolution of the digital terrain model (DTM) is the main problem to solve as it will dictate the uncertainty in the potential astronomical links of the sites investigated (see [12] for a recent use of this technique).

Of course, a kind of data that can be used is that derived from Geo-Radar and other geo-physical explorations of potential sites. In this case, the data can also be put into context of the landscape inspection and then derive orientation measurements and potential astronomical connections to those sites. One example of this kind of research has been carried out with a series of the circular enclosures in southern Portugal [21].

Finally, we should not forget the data acquisition techniques where we can take into account the 3D reconstruction of built structures. Those can then be incorporated into DTMs and planetarium models to investigate, verify and discover potential astronomical relations of light and shadow effects ([13,22,23], see Figure 4).

One key issue with the sustainability of data acquisition is the replicability of the research. This implies taking care of performing reliable estimates for our data (including errors), and a comprehensive description of the methodology employed and the sites measured.

One fundamental question we must carry out from this enumeration, regarding the sustainability of this kind of research, is that as the methodology does not imply altering the sites, this kind of measurements can be performed without (much) impact to already existing heritage sites. Besides it provides new data that normally have not been taken into account before and thus complements and expands our understanding of these sites and helps to bring new heritage into knowledge.

Figure 4. A 3D reconstruction of the dolmen of Dombate (A Coruña, Spain). Top: The 3D rendering has been included within a digital terrain model into the Stellarium software allowing investigation of the illumination of the paintings inside the megalithic chamber today precluded by the preservation measures of the dolmen. Bottom: The backstone of the dolmen, possibly the first to be erected at the time of construction, displays an elaborate program of decoration with sophisticated geometric forms, which are directly illuminated by the first sunbeams of the winter solstice. The 3D reconstruction allows verifying that the illuminated part never reaches further up than the area with red paintings. © A.C. González-García.

Nowadays, there is a rather well established methodology that shows that this discipline can be fully incorporated into the humanities as a useful tool to understand human past. In parallel there is an on-going substantial effort to upgrade the methodology by incorporating new techniques as well as new epistemological discourses.

In this sense, cultural astronomy studies in general and in particular those of archaeoastronomy are closely connected to the spatial and landscape dimension. Such connection links archaeoastronomy to the social aspect of the landscape. Thus, space is socially and culturally built. A number of variegated disciplines highlight that a 'landscape' is a created reality by social action. This can be found in different social disciplines such as human geography [24,25], anthropology [26–29], archaeology [30,31] and the history of religions [32–34].

From this point of view, archaeoastronomy is a key element for understanding and complementing our knowledge on the social construction of space and, given the recurrent and cyclical nature of the astronomical events, also the time and temporal dimension of these landscapes.

In a recent paper, Kristiansen [35] argues that while the objectivist ideal of procesualism has been long time set aside, it is also evident that the subjectivist version of post-procesualism or the hermeneutic liberalism are not the solution either. The enhancement of critic reflexivity as a central component of all theories and practices, and the consolidation of a weak or soft model of science (with rigorous data, robust methodologies and reflexive theories: R3 science), paves the way to new proposals that Kristiansen exemplifies with his "new paradigm" for archaeology.

His proposal combines the science-based potential of the archaeological investigation (big data, profound archaeometries, information sciences and visualization techniques) with solid theories to produce significant narratives, as a way to merge the point of view of anthropology, and humanities in general, with that of the 'hard' sciences.

In our understanding, cultural astronomy, given the characteristics presented above, is well positioned to face such integration also taking care of the sustainability of the research. The use of massive empirical datasets, together with the use of innovative visualization techniques (like 3D modeling of buildings inserted in a planetarium software) integrated in more solid theories, allow constructing narratives that provide an integrative and holistic sense to the social study of the sky.

In the following pages we will show how such an endeavor can be exemplified in one case study, the group of Romanesque Churches of the valley of Bohí (Lleida, Spain). Finally, we will present how this endeavor can help in the promotion of heritage sites, taking the high mountain sanctuaries and complexes in the central part of the island of Gran Canaria as an example.

3. Cultural Astronomy in Action: Vall Bohí Medieval Churches

It should be useful to include the description of a case study, where a series of sites can be sampled, in order to illustrate the aforementioned issues. In this sense, we present in this section a site where the local landscape, the distant horizon and the skyscape could interact, but at the same time would allow preserving the spirit of cultural astronomy as a sustainable discipline which respects the environment.

Located at the foothill of the Pyrenees, the secluded and beautiful Bohí Valley gained a significant relevance from the 9th century onwards. After the Muslim conquest of the Iberian Peninsula, some Christian counties were formed at the valleys of the Pyrenees with the support of the Carolingian kings. Apparently, the local feudal lords of this area received a significant amount of money to support the conquest of Saragossa and other nearby populations and they used such income for building a number of churches between the 11th and 12th centuries [36].

Nine of them remain in the area, all of them built in a Romanesque style with strong Lombard influences. They present elegant mason work and tall and stylized towers and the interior is frequently decorated with elaborate paintings (see Figure 5). These characteristics and their excellent state of preservation granted their inclusion in the UNESCO's World Heritage List in 2000.

According to the canonical medieval sources, the orientation of a church should be such that the apse must be facing towards east, and in particular at the equinox as defined in the Nicaea's Concilium [37]. In previous studies we have verified that, for the Iberian Peninsula, and for the period right before the introduction of the Romanesque style, such a definition is somewhat ambiguous. Pre-Romanesque churches in the Iberian Peninsula are vastly located in the northern half of the Peninsula and are normally facing a few degrees north of equinoctial east [38]. This can hardly be explained by an orienting method where the shadow of a gnomon is used to obtain the orientation of the church. However, a direct observation of sunrise on a given day and the use of the Julian calendar could be a likely explanation. Both the alleged day of the equinox, March 21st, or March 25th, a prominent Christian feast, would move with time as the Julian calendar shifted with respect to the seasons. One interesting exception to the norm for the Iberian Peninsula was found in the Serrablo Valley, where most churches (dating from the 10th century) are facing east but slightly south, instead of north, of the equinox [38]. Serrablo is not far from Bohí (c. 90 km as the crow flies), and those churches could perhaps be understood as the direct precedents of these ones in Bohí.

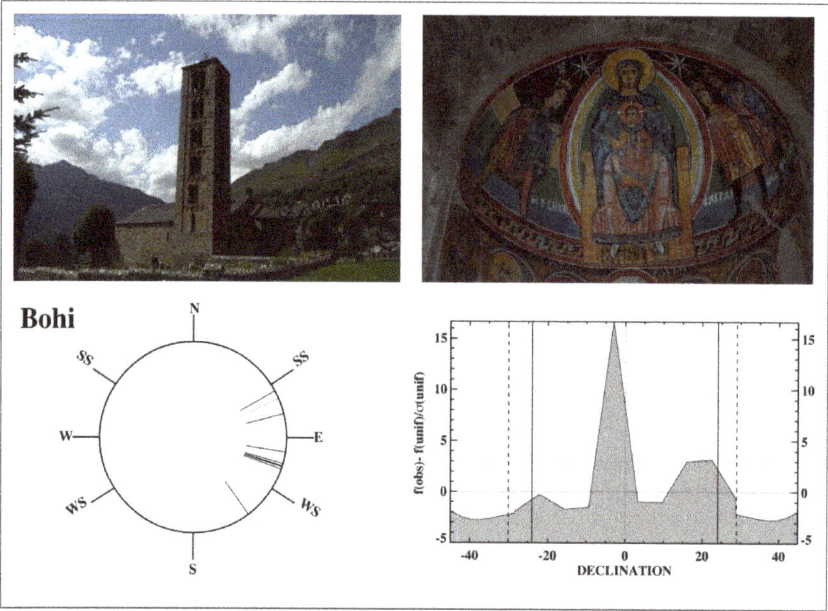

Figure 5. Top-left, Sant Climent de Taül, at Vall Bohí, is a wonderful example of the Romanesque churches in this mountainous area. Frequently the interior is decorated with magnificent paintings, often depicting astronomical symbolism associated to the Christian cult, such as in Santa María de Taül (top-right). The bottom-left panel shows the orientation diagram for the 10 churches presented in Table 1. Each measurement is plotted as a short stroke inside the circle. The strokes outside the circle indicate the cardinal points and the rise and set of the sun at the solstices for the latitude of the valley. However, this diagram does not take into account the altitude of the horizon, which at this alpine area is quite significant. To account for it, the declination histogram is presented in the bottom-right panel. It shows that, despite the azimuths being mostly towards several degrees south of east, the altitude of the horizon renders the orientations of a significant set of churches closer to equinoctial dates. © J.A. Belmonte and A.C. González-García.

Table 1. Orientation data for the nine churches in Vall Bohí, plus one in Andorra (Sta. Coloma). Columns indicate the church name, azimuth as measured towards the church apse, the altitude of the horizon in that direction and the astronomical declination. The mean latitude of the valley is 42°30′ North.

Church	a (°)	h (°)	δ (°)
Sta. María de Taüll	107.7	15	−2.4
S. Climent de Taüll	107.7	14	−3.1
S. Joan de Bohí	143.7	14	−24.4
Sta. Eulalia de Erill la Vall	107.2	10.5	−5.3
La Nativitat de Durro	109.2	19	−0.6
S. Feliu de Barruera	98.7	19	6.6
Sta. María de Cardet	105.75	14.5	−3.6
La Assumpció de Cóll	60.75	14.5	31.1
S. Joan de Caselles	75.5	7	15.3
Sta Coloma, Andorra	66	6.5	21.9

In summer 2011, nine churches in the area included in the UNESCO list (plus Santa Coloma in Andorra la Vella) were measured in a field campaign. Their data are included in Table 1. Measures were obtained using a Silva Survey Master tandem with a professional compass plus a clinometer. Thus, the accuracy of a single measurement can be estimated to be of $\frac{1}{4}°$ in azimuth and $\frac{1}{2}°$ in altitude of the horizon, this translates into roughly $\frac{3}{4}°$ in astronomical declination. No single rock or terrain area was moved when taking the measurements. These where corrected for magnetic declination employing the Enhanced Magnetic Model (EMM) found on the National Geophysical Data Center (NGDC) website (https://www.ngdc.noaa.gov/geomag-web/), the latitude was obtained with a hand-held GPS and astronomical declination was calculated employing our own software including a correction for atmospheric refraction [39].

Figure 5 shows that all churches were facing the east horizon, and all but one were inside the solar limits. In particular, the inspection of Table 1 and the declination histogram in Figure 5 shows that a good number of them were facing slightly south of the equinox, by averaging −3° in declination (above 15° in azimuth). Interestingly, the 14° of altitude of the close horizon at Sant Joan de Bohí corrects for the $143\frac{3}{4}°$ azimuth permitting, within the errors, a winter solstice alignment for this church, devoted to Saint John the Evangelist whose feast ranks close to the winter solstice, precisely. This was a rare phenomenon in the Iberian Peninsula in those distant epochs.

Another interesting consequence is that these churches seemed to comply with the norm found in the Serrablo valley, as explained early on. This could be due to the difficulty in building and properly orienting the churches in these alpine valleys with very close and abrupt horizons. However, the fact that both groups were close but not in direct contact (c. 90 km of rough land straddle them) poses an interesting problem and opens the possibility that for both groups the persistence of such orientation, slightly south of due east, may point to intentionality (perhaps similar dates for sunrise along the year).

Despite being part of different feudal territories (Counties of Aragon and Ribagorza, respectively) both were at the time dependent on the Frank kings and perhaps, given the difference with the orientation found in the contemporary churches of other parts of the Iberian Peninsula, the link should perhaps be sought for at the other side of the Pyrenees. This is something for further exploration in future research.

4. The Canarian Paradigm

Cultural astronomy studies, including both ethno- and archaeoastronomy, have now a tradition of a quarter century in the Canary Islands. The statistical relevance of astronomical implications for a certain number of sites and the spectacular character of some of the findings clearly points out to intentionality in the astronomical relationships discovered so far. There are two paradigms where this relationship is, besides, clearly related to sustainability issues: The mountain of Tindaya, in the island of Fuerteventura, and the 'Risco Caído and the sacred mountains of Gran Canaria Cultural Landscape'.

One of the earliest substantial discoveries in the islands was the probable astronomical connection of the aboriginal footprint engravings (usually called podomorphs) of Montaña Tindaya, a singular mountain dominating the northern plains of the semi-desert island of Fuerteventura (see Figure 6). Tindaya was a local reference of certain importance for ethnographic studies and also an important archaeological site where the largest sample of podomorphs in the world could be found. The mountain was also the centre of a series of aboriginal archaeological sites with obvious religious implications (Figure 6).

However, all these important issues were not enough to save the mountain from wild modern building speculation. Tindaya actually is the nucleus of an eroded volcanic cone formed by an appealing and beautiful stone called trachyte. Hence, a series of quarries were open in the skirts of the mountain, which could, eventually, have destroyed it completely (see e.g., Figure 1). In the mid 1990s, a team led by the archaeologist María Antonia Perera decided to include Tindaya and its podomorphs in a project where cultural astronomy would play the most relevant role to revalue the mountain and its environment. Despite some misfortune initiatives which tried to make a putative use of the results, such as the bizarre idea of making an 'insculpture' inside the mountain, the project was certainly successful [40].

The data showed that these petroglyphs, counted by hundreds, do not follow a random pattern but rather they have a clear custom of orientation with a concentration in the W-SW sector of the horizon (see Figure 7). Different hypotheses have been offered for this pattern and the visibility of the Peak of Teide in the distant island of Tenerife is perhaps among the most suggestive (see Figure 8). Teide, a huge volcanic cone 3714 m high, has been interpreted as a sort of axis mundi for the aboriginal populations of the Canary Islands because it is visible from all of them and it should have been really impressive either at daylight when covered by snows or at night when in an eruptive process. However, it cannot explain the whole sample alone.

Figure 6. Montaña Tindaya, in Fuerteventura, as seen from the ancient 'esequen' (stone circle) of Llano del Esquinzo. The southern (right) peak is virtually full of footprint engravings (podomorphs) with a non-random orientation pattern. Researchers are seated in cyclopean chairs facing the mountain. © J.A. Belmonte.

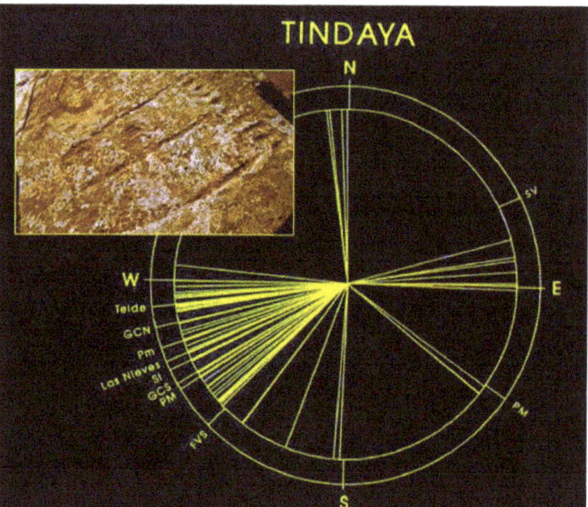

Figure 7. Orientation diagram of the podomorphs of Tindaya (see inset of a pair of footprint engraving), showing several topographical and astronomical references, including the winter solstice sunset (SI) and southern moon standstills (PM for Major Standstill and Pm for Minor Standstill). © M.A. Perera, J.A. Belmonte and C. Esteban.

Figure 8. Teide peak in faraway Tenerife Island, the axis mundi of many populations of the archipelago, as seen from Tindaya in Fuerteventura. This could be a topographic reference for orienting the podomorphs. © Gianfranco Costa.

Currently, the most suggestive hypothesis of all is the one postulating a relationship with the epoch of maximum rainfall in the island in the period from late November to late December and the vision of Venus as the evening star in combination with the crescent moon closest to the winter solstice [41]. Present day peasants and goatherds in Fuerteventura and other islands have interpreted the visibility and behavior of Venus as the evening star in this period of the year as a water carrier. Hence it could have been related to fertility cults in pre-Hispanic times [42]. As a matter of fact, cultural astronomy studies, without the need to alter the terrain and only looking at the horizon and the sky, have been able to reinforce the value of Tindaya, preserving the mountain for future generations in a sustainable way.

On the other hand, the island of Gran Canaria (see Figure 9) presented the most evolved and richest pre-Hispanic culture of the Canary Archipelago. The population had, as for each of the other islands, an undoubted proto-Berber ancestry, which lasted from the turn of the era to the Castilian conquest in 1483 AD. The social structure was complex and hierarchical, similar to a proto-state. This island is characterized by the presence of sanctuaries at the top of significant mountains and on the scarps of the huge volcanic calderas of the island, which are often called 'almogaren' thanks to the early chronicles, where particular rituals took place at precise moments of the year [43].

In particular, the area of the Caldera de Tejeda (Figure 9) presents a paradigmatic example of an adaptive process to a harsh but attractive environment, offering an excellent horizon. This includes impressive natural monuments such as the Roques Bentayga and Nublo, acting as reference landmarks where land- and skyscapes could be in close contact and permanent interaction (see Figure 10). This chain of facts suggested it as the perfect location for a cultural landscape which might be defended within the framework of UNESCO and the International Astronomical Union (IAU) Astronomy and World Heritage Initiative [6] under the name: 'Risco Caído and the sacred mountains of Gran Canaria' [44]. Risco Caído was the name of one of the most representative sites used as a sort of brand.

The area of the cultural landscape is impressive, including the Tejeda Basin and the Ravine of Barranco Hondo (where Risco Caído is located; see Figure 11). These are areas of the summit of Gran Canaria that have been inhabited since antiquity to the present day and where the local population had to make a huge adaptive effort to a not easy environment, applying sustainable practices before the word was even defined.

The relatively high cultural level of the ancient Canarians or 'Canarios', a name later extended to the rest of the archipelago, is clearly illustrated by the existence of irrigated land agriculture, with the stock of the products in communal granaries such as the one of El Alamo in Mesa de Acusa, in the western border of the Caldera. Indeed, a lunisolar calendar ought to be developed to control time and the productive cycle [45]. A large number of petroglyph stations, including alphabetic inscriptions, have been reported in Gran Canaria. The examples of the artificial sanctuary caves at Risco Chapín and Risco Caido (inside the cultural landscape limits, Figure 9; Figure 10) with the largest collection of pubic triangles in the world, which could be interpreted within a fertility cult, are among the most relevant.

The recently discovered light and shadow effects at Risco Caído [46] are indeed a highlight within this particular context. In Gran Canaria, dedicated fieldwork strongly suggests that most of the high-mountain sanctuaries, often located at high spots dominating a wide, and sometimes, impressive panorama, could be related with solar and lunar observations and, certainly astral cults [47] since: "the Canarians are idolatrous, worshiping the Sun, the Moon and other Planets" (Alvise de Cadamosto, 15th century AD [48]), or " ... they worshiped fire, the Sun and the Moon, and the Dog Star, when they started the year with grand feasting ... " (Tomas Arias Marín de Cubas, 1694 [49]). The cultural landscape under examination within the property area includes a good sample of these relationships.

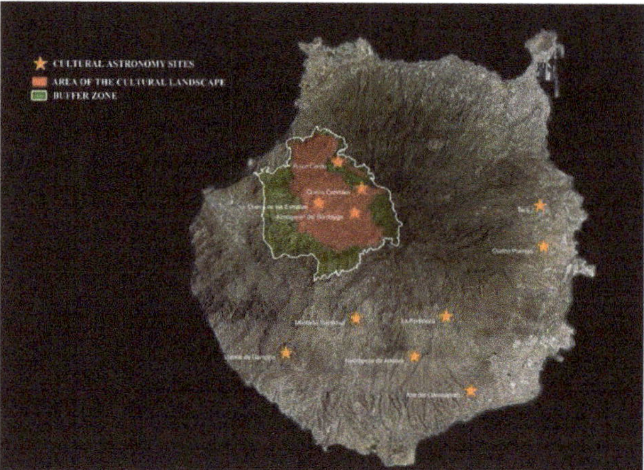

Figure 9. Star depicts settlements in Gran Canaria: Ancient Canarian sites where astronomy did play a most relevant role. Four of them, including some of the most important, are within the limits of the cultural landscape 'Risco Caído and the sacred mountains of Gran Canaria' (red color). This property corresponds to the best preserved and most sustainable area of the island. © Cabildo de Gran Canaria.

Figure 10. A general view of the Tejeda Basin, the nucleus of the cultural landscape, as seen from the base of Roque Nublo (Figure 12). It is dominated by Roque Bentayga (centre left) and Risco Chapín to the right. This is a nearly perfect example of an interacting land- and skyscape. Tenerife, with Teide, is seen in the distance, to the left above the clouds. © Cabildo de Gran Canaria.

Figure 11. Barranco Hondo Ravine, the ancient sector presumably named Artevigua in the chronicles, where the sanctuary of Risco Caido is located. This is a perfect example of sustainability applied to a harsh environment. Roque Nublo is seen in the distance. © Cabildo de Gran Canaria.

To handle with the initiative and defend the importance of the property in front of UNESCO evaluating panels (notably the International Council of Monuments and Sites, ICOMOS), a multidisciplinary international team of specialists was created, including archaeologists, anthropologists, botanists, architects, geologists, environmental experts, developing managers and, indeed, cultural astronomers (see Figure 12). Within this framework, 'Risco Caído and the sacred mountains of Gran Canaria' initiative has been considered as an excellent laboratory where the close relationship between land- and skyscapes in human culture can be illustrated. The idea is to prove that the area selected within the island (Figure 8) is a paradigm as a marvelous example of a cultural landscape worth being declared as a World Heritage Site. In this line of argument, the following outstanding universal value (or OUV in UNESCO jargon) criteria have been settled in the proposal (Astronomically relevant issues are highlighted in bold face):

Criterion (iii): "The ensemble of archaeological sites constitutes a unique and exceptional testimony to an extinct island culture that evolved in isolation for a period of more than one thousand five hundred years. Archaeological evidence and documentary sources relative to the proposed property bear witness to the fact that this culture dates back to the first settlers that arrived to these shores from the Berber Maghreb, which in itself is outstanding, as this thus constitutes a unique case of an island culture with roots in the pre-Islamic Amazigh world, manifestations of which are few and far between. The place expresses a very strong and very original relationship of human beings with nature (both Earth and Sky). **The proposed property constitutes an outstanding testimony to an island culture that integrates the skyscape as a fundamental part of its worldview, its rituals and beliefs, and which has also developed its own astronomical culture in close harmony with the natural environment and the surrounding landscape.** Clear evidence of this is the astronomical sanctuaries, such as Risco Caído that represent the pinnacle of the evolution of this knowledge and practices. This legacy illustrates the odyssey of the indigenous island cultures of the planet that have evolved over long periods without outside influence, ultimately creating their own cosmology and a unique world of knowledge and beliefs". And:

Criterion (v): "The indigenous troglodyte settlements of Caldera de Tejeda and the surrounding area constitute an inimitable example of this type of human habitat in ancient island cultures, illustrating a level of organization of space and adaptive management of resources that is highly efficient and complex. The extensive geological backdrop and the natural landscapes fuse with the cave settlements, sanctuaries, sites and terraces, developing a unique culture that still maintains its principal references, as well as its symbolic and cosmological connotations. This type of human settlement has survived through history, creating new ways of occupying the space that express the syncretism between the indigenous culture and the new culture established after the Spanish conquest. The survival of techniques and ancestral land uses, such as transhumance, should also be added to this. **The orientation and alignment of certain sanctuaries and artificial caves also indicate the intimate relationship that this type of settlement has with the skyscape and the principal symbolic elements of the landscape**. Spatial distribution of the settlements and the archaeological finds give us an in-depth understanding of how indigenous people used the territory of the sacred mountains. The areas surrounding the sites contain habitats and species of flora and fauna that also inform us on the way of life of the early settlers. Knowledge of the skills and cultural traditions of the indigenous settlers on the island has been definitively changed as a direct consequence of the new evidence provided by this incomparable territory".

Hence in Gran Canaria, and notably in the area of the property, a paradigmatic example of the interaction between topography and the sky is found; a paradigm which cultural astronomy has helped to disentangle. The comparative analysis made for the memory [50] suggests that there is no other site in the world where such a combination is so clearly illustrated although parallelisms could be found with Rapa Nui or Thebes in Egypt, but not in such a gigantic and outstanding scale as here.

Hence aspects like relations between astral divinities, seasonal cycles, the need to measure time or fertility worship with its rock-art symbols, is highlighted in the different pieces of evidence that illustrate the cosmology held by the ancient inhabitants of the Canary Islands.

Although it is true that an intrinsic relationship cannot be proven entirely, the evidence is suggestive and talks about a people who were able to adapt to a harsh environment that we researchers are obliged to keep today. In any event, it is not a question of interpreting an entire cultural context in accordance with the sky, just as it would be a serious mistake to forget about it.

Figure 12. Part of the team of scholars supporting the initiative of a UNESCO cultural landscape in Gran Canaria summit in front of Roque Nublo. The group is formed by archaeologists, anthropologists, botanists, architects, geologists, environmental experts and, indeed, cultural astronomers. © Cabildo de Gran Canaria.

5. Conclusions

To conclude, we would like to highlight once again that cultural astronomy is thus a discipline with a sustainable methodology that helps identifying new heritage or provides new value to already existing heritage or, finally, helps identify heritage at risk.

As a relatively new discipline, cultural astronomy accumulates new observations and points of view on different geographic areas, historic periods and subjects. Cultural astronomy allows us to grasp the worldview of past and ancient societies. Besides, it gives new sense to the archaeological and cultural heritage.

A further point that should be mentionehere is that cultural astronomy has a clear social educative value both to present the value of astronomy in ancient societies [51] or for actually teaching astronomy to present day students [52].

Finally, and as far as it engages with society to promote new initiatives, we can ensure the sustainability of the research. However, cultural astronomy also makes society aware of the importance of the sky for our past. Hence, it can help in promoting a sustainable use of the sky, for instance by promoting the use of particular luminaries at heritage sites where the sky is important for the apprehension of the past intent of the builders of ancient sites. The Starlight Initiative [53], and its strong cultural aspects, would be a good example of this.

Author Contributions: Conceptualization, A.C.G.-G. and J.A.B.; Writing—original draft preparation, A.C.G.-G. and J.A.B.; Writing—review and editing, A.C.G.-G. and J.A.B.

Funding: This work was partially supported by the projects P/310793 'Arqueoastronomía' of the IAC, and AYA2015-66787 'Orientatio ad Sidera IV' of the Spanish MICIU.

Acknowledgments: We would like to thank Prof. Ioannis Liritzis for the invitation to produce this work. The authors would especially like to thank Gianfranco Costa, Cipriano Marín and the Gran Canaria Council for some of the beautiful images illustrating this paper

Conflicts of Interest: The authors declare no conflict of interest.

References

1. Iwaniszewski, S. Por una astronomía cultural renovada. *Complutum* **2009**, *20*, 23–38.
2. Krupp, E.C. Archaeoastronomy. In *History of Astronomy, an Encyclopedia*; Lankford, J., Ed.; Routledge: New York, NY, USA, 1997; pp. 21–30.
3. García Quintela, M.V.; González-García, A.C. Arqueoastronomía, Arqueología y Paisaje. *Complutum* **2009**, *20*, 39–54.
4. Criado, F. *Arqueológicas: En Busca de la Razón Perdida*; Bellaterra: Barcelona, Spain, 2012; p. 23.
5. Castro, B. A historical review of the Egyptian calendars: The development of time measurement in ancient Egypt from Nabta Playa to the Ptolomies. *Sci. Cult.* **2015**, *1*, 15–27.
6. Ruggles, C.L.N.; Cotte, M. (Eds.) *Heritage Sites of Astronomy and Archaeoastronomy in the Context of the UNESCO World Heritage Convention: A Thematic Study*; ICOMOS-IAU: Paris, France, 2010.
7. Ruggles, C.L.N. (Ed.) *Heritage Sites of Astronomy and Archaeoastronomy in the Context of the UNESCO World Heritage Convention: Thematic Study no. 2*; ICOMOS-IAU: Paris, France, 2017.
8. Belmonte, J.A.; González García, A.C.; Polcaro, A. On the orientation of megalithic monuments of the Transjordan plateau: New clues for an astronomical interpretation. *J. Hist. Astron.* **2013**, *44*, 429–455. [CrossRef]
9. Nature on Sustainability. Available online: https//www.nature.com/subjects/sustainability (accessed on 6 March 2019).
10. Prendergast, F. Techniques of Field Survey. In *Handbook of Archaeoastronomy and Ethnoastronomy*; Ruggles, C.L.N., Ed.; Springer: New York, NY, USA, 2015; pp. 389–410.
11. Ruggles, C.L.N.; Martlew, R.D. The North Mull Project (3): Prominent Hill Summits and their Astronomical Potential. *Archaoastronomy* **1992**, *17*, S1–S13. [CrossRef]
12. Higginbotton, G.; Clay, R. Origins of Standing Stone Astronomy in Britain: New Quantitaive Techniques for the Study of Archaoastronomy. *J. Archaol. Sci. Rep.* **2016**, *9*, 249–258.

13. González-García, A.C.; Criado-Boado, F.; Vilas Estévez, B. Megalithic Skyscapes in Galicia. *Cult. Cosmos* **2017**, *21*, 87–103.
14. González-García, A.C.; Belmonte, J.A. Thinking Hattusha: Astronomy and landscape in the Hittite lands. *J. Hist. Astron.* **2011**, *42*, 461–494. [CrossRef]
15. Belmonte, J.A. *Pirámides, Templos y Estrellas*; Crítica: Madrid, Spain, 2012.
16. Liritzis, I.; Bousoulegka, E.; Nyquist, A.; Castro, B.; Alotaibi, F.M.; Drivaliari, A. New Evidence from archaeoastronomy on Apollo oracles and Apollo-Asclepius related cult. *J. Cult. Herit.* **2017**, *26*, 129–143. [CrossRef]
17. Rodríguez Antón, A.; González-García, A.C.; Belmonte, J.A. Astronomy in Roman Urbanism: A Statistical Analysis of the Orientation of Roman Towns in the Iberian Peninsula. *J. Hist. Astron.* **2018**, *49*, 363–387. [CrossRef]
18. Ruggles, C.L.N. *Astronomy in Prehistoric Britain and Ireland*; Yale University Press: New Haven, CT, USA, 1999.
19. Hoskin, M. *Tombs, Temples and their Orientation*; Ocarina Books: Bognor Regis, UK, 2001.
20. Zotti, G. Open-source virtual archaeoastronomy. *Mediterr. Archaeol. Archaeom.* **2016**, *16*, 17–24.
21. Gómez, J.; Mejuto, M.; Rodríguez, G. Generación de una capa astronómica para la IDE arqueológica. Arqueoastronomía en el sur de Portugal. *Rev. Catalana Geogr.* **2012**, *17*, 44.
22. González-García, A.C.; Vilas-Estévez, B.; López-Romero, E.; Mañana-Borrazás, P. Domesticating Light and Shadows in the Neolithic: The Dombate Passage Grave (A Coruña, Spain). *Camb. Archaeol. J.* **2019**, *29*, 327–343. [CrossRef]
23. Vlachos, A.; Liritzis, I.; Georgopoulos, A. The lightning of the god's face during solar stands in the Apollo temple Delphi. *Mediterr. Archaeol. Archaeom.* **2018**, *18*, 225–246.
24. Sitwell, O.F.G.; Latham, G.R. Behavioural Geography and the Cultural Landscape. *Geogr. Annaler. Ser. B Hum. Geogr.* **1979**, *61*, 51–63. [CrossRef]
25. Tuan, Y. Language and the Making of Place: A Narrative-Descriptive Approach. *Ann. Assoc. Am. Geogr.* **1991**, *81*, 684–696. [CrossRef]
26. Santos-Granero, F. Writing History into the Landscape: Space, Myth, and Ritual in Contemporary Amazonia. *Am. Ethnol.* **1998**, *25*, 128–148. [CrossRef]
27. Raffles, H. Local Theory: Nature and the Making of an Amazonian Place. *Cult. Anthropol.* **1999**, *14*, 323–360. [CrossRef]
28. Hayden, R.M. Antagonistic Tolerance. Competitive Sharing of Religious Sites in South Asia and the Balkans. *Curr. Anthropol.* **2002**, *43*, 205–231. [CrossRef]
29. Holtzman, J. The Local in the Local. Models of Time and Space in Samburu District, Northern Kenya. *Curr. Anthropol.* **2004**, *45*, 61–84. [CrossRef]
30. Ingold, T. The Temporality of the Landscape. *World Archaeol.* **1993**, *25*, 152–174.
31. Criado Boado, F. *Del Terreno al Espacio*; GIAP-U. de Santiago: Santiago, Spain, 1999.
32. Durkheim, É. *Las Formas Elementales de la Vida Religiosa*; Akal: Madrid, Spain, 1982.
33. Smith, J.Z. *To Take Place*; U. of Chicago Press: Chicago, IL, USA; London, UK, 1987.
34. Woodard, R.D. *Indo-European Sacred Space*; U. of Illinois Press: Urbana, IL, USA; Chicago, IL, USA, 2006.
35. Kristiansen, K. Towards a New Paradigm? The third Science revolution and its possible consequences in Archaeology. *Curr. Swed. Archaeol.* **2014**, *22*, 11–34.
36. Bellmunt Figueras, J. *La Vall de Boí*; Pagès Editors: Barcelona, Spain, 2017.
37. González-García, A.C. A voyage of Christian medieval astronomy: Symbolic, ritual and political orientation of churches. In *Stars and Stones: Voyages in Archaeoastronomy and Cultural Astronomy*; Pimenta, F., Ed.; BAR International Series: Oxford, UK, 2015; pp. 268–275.
38. González-García, A.C.; Belmonte, J.A. The Orientation of Pre-Romanesque Churches in the Iberian Peninsula. *Nexus Netw. J.* **2015**, *17*, 353–377. [CrossRef]
39. Schaeffer, B. Astronomy and the Limits of Vision. *Vistas Astron.* **1993**, *36*, 311–361. [CrossRef]
40. Perera Betancort, M.A.; Belmonte, J.A.; Esteban, C.; Tejera Gaspar, A. Tindaya: Un acercamiento arqueoastronómico a la sociedad prehispánica de Fuerteventura. *Tabona* **1996**, *9*, 163–193.
41. Belmonte, J.A.; Hoskin, M. *Reflejo del Cosmos*; Equipo Sirius: Madrid, Spain, 2002.
42. Belmonte, J.A.; Sanz de Lara, M. *El Cielo de los Magos*; La Marea: La Laguna, Spain, 2001.
43. Esteban, C.; Belmonte, J.A.; Schlueter, R.; González, O. Equinoctial markers in Gran Canaria Island, Parts I & II. *Archaeoastronomy* **1996**, *21*, S73–S79.

44. Belmonte, J.A.; Cuenca Sanabria, J.; Gil, J.C.; Marín, C.; de León, J.; Ruggles, C.L. The cultural landscape 'Risco caído and the sacred mountains of Gran Canaria: A paradigmatic proposal within UNESCO Astronomy and Wolrd Heritage initiative. *Mediterr. Archaeol. Archaeom.* **2018**, *18*, 377–385.
45. Belmonte, J.A.; Perera Betancort, M.A.; González-García, A.C. Calendario, signo y símbolo: Tres claves para una aproximación al poblamiento del Archipiélago Canario. In *Homenaje a Antonio Tejera*; Chaves, E., Ed.; Universidad de La Laguna: La laguna, Spain, 2019; pp. 118–141.
46. Cuenca Sanabria, J.; de León, J.; Marín, C.; Gil, J.C.; Belmonte, J.A.; Gil Sarmiento, C.; Márquez Zárate, J.M. The almogaren of Risco Caído: A singular astronomical sanctuary of the ancient Canarians. *Mediterr. Archaeol. Archaeom.* **2018**, *18*, 11–18.
47. Belmonte, J.A. Pre-Hispanic sanctuaries in the Canary Islands. In *Handbook of Archaeoastronomy and Ethnoastronomy*; Ruggles, C.L.N., Ed.; Springer: New York, NY, USA, 2015; pp. 1115–1124.
48. Cadamosto, A. *Viagens de Alvise de Cadamosto e de Pedro de Sintra*; Academia Portuguesa da Historia: Lisbon, Portugal, 1948.
49. Marín de Cubas, T.A. *Historia de las Siete Islas de Canaria*; Real Sociedad de Amigos del País de Las Palmas: Las Palmas de Gran Canaria, Spain, 1986.
50. de León, J.; Marín, C. (Eds.) *Risco Caído and the Sacred Mountains of Gran Canaria Cultural Landscape*; Cabildo de Gran Canaria: Las Palmas, Spain, 2018.
51. Hannah, R. The roles of observational astronomy in ancient Greece. *Sci. Cult.* **2015**, *1*, 47–56.
52. Pannou, E.; Violetis, A. Teaching astronomy using monuments of cultural heritage: The educational example of "horologion of Andronikos Kyrrhestes". *Sci. Cult.* **2018**, *4*, 77–83.
53. Starlight Initiative. Available online: https://fundacionstarlight.org/ (accessed on 6 March 2019).

© 2019 by the authors. Licensee MDPI, Basel, Switzerland. This article is an open access article distributed under the terms and conditions of the Creative Commons Attribution (CC BY) license (http://creativecommons.org/licenses/by/4.0/).

Article

Archaeometric Approaches to Defining Sustainable Governance: Wari Brewing Traditions and the Building of Political Relationships in Ancient Peru

Patrick Ryan Williams [1,*], Donna J. Nash [2], Joshua M. Henkin [3] and Ruth Ann Armitage [4]

1 Field Museum, Chicago, IL 60605, USA
2 Department of Anthropology, University of North Carolina at Greensboro, Greensboro, NC 27412, USA; djnash@uncg.edu
3 Medicinal Chemistry and Pharmacognosy, University of Illinois at Chicago, Chicago, IL 60607, USA; henkinj@gmail.com
4 Department of Chemistry, Eastern Michigan University, Ypsilanti, MI 48197, USA; rarmitage@emich.edu
* Correspondence: rwilliams@fieldmuseum.org; Tel.: +1-312-665-7008

Received: 28 February 2019; Accepted: 2 April 2019; Published: 18 April 2019

Abstract: Utilizing archaeometric methods, we evaluate the nature of production of feasting events in the ancient Wari state (600–1000 CE). Specifically, we focus on the fabrication of ceramic serving and brewing wares for the alcoholic beverage *chicha de molle*. We examine the source materials used in the creation of these vessels with elemental analysis techniques (INAA and LA-ICP-MS). We then assess the chemical traces of the residues present in the ceramic pores of the vessels to detect compounds indicative of the plants used in *chicha* production (DART-MS).While previous research has identified circumstantial evidence for the use of *Schinus molle* in the production process, this research presents direct evidence of its existence in the pores of the ceramic vessels. We also assess what this material evidence suggests about the sustainability of the feasting events as a mode of political interaction in the Wari sphere. Our evaluation indicates that regional resource use in the production of the ceramic vessels promoted locally sustainable raw material procurement for the making of the festivities. Likewise, drought resistant crops became the key ingredients in the beverages produced and provided a resilient harvest for *chicha* production that was adopted by successor groups.

Keywords: elemental analysis; archaeological chemistry; organic residue analysis; phytochemistry; ethnobotany; Andean Middle Horizon

---

1. Introduction

Sustainable governance requires a shared set of values by which political elites affirm their allegiance to a set of ideals. They may differ in their interpretation of how they arrive at those ideals, but in successful states they ultimately cooperate for their political futures [1]. That shared set of values requires an understanding of their common past and of their collaborative potential and those understandings are forged in political discourse. In ancient Peru, and in many other societies, those shared interests were built in ritual acts consummated in special accords.

In the Wari empire (600–1000 CE), those agreements were formed, in part, during elite festivals, involving the consumption of *chicha de molle*, an alcoholic beverage of superb potency. Served in ceramic vessels that invoked the supernatural and or communicated elite allegiances [2,3], *chicha* consumed during ritual drinking sessions fomented political alliances and reified Wari ideology [4]. The most elaborately decorated ceramic cups may have established a link between vessels and solidified the relationships between the Wari elite who drank from them and the supernatural beings that controlled water availability and fecundity that were presumably represented on these cups. The practice of drinking *chicha* invoked the flow of mountain water and represented the shared desire to exert supernatural control over the most precious resource—water [5].

These rites of incorporation relied on both specialized knowledge on how to produce them, as well as the material means to enact them at the various venues throughout the empire. Specialized knowledge involved the harvesting of clays to build the ceramic bodies and the acquisition of pigments to create the highly ritualized iconography painted on the vessels. It required the esoteric knowledge of the geometric designs and the graphic and vivid supernatural beings painted on the vessels and the means to execute them [6]. It also required a knowledge of the art of brewing and an understanding of the ethnobotanical materials that were the ingredients in a successful beer [7].

The material means to produce these events, in the Wari case, revolved around highly decorated and ritually charged iconography on ceramic brewing and serving wares [8]. Use of renewable local resources may have ensured that the various regional venues for these events were not dependent on resources beyond the regional governor's immediate control. That allowed for these festive events to be independent of disruption to trade routes, political bickering outside the local area, or interference from external adversaries. In other cases, ornate imported ceramic vessels might signal stronger affinity with the imperial center. However, their replenishment would be dependent on external producers distant from the brewing and feasting locales. Interference in the delivery of these goods could impact the ability to carry out the festivals that reaffirmed political ties and alliances.

The other important component for producing a Wari festival was the raw material for the brew itself. *Chicha* can be made of many different ingredients [9]. The most common base for *chicha* in the late prehispanic period was maize, and it remains to this day the most popular ingredient in the Andes. Other *chichas* from distinct areas of the Andes were reported by Spanish chroniclers to be made of tubers, peanuts, strawberries (*Fragaria chiloensis*) and other fruits, quinoa, and the berry of the Peruvian pepper tree, *Schinus molle*. This latter ingredient is especially interesting since finds of large quantities of desiccated *molle* seeds are often found in Wari sites, and provide a compelling indication of the type of *chicha* favored by the Wari [10]. We argue that both the ceramic vessels from which Wari beer was served, as well as the composition of the beer itself, were critical to making these events uniquely Wari. And it was only in these special circumstances that political allegiances sanctioned by Wari customs could be formed.

These accords were materialized both in the media from which the sacred beverage was consumed and in the liquid itself. The ceramic vessels in which the *chicha* was served were especially created for the elite event, and the brew which cemented the agreements conceived was also an extraordinary concoction focused on Wari culinary traditions rooted in local-imperial relations created for the occasion. Likewise, the events themselves were held in Wari architectural frames implanted within administrative centers throughout the imperial realm. These architectural complexes included features like platforms that highlighted the role of the patron of the feast in asymmetrical power relations with the attendees [11].

In this contribution, we utilize archaeometric methods to elucidate the technologies for producing and serving the political elixir for cementing relationships between elites in the Wari realm: *chicha*. In particular, we evaluate the special raw material sources used in the production of ritually decorated ceramics in which the brew was provided, as well as the essential constituents of the brew itself as preserved in the pores of the ceramics from which it was made and served.

1.1. Cerro Baúl Brewery as Study Focus

The materials on which this research is based were recovered from an ancient Wari brewery discovered at the site of Cerro Baúl [10,12]. Cerro Baúl was the southernmost administrative center in the Wari realm. Located in the Moquegua region of southern Peru, it occupied the summit of a unique mountaintop locale on the frontier with Wari's imperial rivals, the colonies of the Tiwanaku polity [13,14] (Figure 1). It was thus both a cosmopolitan political center and an embassy to a rival polity. The nature of political interaction on this frontier was a critical example of Wari state practice, and at its heart was a large-scale brewing facility dedicated to producing a unique Wari brew served in highly decorated drinking vessels made on site [6]. Cerro Baúl provides evidence that Wari colonial governance in Moquegua was maintained in part through a ritual feasting center around *chicha* production. This pattern of governance persisted for four centuries on the southern Wari frontier, and implies the practice, as well as the governmental system it supported, were remarkably sustainable, including during periods of more arid climate.

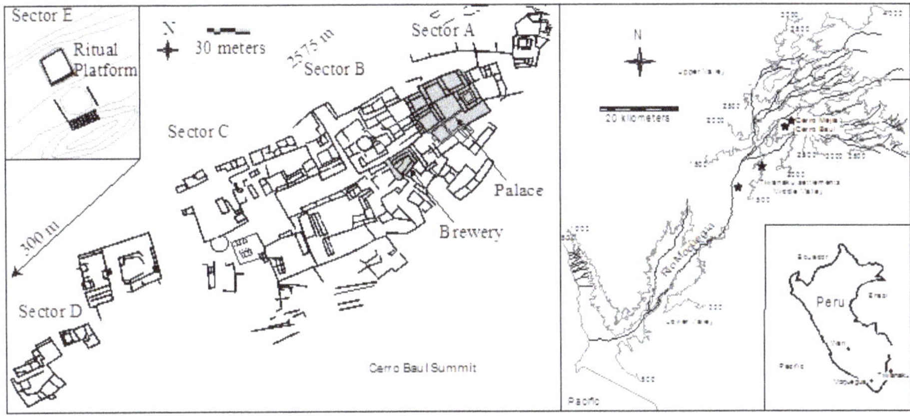

Figure 1. Map of the Cerro Baúl site in the Moquegua research area.

Previous research has highlighted the contexts of the brewing traditions and the material remains of production debris [15–17] as well as the nature of Wari political economy [18]. Here we focus on archaeometric analysis of the materials used to create this event, arguing that it was the special contexts of production that made these alliances unique. We assess the nature of the material ingredients that formed the ceramic vessels and the potent brew they contained (Figure 2). This paper uses the results of ceramic sourcing studies and residue analysis of the ceramics from the *chicha* brewery at Cerro Baúl to understand how this practice sustained political interactions over the latter half of the first millennium CE.

Figure 2. *Cont.*

Figure 2. Example vessels from the Cerro Baúl brewery identified as part of the Baúl chemical group by INAA. They represent cups, fermentation and serving vats, and a *kero* or drinking mug.

1.2. The Brewery Context

The excavated area of the brewery constitutes approximately 500 square meters encompassing four distinct spaces (Rooms 1–4) as well as an additional set of smaller rooms to the southwest (Figure 3). Distinct rooms were dedicated to grinding maize, preparing comestibles, boiling the mash, and fermenting and storing the brew. The boiling room contained the remains of between eight and twelve hearths along the northwest wall, each framed by two upright stones placed 50 cm from the wall to support a ceramic boiling vat. The hearths contained the burnt remains of fuel ash, identified principally as dung, including guinea pig as well as camelid sources. Trash pits in the floor of the boiling room contained the remains of large numbers of the spent drupes of the Peruvian pepper berry, *S. molle* [16]. The fragmented remains of large cooking vessels were also recovered from this space, with sooted base fragments and clear evidence of use. One of these fragments was collected in situ and wrapped in foil for residue analysis, reported below.

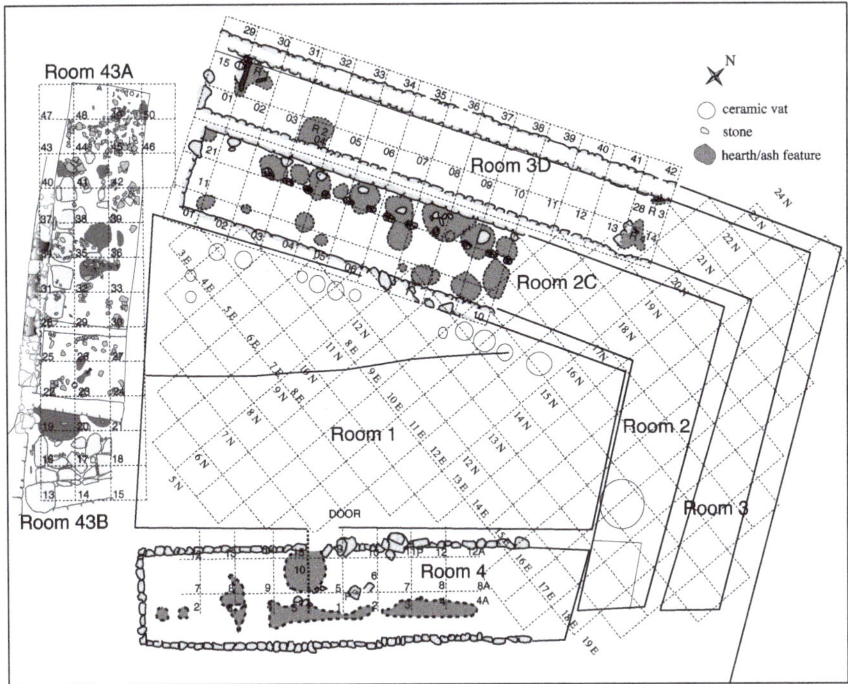

Figure 3. Map of the brewery at Cerro Baúl, illustrating the fermentation/storage area (Room 1), boiling area (Room 2C), and the grinding and food preparation rooms (Rooms 4, 43A). Other rooms may have been used for storage or access to brewing facilities.

The adjacent fermentation or storage room contained the remains of several oversize jars, originally lined up along the northwest wall. These jars held up to 150 L of brew each and were decorated as Wari personages, with a face on the neck of the jar, or with a chevron headband painted on the rim of the neck as shown in Figure 2. Another large vat found in situ embedded in the floor was over a meter in diameter and may have held up to 1000 L of liquid. The fermentation room and the adjacent grinding rooms were the areas where most of the fineware fermentation vats and drinking and serving vessels were found smashed on the floors [10]. These serving vessels include cups of ca. 300 mL volume, keros (drinking mugs) of upwards of 2-L volumes, open bowls, and restricted jars. Four of the sherds from the fermentation/storage vessels were exported for residue sampling as described later. The highly decorated nature of the assemblage, the comparative high quality of the ceramic materials, and the highly privileged location of the brewery indicates it was an elite production space. While evidence of smaller scale brewing has been identified in intermediate elite contexts at the adjacent site of Cerro Mejia and Cerro Trapiche and in the houses of Tiwanaku leaders, we have not seen contemporary commoner houses with these production contexts.

Radiocarbon dates and evidence of modification of space in the brewery indicate a long occupation, with at least two phases of construction. Radiocarbon dates range from 1400 ± 60 BP to 900 ± 40 BP uncalibrated [10], which span the range of occupation of the Wari settlement as a whole. When the site was abandoned ca 1050 CE, the brewery was intentionally burnt to the ground and the fineware ceramic drinking vessels were smashed in the smoldering flames. As the fire extinguished, seven semi-precious stone and shell bead necklaces were placed atop the ashes and covered with sediment to cap the ritual closure of the facility. The large fermentation jars in the central storage room were also broken and pieces were dispersed throughout the brewery. These drinking cups, serving wares, and fermentation vats were the samples for the brewery ceramic paste sourcing analysis discussed next, some of which are displayed in Figure 2.

2. Materials and Methods

2.1. Ceramic Paste Analysis Methods

A sample of 20 ceramic sherds from the Baúl brewery were analyzed by instrumental neutron activation analysis (INAA) at the Missouri University Research Reactor (MURR) and another 19 samples from nearby consumption contexts were analyzed by both INAA at MURR and by laser ablation inductively coupled plasma mass spectrometry (LA-ICP-MS) at the Elemental Analysis Facility (EAF) at The Field Museum [19–21]. We also analyzed an additional 41 sherds from other summit contexts on Cerro Baúl for a total of 80 ceramic samples from the top of the mountain citadel (Table 1).The INAA dataset also contains comparative ceramic samples of 10 sherds from Formative sites in the local Moquegua region, 51 sherds from the Wari site of Cerro Mejia, 44 samples from Wari and Tiwanaku sites on the slopes of Cerro Baúl, as well as 29 ceramic sherds from the Wari heartland [19].

Table 1. Summary table of ceramic sourcing samples from Cerro Baúl and comparative sites.

Context	N	Lab Numbers	Method	Group(s) Assigned
Baúl brewery (elite)	20	PRW002,4–8,11, 13–14, 16, 20, 27, 29–32, 36, 38–40	INAA	Baúl group (90%) Mejia A (5%) Unassigned (5%)
Baúl palace and consumption (elite)	19	PRW101–102, 104, 106, 109, 112–113, 118, 120, 123, 125, 131, 133, 136, 144–145, 147–148, 150	INAA & LA-ICP-MS	Baúl group (100%)
Other Baúl summit (elite)	41	PRW001, 3, 9–10, 12, 15, 17–19, 21–26, 28, 33–35, 37,103, 105, 107–108, 110–111, 114–117, 119, 121–122, 124, 126, 130, 132, 134, 138, 143, 149	INAA	Baúl group (46%) Mejia A, B, D, E, or G (29%) Unassigned (25%)
Formative sites (local)	10	PRW183–192	INAA	Mejia A (40%) Mejia D (30%) Unassigned (30%)
Cerro Mejia (local)	51	PRW046–096	INAA	Mejia A (24%) Mejia B (31%) Mejia D, G, or F (12%) Baúl Group (6%) Unassigned (27%)
Baúl slopes (local)	44	PRW099–100, 127–129, 135, 137, 139-142, 146, 151–182	INAA	Mejia A (52%) Mejia B, C, D, E, or F (32%) Baúl Group (2%) Unassigned (14%)
Tiwanaku sites in Moquegua	29	TW001–029	LA-ICP-MS	Local Tiwanaku (86%) Unassigned (14%)
Wari heartland	29	PRW195–223	INAA & LA-ICP-MS	Wari-1 (55%) Wari-2 (21%) Wari-3 (10%) Mejia A (4%) Unassigned (10%)

Sharratt et al. [22] conducted a clay survey in 2005, providing 50 clay locales that were analyzed via LA-ICP-MS. In conjunction with this analysis, they analyzed an additional 29 Tiwanaku sherds from the local region and compared them with 19 of the 29 sherds from the Wari heartland to provide the link between local clay sources and ceramic production workshops [20–22]. This analysis complements the INAA data and the 19 samples from Baúl consumption contexts outside the brewery mentioned earlier.

INAA analytical methods are detailed in Glascock [23], while we describe ICP-MS methods here. A Varian inductively-coupled plasma-mass spectrometer (ICP-MS), equivalent to the Varian 810 instrument, was used at the EAF (quadrupole, auto-optimized spectrometer on 55 selected isotopes). The facility uses a New Wave UP213 (helium carrier gas, 213 nm laser operated at 0.2 mJ and a pulse frequency of 15 Hz) laser in conjunction with the LA-ICP-MS to introduce solid samples. Methodology follows that presented previously [20,22].

Ceramic samples were laser-ablated with a spot size of 150 microns and a dwell time of ninety seconds. Each sample was ablated ten times and a total of 55 elements was measured, using 29Si as an internal standard to control for time variability in ablation efficiency and resulting signal strength. Concentrations were calculated using NIST standards n610, n612, and Brick Clay (n679) via the approach first presented by Gratuze, Blet-Lemarquand, and Barrandon [24]. Large temper grains in the ceramic were avoided when positioning the laser so as to focus on the clays used in production. Ablation took place on sherd cross-sections to avoid pigments and treatments on ceramic surfaces.

Statistical analysis was performed using SPSS and Gauss Runtime statistical routines developed by MURR.A principal components analysis was carried out on the 16 elements that best produced group separation in the INAA and ICP-MS data [20]. We converted concentration values to base-10 logarithms to improve normality for trace element data and to minimize scaling differences of trace elements. We conducted a principal components analysis using elements from the variance–covariance matrix in order to examine the multivariate patterns in the data.

2.2. Residue Analysis Methods

In order to provide a baseline for residues within Wari ceramic vessels, Donna Nash conducted an experimental archaeology process for the production of *chicha de molle*. In collaboration with ethnographic informants, Nash prepared three types of *chicha* in newly fabricated clay vessels: *chicha de jora* (*maíz*), *chicha de molle*, and *chicha de maíz y molle*. Maize was germinated, then ground and boiled. *Molle* was winnowed and the seeds steeped in hot water for a short period of time. The liquid was then left to repose in clay fermentation vessels for five days before tasting. The *chicha* was left in the vessels for another five days, its maximum shelf life before consumption, and the empty clay vessels for each type of *chicha* were broken into sherds. Several years later, the clay vessel sherds were subjected to chemical analysis of residues embedded in the ceramic pores, as described herein. Likewise, five fragments from clay vessels from the ancient brewery at Cerro Baúl were subjected to the same residue analysis to ascertain which brew was chemically most similar to the archaeological residues embedded in the ceramic vessels.

Direct analysis in real time mass spectrometry (DART-MS) was employed to characterize potsherd residues and spent *molle* drupes recovered from the Cerro Baúl brewery, which were compared to experimental archaeology controls (produced by Nash and her ethnographic informants as described above).Analyses were carried out using a DART ionization source (IonSense, Saugus, MA) on a JEOL AccuTOF mass spectrometer, additional background on the DART-MS may be found elsewhere [25]. Spectra were collected in negative ion mode with the DART ion source at 500 °C using helium as the DART gas. Orifice 1 on the AccuTOF was set to −30 cV to minimize fragmentation; the ring lens voltage and orifice 2 were set to −5 V. The temperature at orifice 1 was held at 150 °C for analysis. The DART grid voltage was kept at the default value of −530 V. The mass spectrometer RF ion guide ("peaks voltage") was set to 800 V to maximize sensitivity above m/z 80. Mass calibration was carried out using PEG-600 in methanol during each acquisition. The mass resolving power was approximately 6000.

Individual *molle* drupes were held in a pair of tweezers and introduced into the space between the DART ion source and the AccuTOF mass spectrometer orifice. No sample preparation was required. For the ceramic materials, a small amount of the surface was removed using a rotary grinding tool and placed into a small autosampler vial with a few drops of HPLC grade methanol. This ceramic slurry was introduced into the space between the DART ion source and the mass spectrometer orifice on the closed end of a capillary melting point tube. The methanol served primarily to allow a significant quantity of ceramic slurry to be subjected to ionization rather than as a solvent for the residue [26]. DART-MS has a distinct advantage over the traditional residue analysis methods like gas chromatography-mass spectrometry (GC-MS) and liquid chromatography-mass spectrometry (LC-MS) in that it requires no sample preparation, or as in the case here, very little. The speed of analysis—just a few seconds in the ion source—comes at a cost: molecules can only be identified based on their molecular mass. Fragmentation can be accomplished by using a higher voltage on Orifice 1, but interpretation is complicated by the sheer number of ions produced from the mixture being ionized. This analysis was an initial attempt to determine if the pertinent residues could be distinguished with DART-MS, an approach that uses minimal solvent, no additional reagents, and requires very little time. It is, one could argue, a sustainable initial approach to screening ceramics prior to further study with more traditional methods like GC-MS or LC-MS.

The sensitivity of the DART-MS method is dependent upon the amount of material presented for ionization. For the ethnographic samples, the control ceramic and *chicha de jora* vessels were sampled both before and after being scrubbed with a brush under running water to simulate the process of field cleaning of ceramic samples. Each sample was run in triplicate unless otherwise noted. Compounds were identified in the spectra using modified versions of the databases compiled by one of us (JMH) of compounds that have been previously identified and isolated in *Schinus* spp. and in maize, from Reaxys and/or NAPRALERT web-based tools [27]. The modified databases removed compounds present as glycosides (or linked to other sugars) as these compounds do not readily ionize by DART-MS.

Identifications are based on the observed peaks being within 10 millimass units (± 1 in the second decimal place) of the calculated exact mass of the compounds in the database. Each compound is identified based on only one mass, as the deprotonated species (molecule less one hydrogen atom, or M–H anion); as multiple compounds can share the same formula and thus the same mass, identifications must also be based on what compounds are present from this small subset derived from web-based tools and the analysts' (RAA and JMH) discernment in relation to the mass spectra.

3. Results

3.1. Ceramic Paste Sourcing Results

We discovered that Wari ceramics from Cerro Baúl were chemically distinct from contemporary materials in the local region, including both Tiwanaku and local sherds.They were also distinct from a control group of ceramics excavated at the Wari capital in Ayacucho. Statistical analysis of the INAA data resulted in the definition of three chemical groups from the Wari heartland (Wari-1, 2, and 3), one chemical group concentrated on the Cerro Baúl site summit (Baúl Ref.), and seven groups of local ceramic production (Mejia A-G) [19].In the INAA data, a biplot on the first two principal components calculated from the variance–covariance matrix of the ceramic samples shows the brewery ceramics can be distinguished from both Wari heartland materials and other local ceramics on an axis of variability associated with the second principal component (Figure 4).

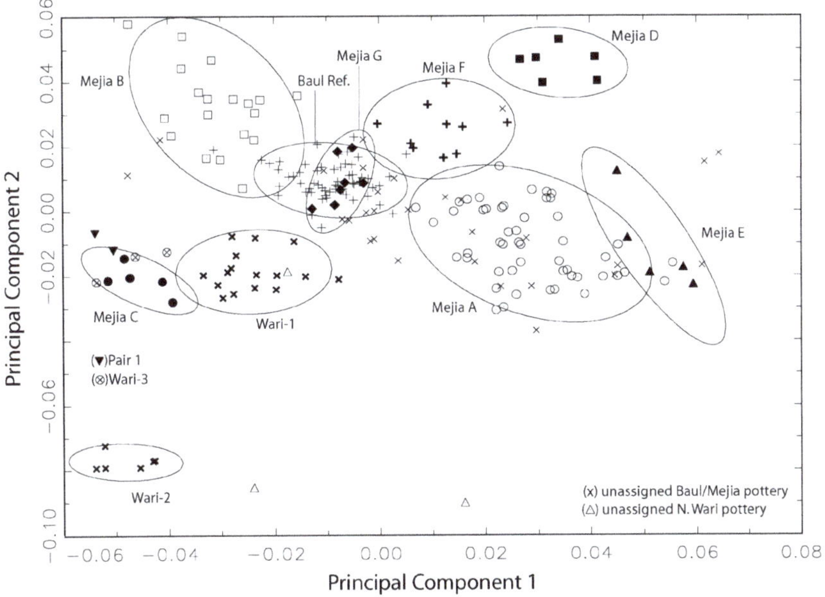

Figure 4. PC1 v PC2 for INAA data from Baúl, Mejia (local) and Wari heartland ceramics [19].

This component is positively loaded on Na, Sr, Ca, and Th and negatively loaded on Fe, V, Co, Sc, Ti, Sb, Cr, As, and Zn. A biplot of Cr and La distinguishes the Baúl group from the local Mejia pastes (Figure 5).

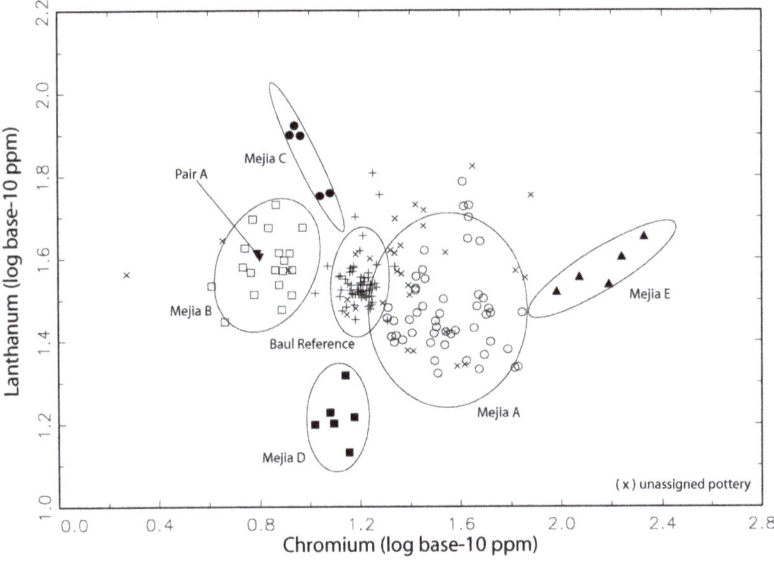

Figure 5. Chromium v Lanthanum plot for INAA data [19].

In the ICP-MS data, the distinction of the Cerro Baúl brewery ceramics is also pronounced, but this time the second principal component, which together with PC1 accounts for 47% of the total variance, is loaded positively on Be, K, Nb, Rb, U, and Th. It is negatively loaded on Fe, Cu, Sr, V, Co, Mg, and Ca [22]. Analysis of local clay sources indicate that the Baúl ceramics match most closely with a clay source across the valley from the site. The local Tiwanaku and Mejia ceramics best match clay sources from geological formations in the Torata and Moquegua valleys themselves [22]. None of the archaeological ceramics from the Moquegua region were chemically similar to the three sources identified from the Wari heartland.

Ceramic samples from the brewery (n = 20) represent 25% of the collection of pottery sourced by INAA on the summit of Cerro Baúl (n = 80). While the Baúl chemical group accounts for about 70% of the summit assemblage, it accounts for all but two samples from the brewery. Thus, 90% (18/20) of the brewery ceramic vessels are from the Baúl ceramic chemical group (Table 1). Most of those are serving wares constituted by bowls and drinking cups. It is apparent that the brewery was a key consumer of the special ceramic workshop production represented by the Baúl chemical group. The palace and adjacent consumption contexts on the summit of Cerro Baúl near the brewery were also key consumers of the Baúl workshop ceramic wares. Nash [6] identified the probable location of a ceramic workshop in the palace on the summit of the mountain that is the likely locale of production for these specialized ceramic vessels.

Off the summit of Cerro Baúl, the finely made and decorated Baúl group ceramics are very rare [6]. At the adjacent Wari site of Cerro Mejia, only 6% of the samples match the Baúl group, while in local settlements on the slopes of Cerro Baúl, only 2% (a single vessel) of the ceramic assemblage is part of the Baúl group [19]. Outside of these sites immediately adjacent to the Wari administrative center at Cerro Baúl, none of the other Moquegua valley samples match the Baúl chemical group. In fact, greater than 90% of all these ceramic assemblages were produced from Moquegua or Torata valley clay sources with the exception of the elite wares that are part of the Baúl group from the summit of the mountain.

Thus, the archaeometric data indicate that the brewery ceramics were produced from a specialized set of raw materials, dedicated to political consumption events in ritually charged monumental architecture on the summit of Cerro Baúl. We previously identified the likely clay source for that ceramic production on a mountain across the valley from the site [22]. Interestingly, that clay source is uniquely used by the ceramicists at Cerro Baúl and is not used by any other contemporary groups, despite the high quality of the clays. This likely demonstrates an exclusionary control over the access to the clay sources by the Wari elite.

3.2. Residue Analysis Results

The results of the residue analysis of the ethnographic ceramics by DART-MS are presented in Table 2. Several compounds from the *Schinus* database were found in a control vessel that contained none of the *chicha* recipes and were thus excluded from further results as being not indicative of the presence of *Schinus molle* residues. The base peak of all of the *chicha de jora* samples (even after washing) is observed at m/z 341.108. Of the possible formulas calculated for this mass was $C_{19}H_{17}O_6$, which would correspond to a number of tetrahydroxyflavone compounds, or $C_{12}H_{22}O_{11}$, which would correspond to a disaccharide like maltose. While the calculated masses for each of these compounds is different, it differs only in the third decimal place, meaning that within the mass resolution of the AccuTOF mass spectrometer, we cannot differentiate between these compounds. Maltose is the most likely source of that peak.

The *chicha de molle* experimental samples contained few of the compounds that were both found in the literature for *Schinus* species and could be ionized by DART-MS. Fewer yet were identified in the *chicha* with both corn and *molle*. The potentially relevant compounds that could pertain to *molle* which seem to have been identified in ethnographic controls of (1) *chicha de molle* alone and (2) *chicha de molle* with maize were respectively: (1) terebanene, teredenene, or β-spathulene (a sesquiterpenoid); safrole (a phenylpropanoid); and gallic acid (a phenolic acid) and (2) terebanene, teredenene, or β-spathulene (a sesquiterpenoid). None of these compounds alone or altogether as an assemblage affords a firm identification of *molle* to these ethnographic controls, but all are consistent with *molle*. Indeed, the DART mass spectra of the *molle* drupes showed all of these compounds, as well as a compound with m/z 453.332, which corresponds to the $[M-H]^-$ ion for the formula $C_{30}H_{46}O_3$. This is most likely either (1) masticadienoic acid (also known as terebinthone), (2) a positional isomer of that compound such as isomasticadienoic acid, or otherwise (3) moronic acid; these compounds are indistinguishable based on the molecular mass. The DART mass spectrum of the drupes from the Cerro Baúl excavation also showed an additional signal most likely arising from β-sitosterol (m/z 413.380, $C_{29}H_{50}O$ $[M-H]^-$).

Table 2. Experimental ceramic residues on vessels from DART-MS.

Chicha de jora (maize)				
Compound	**Formula**	**Measured Mass**	**Calculated Mass**	**Intensity**
Methyl gallate	$C_8H_8O_5$	183.024	183.029	trace
Shikimic acid	$C_7H_{10}O_5$	173.041	173.045	trace
Chicha de molle				
Compound	**Formula**	**Measured Mass**	**Calculated Mass**	**Intensity**
Safrole	$C_{10}H_{10}O_2$	161.067	161.060	low
Gallic acid	$C_7H_6O_5$	169.016	169.014	trace
Terebanene, teredenene, or β-spathulene	$C_{15}H_{22}$	201.156	201.164	trace
Palustrol/rosifoliol/shyobunol	$C_{15}H_{26}O$	221.192	221.191	low

Identifications of triterpenoids that are chemotaxonomically characteristic especially of certain anacardiaceous resins (*Pistacia* spp., *Rhus* spp., and *Schinus* spp.) are particularly heartening. In J. Henkin's view these triterpenoids are, where present in the DART-MS spectra, the best phytochemical evidence for *chicha de molle* production from the archaeologically recovered ceramics and *molle* drupes analyzed (Figure 6; Table 3). Hydroxymasticadienoic acid, (iso)masticadienoic acid, moronic acid, schinol, and/or simiarenol together as this relevant triterpenoid group underlie a solid chemotaxonomic argument for *molle* processing and *chicha de molle* production at Cerro Baúl. These triterpenoids should be regarded as *molle* biomarkers in further research along these lines (Figure 7). Sesquiterpenoid and triterpenoid biomarkers identified, plus the presence of the phenylpropanoid safrole, support the archaeological evidence for *chicha de molle* production at Cerro Baúl.

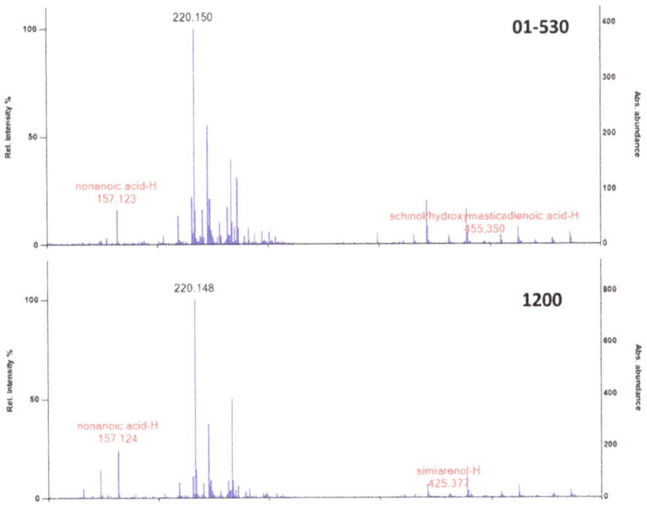

Figure 6. *Cont.*

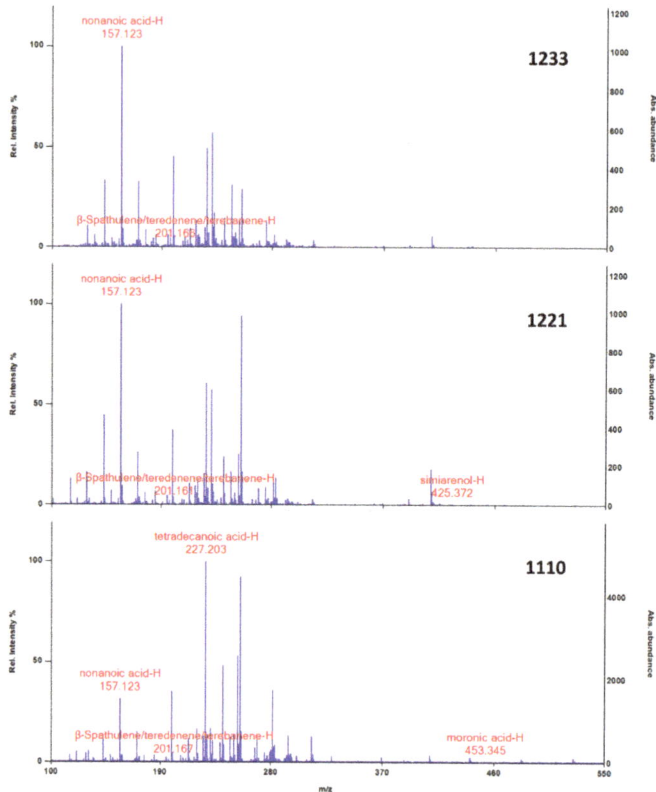

Figure 6. Example DART-MS spectra for one replicate each of the Cerro Baúl brewery ceramic residue samples.

Table 3. Ceramic residues on archaeological sherds from DART-MS (nd = not detected). Each + indicates a positive identification (based on mass for the $[M–H]^-$ ion) in one replicate. Thus ++ means that the compound was identified in two replicates. For CB89-1221, only two replicate analyses were carried out.

Compound	Formula	Measured mass, $[M-H]^-$	Calculated Mass, $[M-H]^-$	CB04-01-0530	CB89-1110	CB89-1200	CB89-1221	CB89-1233
simiarenol	$C_{30}H_{50}O$	425.376 ± 0.002	425.378	nd	nd	+	+	nd
masticadienoic acid (terebinthone) or moronic acid	$C_{30}H_{46}O$	453.345± 0.000	453.337	nd	+	nd	nd	+
hydroxymasticadienoic acid	$C_{30}H_{48}O_3$	455.348 ± 0.002	455.353	nd	nd	+	nd	nd
terebanene, teredenene, or β-spathulene	$C_{15}H_{22}$	201.163 ± 0.003	201.164	nd	+	nd	+	++
safrole	$C_{10}H_{10}O_2$	161.061 ± 0.009	161.060	nd	nd	nd	+	+

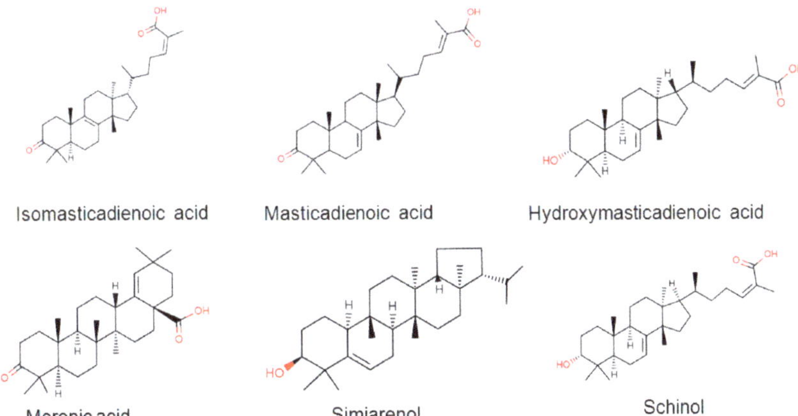

Figure 7. Triterpenoid compounds identified by DART-MS in *molle* samples, diagnostic especially for *Schinus* spp. and its close relatives [26].

The results in Table 3 show only the compounds that, based on literature sources, are indicative of *Schinus molle* and that are not found in the control samples that contained only corn-based *chicha*. In none of the spectra were the identified *Schinus* compounds the most intense signal, referred to as the base peak of the spectrum (Figure 6).For two of the sherds the observed base peak was at m/z 220.150 (CB04-01-0530 and CB89-1200), a compound that we were unable to identify, but appears to give primary fragments at m/z148 and 205 when Orifice 1 on the AccuTOF is set to −90V.The intensity of this signal may indicate it is a contaminant from the environment or handling. For the remaining three sherds, the base peak was most likely a fatty acid. These include myristic acid (tetradecanoic acid, m/z 227.201) in replicates 1 and 2 of CB89-1110, palmitic acid (hexadecanoic acid, m/z 255.232) in the third replicate of both CB89-1110 and -1233, and the shorter-chain pelargonic acid (nonanoic acid, m/z 157.123) in both replicates of CB89-1221 and the second of -1223. None of these fatty acids are observed in significant quantities (based on signal intensity) in the control vessels.

Interestingly, the medium-chain fatty acids are absent in the ceramic vessel that contained the ethnographic *chicha de jora* but dominate the spectra of the ones containing *chicha de molle*, even when corn was also present. Caproic, caprylic, and pelargonic acids (hexanoic, octanoic, and nonanoic) are the top three intensity peaks in the DART-MS spectra of all of vessels that contained the ethnographic *chicha* made with *Schinus molle* berries. Medium-chain fatty acids, while not anticipated by the initial chemoinformatic search that has driven the residue analysis aspect of this research into the material culture of Cerro Baúl, do appear prominently in ethnographic and archaeological *chicha de molle* DART-MS spectra. Truly accounting for the fatty acids' prominence in terms of the phytochemistry of *S. molle* itself, the beverage processing, and/or other environmental inputs would be a sensible focus for any follow-up research. These fatty acids, specifically their isotopic composition, are an excellent target for further analyses. Confirmation of the exact identities of the terpenoids and phenolics would further increase confidence in the identification of the *molle* residues.

4. Discussion

How do unique or rare materials contribute to sustainable governance? In reality, it is the know-how in this case that creates the unique experience, not the material itself. Clay sources utilized to produce brewing vessels may be a limited resource or may be distant or difficult to procure. But in the case of Cerro Baúl, the clay source identified by the chemical data was abundant and located less than half a day's walk from the brewery. Wari ceramic specialists did establish a privileged access to this source and created a recipe for working the material. Yet, they did not overexploit the source nor did they focus on incorporating materials that were not readily available in the local area.

The clay itself was a local resource, and it was particularly reserved by the Wari for their specialized workshop in the region. Wari ceramic specialists did not rely on extra-local materials (except perhaps for some pigments) for their ceramic production, nor did they rely on imported ceramic vessels with their ritually charged iconography. They decorated these vessels themselves with highly significant ritual iconography, representing Wari supernatural beings and geometric motifs. The means of forming and decorating these vessels replicated ways from the Wari heartland and did require specialized knowledge. That knowledge was reproduced locally to create the special vessels for the feasting events. Having a local source for the ceramics made the Baúl brewery a sustainable producer of both the containers for production and consumption of the beer itself since they were not reliant on foreign materials.

Yet, these ornate, regionally produced ceramic vessels did signal strong affinity with the imperial center. They were not dependent on external producers distant from the brewing and feasting locales in order to provide the vessels for these political events. They were not subject to disruptions in long distance supply chains, external political interference, nor disasters in distant production locales for these important constituents of the Wari feast. They were, however, dependent on the Wari specialist potters who knew how to create the Wari vessel forms and to decorate them with the highly stylized ritual iconography. These specialists had to experiment with local resources: tempers, clays, and pigments, in order to transform those raw materials into legitimate and authentic Wari ceramic wares. This local production of highly specialized containers provided a high degree of locally sustainable fabrication of genuine Wari ritual receptacles. That was critical to the reproduction of the Wari political economy, not only at Cerro Baúl, but likely in every major Wari center throughout the empire.

The other material components of crucial importance to the brewery's production were of course the grains and fruits fermented to make the political beverage, selected by the Wari specialist brewers for their recipes. Corn is a fairly water intensive crop to produce, though yields are quite high. Other native grains like quinoa have lower yields (500 kg per acre), but may be less water intensive. In modern farming, corn generally requires more water per field than many other native Andean crops, including other grains, tree fruits, potatoes, and other potential *chicha* producing ingredients. Modern corn, though, does provide high calories per unit area (up to 15 million calories per acre) and thus produces significantly more energy despite its higher water requirements.

The pepper berries are a renewable resource as well, available on trees that naturally form hedgerows and are commonly available on the edges of cultivated fields. Growing to heights up to 15 m, they are also hearty and resistant to drought conditions. In an environment of climatic variability, they are a raw material resource that is resilient in its availability. While the fruits are especially prevalent after extended periods of heavy rain, they are present on trees in the study area year round [9]. Thus, while maize is a fairly water intensive crop, *molle* is a drought resistant tree that consistently produces fruit. In addition to the cultural properties *molle* contributes, it also provides a consistent ingredient for *chicha* that is drought tolerant.

The Middle Horizon (600–1000 CE) was generally a period of climatic stability with some variability in precipitation pronounced in later times. Wari terracing technology provided some mitigation of drought conditions through more efficient water use [28].Towards the end of the period, increased variability in precipitation lead to localized conditions of water deficiency, which in the succeeding centuries became severe and prolonged droughts [29].The use of *molle* as a *chicha* product seems to be pronounced in the final years of the Baúl brewery's production, slightly after 1000 CE, evidenced by the fact that the residue results presented here come from ceramics used in the ceremonial closure of the brewery and that *molle* dumps are associated with the final use of the floor in the brewery. This may reflect the reliability of *molle* as a resource, even in times of drought stress.

Thus, it was both local resourcing of ceramic vessels for brewing and serving the alcoholic beverage and a focus on a set of brewing ingredients that were locally produced and resilient to climatic variations in production that created a sustainable brewing operation for four centuries.

These variables were incredibly important to the Wari political economy and to the building of local allegiances that sustained imperial relationships over decades. Archaeometric data confirms that local source material was utilized in both ceramic production and beverage fabrication. Yet, it was the ceramicists' specialized knowledge on how to construct and decorate the elaborate drinking vessels that made these events quintessentially Wari. It was also the brewers' knowledge of the culinary recipes of different *chicha* production strategies across dispersed areas of the Andes mountains that brought that shared political identity to strangers in many lands. Local resourcing joined with shared knowledge over vast areas to create the political unity that Wari represented as the Andes' first empire.

Author Contributions: Conceptualization, P.R.W. and D.J.N.; Formal analysis, P.R.W., D.J.N., J.M.H., and R.A.A.; Funding acquisition, P.R.W., D.N., and R.A.A.; Investigation, P.R.W., D.J.N., J.M.H., and R.A.A.; Methodology, P.R.W., J.M.H., and R.A.A.; Writing—original draft, P.R.W., D.J.N., and J.M.H.; Writing—review & editing, J.M.H. and R.A.A.

Funding: Field research on the Cerro Baúl brewery was funded by the National Science Foundation (BCS-0074410, BCS-0226791), the National Endowment for the Humanities (RZ-50098), the G. A. Bruno Foundation, and the Grainger Foundation. DART-MS analyses were carried out on EMU's AccuTOF mass spectrometer, funded under NSF MRI-R^2 (0959621).LA-ICP-MS analyses were carried out on TFM's Varian inductively-coupled plasma-mass spectrometer funded under NSF MRI (0320903).

Acknowledgments: INAA was conducted at MURR under the direction of Michael Glascock and preliminary interpretation of results was done by Jeff Speakman. LA-ICP-MS was conducted at the Field Museum, with contributions by Nicola Sharratt, Laure Dussubieux, and Mark Golitko. Archaeological materials were exported from Peru under permit from the Ministry of Culture acuerdos 1659/792 on 18 December 2006 and 123 on 2 October 2003.

Conflicts of Interest: The authors declare no conflict of interest.

References

1. Stein, G. Segmentary States and Organizational Variation in Early Complex Societies: A Rural Perspective. In *Archaeological Views from the Countryside: Village Communities in Early Complex Societies*; Schwartz, G., Falconer, S., Eds.; Smithsonian Institution Press: Washington, DC, USA, 1994; pp. 10–18.
2. Cook, A.; Glowacki, M. Pots, Politics, and Power. In *The Archaeology and Politics of Food and Feasting in Early States and Empires*; Bray, T., Ed.; Springer: Boston, MA, USA, 2003; pp. 173–202.
3. Nash, D.J. Art and Elite Political Machinations in the Middle Horizon Andes. In *Images in Action: The Southern Andean Iconographic Series*; Isbell, W., Uribe, M., Tiballi, A., Zegarra, E., Eds.; UCLA-Cotsen Institute of Archaeology Press: Los Angeles, CA, USA, 2018; pp. 481–499.
4. Nash, D.; Williams, P.R. As Wari Weakened: Ritual Transitions in the Terminal Middle Horizon of Moquegua, Peru. In *Ritual during Periods of Decline, Collapse, and Regeneration in Archaic States*; Murphy, J., Ed.; Routledge: London, UK, 2019; under review.
5. Williams, P.R.; Nash, D.J. Sighting the Apu: A GIS Analysis of Wari Imperialism and the Worship of Mountain Peaks. *World Archaeol.* **2006**, *38*, 455–468. [CrossRef]
6. Nash, D. Craft Production as an Empowering Strategy in an Emerging Empire. *J. Anthropol. Res.* **2019**, under review.
7. Goldstein, D.J.; Coleman, R.C.; Williams, P.R. Reconstructing Middle Horizon (500–1000 C.E.) Social Dynamics through Paleoethnobotanical Interpretations of Fermented Beverage Production and Consumption at Cerro Baúl, Moquegua, Perú. In *Drink, Power, and Society in the Andes*; Jennings, J., Bowers, B., Eds.; University of Florida Press: Gainesville, FL, USA, 2009; pp. 133–166.
8. Nash, D. The art of feasting: building an empire with food and drink. In *Wari: Lords of the Ancient Andes*; Bergh, S., Ed.; Cleveland Museum of Art: Cleveland, OH, USA, 2012; pp. 82–99.
9. Goldstein, D.J.; Coleman, R.C. *Schinus molle L.* (Anacardiaceae) *chicha* production in the Central Andes. *Econ. Bot.* **2004**, *58*, 523–529. [CrossRef]
10. Moseley, M.; Nash, D.; Williams, P.; DeFrance, S.; Miranda, A.; Ruales, M. Burning Down the Brewery: Excavation and Evacuation of an Ancient Imperial Colony at Cerro Baúl, Perú. *Proc. Natl. Acad. Sci. USA* **2005**, *102*, 17264–17271. [CrossRef] [PubMed]

11. Nash, D.; Williams, P.R. Architecture and Power on the Wari—Tiwanaku Frontier. *Archaeol. Papers Am. Anthropol. Assoc.* **2005**, *14*, 151–174. [CrossRef]
12. Williams, P.R. Cerro Baúl: A Wari Center on the Tiwanaku Frontier. *Lat. Am. Antiq.* **2001**, *12*, 67–83. [CrossRef]
13. Williams, P.R. Tiwanaku: A Cult of the Masses. In *Visions of Tiwanaku*; Vranich, A., Stanish, C., Eds.; Cotsen Institute of Archaeology at UCLA: Los Angeles, CA, USA, 2013; pp. 27–40.
14. Williams, P.R.; Nash, D. Imperial Interaction in the Andes: Wari and Tiwanaku at Cerro Baúl. In *Andean Archaeology*; Isbell, W., Silverman, H., Eds.; Plenum: New York, NY, USA, 2002; pp. 243–265.
15. Nash, D. Fine Dining and Fabulous Atmosphere: Feasting Facilities and Political Interaction in the Wari Realm. In *Inside Ancient Kitchens: New Directions in the Study of Daily Meals and Feasts*; Klarich, E., Ed.; University Press of Colorado: Boulder, CO, USA, 2010; pp. 83–110.
16. Sayre, M.; Goldstein, D.; Whitehead, W.; Williams, P.R. A Marked Preference: *Chicha de Molle* and Wari State Consumption Practices. *Ñawpa Pacha* **2012**, *32*, 231–258. [CrossRef]
17. Williams, P.R. Wari and Tiwanaku Borderlands. In *Tiwanaku: Papers from the 2005 Mayer Center Symposium*; Young-Sanchez, M., Ed.; Denver Art Museum: Denver, CO, USA, 2009; pp. 211–224.
18. Nash, D.; Williams, P.R. Wari Political Organization: The Southern Periphery. In *Andean Civilization: A Tribute to Michael E. Moseley*; Marcus, J., Williams, P.R., Eds.; Cotsen Institute of Archaeology Press, UCLA: Los Angeles, CA, USA, 2009; pp. 257–276.
19. Williams, P.; Nash, D.; Cook, A.; Isbell, W.; Speakman, R. Wari Ceramic Production in the Heartland and Provinces. In *Ceramics of the Indigenous Cultures of South America: Studies of Production and Exchange through Compositional Analysis*; Glascock, M., Neff, H., Vaughn, K., Eds.; UNM Press: Albuquerque, NM, USA, 2019; pp. 125–133.
20. Dussubieux, L.; Golitko, M.; Williams, P.R.; Speakman, R.J. LA-ICP-MS analysis applied to the characterization of Peruvian Wari ceramics. In *Archaeological Chemistry: Analytical Techniques and Archaeological Interpretation*; Glascock, M., Popelka-Filcoff, R., Speakman, R.J., Eds.; American Chemical Society Symposium Series 968; American Chemical Society Symposium: Washington, DC, USA, 2007; pp. 349–363.
21. Sharratt, N.; Golitko, M.; Williams, P.R. Pottery Production, Regional Exchange and State Collapse during the Middle Horizon: LA-ICP-MS analyses of Tiwanaku Pottery in the Moquegua Valley, Perú. *J. Field Archaeol.* **2015**, *40*, 397–412. [CrossRef]
22. Sharratt, N.; Golitko, M.; Williams, P.R.; Dussubieux, L. Ceramic Production during the Middle Horizon: Wari and Tiwanaku clay procurement in the Moquegua Valley, Peru. *Geoarchaeology* **2009**, *24*, 792–820. [CrossRef]
23. Glascock, M. Compositional Analysis of Archaeological Ceramics. In *Ceramics of the Indigenous Cultures of South America: Studies of Production and Exchange through Compositional Analysis*; Glascock, M., Neff, H., Vaughn, K., Eds.; UNM Press: Albuquerque, NM, USA, 2019.
24. Gratuze, B.; Blet-Lemarquand, M.; Barrandon, J.N. Mass spectrometry with laser sampling: A new tool to characterize archaeological materials. *J. Radioanal. Nucl. Chem.* **2001**, *247*, 645–656. [CrossRef]
25. Cody, R.B.; Dane, A.J. Direct Analysis in Real Time (DART). In *Ambient Ionization Mass Spectrometry*; Domin, M., Cody, R.B., Eds.; Royal Society of Chemistry: Cambridge, UK, 2015.
26. Hopkins, J.; Armitage, R.A. Characterizing Organic Residues on Ceramics by Direct Analysis in Real Time Time-of-Flight Mass Spectrometry. In *Collaborative Endeavors in the Chemical Analysis of Art and Cultural Heritage Materials*; Lang, P.L., Armitage, R.A., Eds.; American Chemical Society: Washington, DC, USA, 2012; Volume 1103, pp. 131–142.
27. Reaxys. *Schinus Molle (Isolation from Natural Product)*; Elsevier: Amsterdam, The Netherlands, 2014.
28. Londoño, A.C.; Williams, P.R.; Hart, M.L. A change in landscape: Lessons learned from abandonment of ancient Wari agricultural terraces in Southern Peru. *J. Environ. Manag.* **2017**, *202*, 532–542. [CrossRef] [PubMed]
29. Williams, P.R. A Re-examination of Disaster Induced Collapse in the Case of the Andean Highland States: Wari and Tiwanaku. *World Archaeol.* **2002**, *33*, 361–374. [CrossRef]

© 2019 by the authors. Licensee MDPI, Basel, Switzerland. This article is an open access article distributed under the terms and conditions of the Creative Commons Attribution (CC BY) license (http://creativecommons.org/licenses/by/4.0/).

Article

3D Digital Heritage Models as Sustainable Scholarly Resources

Erik Champion * and Hafizur Rahaman

Discipline of Theatre, Screen and Digital Media, School of Media Creative Arts and Social Inquiry, Faculty of Humanities, Curtin University, GPO Box U1987, Perth, WA 6845, Australia; hafizur.rahaman@curtin.edu.au

* Correspondence: erik.champion@curtin.edu.au

Received: 28 January 2019; Accepted: 18 April 2019; Published: 24 April 2019

Abstract: If virtual heritage is the application of virtual reality to cultural heritage, then one might assume that virtual heritage (and 3D digital heritage in general) successfully communicates the need to preserve the cultural significance of physical artefacts and intangible heritage. However, digital heritage models are seldom seen outside of conference presentations, one-off museum exhibitions, or digital reconstructions used in films and television programs. To understand why, we surveyed 1483 digital heritage papers published in 14 recent proceedings. Only 264 explicitly mentioned 3D models and related assets; 19 contained links, but none of these links worked. This is clearly not sustainable, neither for scholarly activity nor as a way to engage the public in heritage preservation. To encourage more sustainable research practices, 3D models must be actively promoted as scholarly resources. In this paper, we also recommend ways researchers could better sustain these 3D models and assets both as digital cultural artefacts and as tools to help the public explore the vital but often overlooked relationship between built heritage and the natural world.

Keywords: 3D model; virtual heritage; ecosystem; infrastructure

1. Introduction

Sustainable digital cultural heritage has been considered a serious national issue in America [1]. Sustaining digital libraries are also a crucial issue [2], and these two concepts share common issues, including a problem with securing long-term funding and ensuring that users continually find the heritage collections (and library collections) useful and worthwhile. Digital reconstructions of cultural heritage have been deployed as showcases for cutting-edge technology and to promote tourism to otherwise remote cultures and distant lands [3,4].

Virtual reality, mixed reality, and augmented reality projects also provide tantalizing new ways of engaging the public with the past [5]. As simulations, scholars might modify them to verify or refute historical hypotheses, testing either data or methods. While the original sites may have existed for hundreds or thousands of years, the digital models that underpin these digital projects have a limited shelf-life, and through designed obsolescence, perceived obsolescence, or the limitations of time, training, and resources, they are seldom successfully deployed in the classroom [6].

As with libraries, museums require both long-term funding and public engagement. However, they have severely limited space and facilities for either exhibition or digitalization, let alone continual funding for new technologies or the time to train staff or teach the public how to best utilize new interaction design technology [7]. The field of digital heritage, with its 3D models and 3D projects, has an added sustainability dilemma: 3D digital heritage models can help promote tourism in remote and endangered areas, therefore helping local businesses, but they can also potentially damage fragile historic places and heritage sites through increased visitation [8].

These are profound meta-issues, but a more immediate yet often overlooked problem is how to help scholars support more appropriate, useful, and required research into both digital heritage technologies and user experience design solutions. For example, 3D models, when used in interactive virtual environments or when integrated into augmented and mixed reality environments, may provide immediate and user-directed simulations communicating how even built heritage sites are predicated on natural features, resources, and ecosystems. Archaeological sites are often prepared to take best advantage of dynamic and seasonal natural resources. Monuments are designed to resist (but ultimately succumb to) natural forces. Sacred buildings often frame constellations and cosmic events. The range and nature of architecture is dependent on local or precious materials. Their remains are palimpsests of human encounters, repeated erosion, personal habits, human-caused pollution, and natural calamities. Game engines and interactive virtual reality technologies can show both these relationships plus changes over time and the effects of human visitation, modulated by the decisions of virtual visitors [9]. A more sustainable development of 3D models to promote the aims of cultural heritage may therefore lead to increased public, institutional, and philanthropic interest, engagement, and investment, in both built heritage and its relationship to the natural environment.

Impetus for more sustainable digital heritage models would ideally be generated by the community of scholars dedicated to the study of digital heritage. After all, education is a major reason for the preservation of digital heritage, according to UNESCO's *Charter on the Preservation of the Digital Heritage* [10]:

> "Preservation of the digital heritage requires sustained efforts on the part of governments, creators, publishers, relevant industries and heritage institutions. In the face of the current digital divide, it is necessary to reinforce international cooperation and solidarity to enable all countries to ensure creation, dissemination, preservation and continued accessibility of their digital heritage ... The stimulation of education and training programs, resource-sharing arrangements, and dissemination of research results and best practices will democratize access to digital preservation techniques."

In this charter, UNESCO recommends developers, designers, and publishers to work with heritage organizations (such as libraries, museums, and the private sector), professional associations and institutions, and universities (as well as other research organizations) to preserve digital heritage data and to train and share experience and knowledge in a "sustained" fashion. However, there is a critical problem in the scholarship of 3D digital heritage projects [11]. In our initial investigations into this field of scholarship, we did not find many reports building on, corroborating, or verifying previous digital heritage research. In fact, we could not find many digital heritage models directly linked to research projects and openly accessible both as interactive digital experiences and as scholarly resources.

Admittedly, there are successful portals for acquiring free or purchasable 3D heritage models—notable exceptions include Sketchfab, Smithsonian 3D, Europeana, or the Google Arts and Culture- CyArk Open Heritage Project websites. However, there are still far too few instances of scholarly digital heritage projects that are easily accessible to the public or to scholars that are clearly identifiable as scholarly investigations or carefully delineated research projects. There appear to be even fewer scholarly projects that lend themselves to investigation, pedagogical explanation, scholarly verification, design modification, refinement, or amalgamation into larger or newer projects.

As mentioned above, virtual heritage (VH) is commonly used to describe projects that combine virtual reality (VR) and cultural heritage [12,13]. Stone and Ojika [5] defined virtual heritage as "the use of computer-based interactive technologies to record, preserve, or recreate artefacts, sites, and actors of historic, artistic, religious, and cultural significance and to deliver the results openly to a global audience in such a way as to provide formative educational experiences through electronic manipulation of time and space". Various commentators and charters (London, Seville) have also stated that the success of a VH (Virtual Heritage) project depends on 3D models and associated scholarly content [14,15]. Given the above, our starting hypothesis is that there appears to be a dramatic increase in the number of academic papers on 3D digital heritage (especially virtual heritage), but, conversely, there is a decreasing number of accessible 3D assets [16]. If true, this foretells serious problems in the field of digital heritage as a sustainable scholarly activity, at least if 3D models are considered to be an essential part of scholarly and pedagogical endeavors.

2. Method

We ran a literature survey of the 14 proceedings of the last three consecutive publications of major digital heritage events and conferences (Table 1). These were: *The International Society for Virtual Systems and Multimedia (VSMM), Computer Applications and Quantitative Methods in Archaeology (CAA), International Committee of Architectural Photogrammetry (CIPA), The European Mediterranean Conferences (EuroMed)*, and *The Digital Heritage International Congress* (but not *Digital Heritage 2018*). These conferences were chosen as they are arguably major international conferences in digital heritage, and provided online access to the papers. From a total of 1483 conference papers, 264 were selected (Table 2), and 19 were found to contain explicit links to 3D models and related assets (Table 3). The quality of reporting of meta-analysis method (QUORUM) statement presented by Moher et al. [17] was chosen to help select the papers (Figure 1), using the following steps:

Step 1: Identification and Screening

1. Source selection—Popular and renowned international events such as a conference and symposium on digital heritage and allied domains were preferred as initial sources. Proceedings of the last three consecutive publications of these events were selected. This selection, from 2012 to 2017, covered major events, journals, and conferences, such as *The International Society for Virtual Systems and Multimedia (VSMM), Computer Applications and Quantitative Methods in Archaeology (CAA), International Committee of Architectural Photogrammetry (CIPA), The European Mediterranean Conferences (EuroMed)*, and *The Digital Heritage International Congress.*
2. Retrieval and initial screening—1483 papers from 14 proceedings were collected from their respective digital repositories and publication databases. Articles which contained representative images or references to 3D digital heritage assets were selected for further study. A total of 264 articles were selected at this stage.

Step 2: Final Screening

The selected papers were then reviewed in terms of their abstract and a rapid examination of their structure and content in order to:

1. Exclude irrelevant articles (such as review papers and short survey papers);
2. Eliminate duplicates (or similar papers published in other proceedings with minimum changes);
3. Exclude articles on the digitization of paintings and artworks (32 were excluded at this stage).

Step 3: Review

1. Final sorting—The final selection of articles included for the detailed study were: 31 (out of 173) from *VSMM*, 38 (out of 240) from *CAA*, 79 (out of 305) from *CIPA*, 61 (out of 284) from *EuroMed*, and 55 (out of 481) from *Digital Heritage Congress* (a list is attached as a Supplementary Materials).
2. Study process and analysis—At this phase, the selected articles were studied to find whether they provide any references citing external links or information to:

    (i) Accessible 3D assets (and the degree of their accessibility);
    (ii) Accessible video content;
    (iii) Visual materials (such as VR models, photographs, images of 3D reconstruction, etc.);
    (iv) Other external resources (if any).

A scheme for recording the study's detailed descriptive data into a database was created in MS Excel. For each article reviewed, required information about the criteria mentioned above was inserted into a spreadsheet and presented in a tabular format.

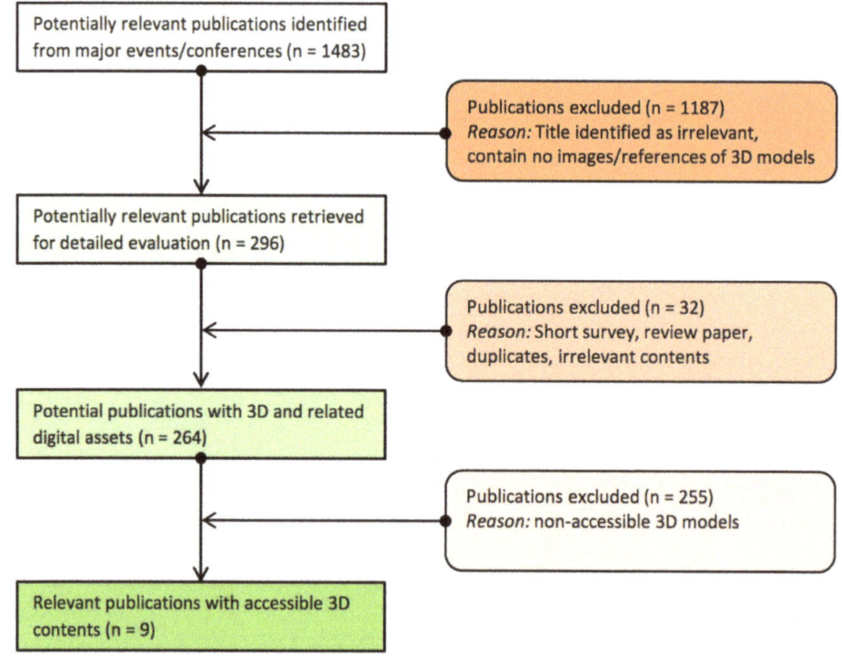

Figure 1. QUORUM process.

Table 1. 3D heritage conference papers.

Conference Publications	2017	2016	2015	2014	2013	2012	Total Papers
VSMM	55	65	53				173
CAA			117	73	50		240
CIPA	111		82		112		305
EuroMed		105		84		95	284
Digital Heritage			270		211		481
TOTAL	166	170	522	157	373	95	1483

Table 2. Total articles containing references to 3D models and heritage assets.

Conference Publications	Total Papers	Mentioning 3D Assets	%
VSMM 2015–2017	173	31	17.9%
CAA 2013–2015	240	38	15.8%
CIPA 2013, 2015, 2017	305	79	25.9%
EuroMed 2012, 2014, 2016	284	61	21.5%
Digital Heritage 2013, 2015	481	55	11.4%
TOTAL	1483	264	17.8%

Table 3. Selected papers included in our study.

Directly Accessible Content	*VSMM*	*CAA*	*CIPA*	*EuroMed*	*Digital Heritage*	Total
3D content	0	1	3	1	4	**9**
Videos	1	2	1	2	6	**12**
Other (VR models, photos, images of 3D models, etc.)	1	4	6	5	17	**33**
3D assets on accessible websites	3	0	5	3	8	**19**

3. Results

From a group of 1483 conference papers, we selected 264, accounting for 17.9% of the total papers published in *VSMM, CAA, CIPA, EuroMed*, and *Digital Heritage Congress* from 2012 to 2017. The results of the study, which have been tabulated in Tables 1 and 2, reveal that a significant number of papers (i.e., 17.9%) referred to and contained images of 3D assets or 3D digital models. Contrary to our initial expectations, accessible 3D assets or 3D models were found in only nine papers, i.e., 3.4% of the selected publications.

Of the 264 selected articles, 12 contained external web-links to video content (4.6%) and 33 articles (12.5%) provided external links for other accessible visual material, including VR models, photographs, and images of 3D models. We found 19 articles with external web-links to 3D models. However, not a single one of the links worked at the time of writing this article (last checked: 1 September 2018).

Of the nine articles that provided external links to accessible 3D assets, they all shared four common locations/repositories. These particular nine articles referred to only five external links for storing their 3D assets. The five external links were http://3dicons-project.eu/ (leading to https://www.europeana.eu/), http://dati.comune.bologna.it/3d, www.cyark.org, https://skfb.ly/DtVq, and https://harvest4d.org/?page_id=1367 (last accessed 7 January 2019).

4. Discussion

In an upcoming conference paper (to be presented at CAADRIA 2019) we will explore the technical solutions to this problem of the "vanishing virtual"—that is, the dilemma of technology superseding itself [16]. There we will propose a component-based 3D model system that is linked to current infrastructure projects. However, in this article we wish to focus on what we propose is a fairly simple yet barely noticed problem: digital technology has compelled us to seize the historical artefact at one point in time (the time of recording, which is not the time of creation or time of use), and then develop hermetically sealed interfaces and interaction mechanisms around these stillborn 3D copies of the found object or the recorded landscape. This may not initially appear to be a problem—after all, faster processes and bigger, high-resolution screens can exhilarate the senses—but we question whether the accumulative, organic, uniquely situated, and highly dynamic built culture of the past is always best served by apparent precision, speed, and scale.

High-resolution scans and photogrammetric models record one slice of historical action, but they do not necessarily communicate how built culture has responded to natural forces, to human change, or to the pressures of time. Secondly, high-technology demonstrations may impress but do not necessarily engage the public (or even scholars) into exploring process and test theories. A photorealistic 3D digital reproduction (born digital or digital surrogate) is not sufficient by itself [18] for the public to interpret and perceive its cultural significance. Thirdly, high-technology showcases necessitate very expensive equipment, specialized resources, and highly-trained staff (who are often trained in research rather than in public engagement).

Fourthly, such advanced equipment can exact a high price not only from the public or private purse, but also from natural resources (in terms of both energy and materials). Merely being in the cloud also has an energy cost: for example, in 2015 Google consumed roughly the same energy as the city of San Francisco [19] and in the same year the Internet was predicted to contribute roughly the same amount to global emissions as air travel [20]. A 2018 *Nature* article warned that by 2030, thanks to an explosion in data centers and increasing shared social media content, Information and Communication Technology could consume up to 21% of total global energy [21]. Nor is moving digital heritage content to increasingly powerful smartphones an ideal situation. By 2020, 5 billion phones could be in circulation, and the rare earth metals they use could run out in 20 to 50 years. Meanwhile, the extraction of iron, aluminum, and copper, not to mention gold and tin, have already resulted in catastrophic spills, deforestation, and toxic poisoning [22].

Digital heritage has a price. While digital heritage projects are likely to contribute only a small percentage to this energy consumption and to the depletion of rare metals and minerals, we suggest that the environmental resources consumed should be explicitly considered in the design of any major project. We also suggest that digital heritage as an educational medium and as a channel for communication and collaboration among scholars across the world, with access to wildly varying resources, be considered.

Moreover, we suggest that digital heritage needs to understand the contribution of 3D models to the field not only as finished products but also as pedagogical and theoretical building blocks. While the design and deployment of high-technology showcases has its place, there should also be room for the design, sharing, and redesign of simpler objects, scripts, and related digital heritage media that can be modified and improved on by not just a single team but also by a community. This pathway may prove to be more sustainable for the digital models themselves, as well as more beneficial to the aims and objectives of the research community and more effective in disseminating and promoting cultural heritage awareness and understanding. In other words, the academic community should put more emphasis on sharing, critiquing, reusing, and improving the elements of virtual heritage projects, rather than relying on overall projects inside proprietary, locked frameworks.

Simple mechanisms to aid the wider sharing of models, infrastructures, scripts, and media might be to design competitions, grants, and prizes to award to digital heritage projects and communities based on their sharing, verification, modification, and improvement of others' original models and data. Secondly, contributions to open access infrastructures, repositories, and tools should be recognized and supported by universities and related research organizations, while tools, projects, and papers that advance these goals could also be specifically recognized. This includes new forms of publications that emphasize collaboration and feedback around 3D models as specific scholarly resources and as components of scholarly arguments. As far as we know, none of the surveyed digital heritage conferences specify awards or recognition for papers and projects that share 3D models as scholarly assets and scholarly arguments, or for projects involving not only the design but also the evaluation and preservation of digital models. We believe this is not only feasible but also likely to increase the direct linking of publicly accessible models.

In terms of scholarly understanding, there is surprisingly little written and debated about the 3D digital heritage model considered as a learning tool or experimental device rather than as a finished (if virtual) object. Simply put, 3D models are not yet fully integrated into scholarly discourse [23]. At an instrumental level, more uptake is required to establish file formats [11,24] that 'travel' and to develop more tools and frameworks (such as http://www.meshlab.net/) in order to allow content to move between different programs. This would help the modification and collaboration of models. Increasing the use of metadata and Linked Open Data tools [25] and frameworks would help increase the visibility and probably also the usage of digital heritage models. However, the single most effective way to increase public access to 3D digital models, we argue, is to develop various levels of copyright specifically for 3D content that allow owners to share various levels of resolution (or precision) of their 3D models and 3D data [26], along with incentives for them to share various levels of resolution and precision of those models.

5. Conclusions

This survey not examined how digital heritage conference papers have addressed the issue of sustainability per se, but it also indicated that the 3D models associated with these papers are not typically seen as worthy of preservation in their own right, which leads us to question both the sustainability of digital heritage as a serious scholarly activity (how can the discipline evolve if we cannot verify each other's data?) and the pedagogical value of these 3D models. However, the problems are so widespread that it appears to be foremost a problem of infrastructure, or more accurately, a problem raised by not having suitable infrastructure. There have been impressive European Union (EU) infrastructure-related projects (ARIADNE, CARARE, 3D–ICONS, Scottish Ten, etc.) and relevant National Endowment of the Humanities projects in the United States, but accessible and well-maintained links to the related 3D assets need to be integrated into academic publication and dissemination systems.

We propose, following the London Charter and others, that 3D models must be recognized as a scholarly resource [27,28]; however, we suggest that there is a key element missing from such charters: a framework or set of guidelines to help create and maintain a robust infrastructure that underpins 3D digital heritage models. There could also be tools and procedures run by digital heritage conferences to provide a framework to view 3D models in relation to the articles (some journals have already begun to explore this, though conferences, to our knowledge, have not).

A further aspect of this article was to suggest that the relationship of built heritage to natural ecosystems has not been fully addressed by digital heritage models. Only when we tackle the challenge of communicating the dynamic and environmentally situated nature of built cultural heritage will we be able to communicate not only the visual effects but also the principles of both the scholarly research underpinning the digital heritage simulation and the sustainability issues of the heritage site itself. Just as the scholarly publication system needs to see itself as more of an evolving scholarly digital ecosystem, which can be continually tested, debated, and updated, so too should the digital heritage project be considered to be not merely a standalone object or a finished product, but a component of process. For how can digital heritage fulfil the noble aims of cultural heritage if it cannot even maintain, preserve, and sustain itself?

Supplementary Materials: The following are available online at http://www.mdpi.com/2071-1050/11/4/2425/s1.

Author Contributions: Conceptualization of the overall project was devised and managed by E.C., but the surveying and analysis of papers and related 3D assets was conducted by H.R. E.C. wrote the initial draft and overall paper, based on the survey study by H.R. who also provided the initial workflow illustration and table, feedback and corrections on the paper draft and also some of the observations and recommendations.

Funding: This research received no external funding.

Acknowledgments: We would like to thank Curtin University for support and internal funding.

Conflicts of Interest: The authors declare no conflict of interest.

References

1. Zorich, D.M. *A Survey of Digital Cultural Heritage Initiatives and Their Sustainability Concerns. Managing Economic Challenges*; Council on Library and Information Resources: Washington, DC, USA, 2003.
2. Hamilton, V. Sustainability for digital libraries. *Lib. Rev.* **2004**, *53*, 392–395. [CrossRef]
3. Zaitceva, O.; Vavulin, M.; Pushkarev, A.; Vodyasov, E. Photogrammetry: From Field Recording to Museum Presentation (Timiryazevo burial site, Western Siberia). *Mediterr. Archaeol. Archaeom.* **2016**, *16*, 97–103. [CrossRef]
4. Georgopoulos, A. 3D virtual reconstruction of archaeological monuments. *Mediterr. Archaeold. Archaeom.* **2014**, *14*, 155–164.
5. Stone, R.; Ojika, T. Virtual heritage: What next? *IEEE Multimed.* **2000**, *7*, 73–74. [CrossRef]
6. Economou, M.; Tost, L.P. Educational tool or expensive toy? Evaluating VR evaluation and its relevance for virtual heritage. In *New Heritage: New Media and Cultural Heritage*; Kalay, Y.E., Kvan, T., Affleck, J., Eds.; Routledge-Taylor and Francis Group: London, UK; New York, NY, USA, 2008; pp. 242–260.

7. Parry, R. *Museums in a Digital Age*; Routledge: London, UK; New York, NY, USA, 2013; p. xviii. 478p.
8. Girard, L.F.; Nijkamp, P. (Eds.) *Cultural Tourism and Sustainable Local Development*; Ashgate Publishing, Ltd.: Farnham, UK, 2009.
9. Vosinakis, S.; Avradinis, N. Virtual Agora: Representation of an ancient Greek Agora in virtual worlds using biologically-inspired motivational agents. *Mediterr. Archaeol. Archaeom.* **2016**, *16*, 29–41.
10. UNESCO. Charter on the Preservation of the Digital Heritage. In Proceedings of the 32nd Session: The General Conference of the United Nations Educational, Scientific and Cultural Organization, Paris, France, 29 September–15 October 2003; pp. 74–76.
11. Tsiafaki, D.; Michailidou, N. Benefits and problems through the application of 3D technologies in archaeology: Recording, visualisation, representation and reconstruction. *Sci. Cult.* **2015**, *1*, 37–45.
12. Addison, A.C. Emerging trends in virtual heritage. *IEEE Multimed.* **2000**, *7*, 22–25. [CrossRef]
13. Roussou, M. Virtual Heritage: From the Research Lab to the Broad Public. In Proceedings of the VAST Euroconference, Arezzo, Italy, 24–25 November 2000; pp. 93–100.
14. Scopigno, R.; Callieri, M.; Dellepiane, M.; Ponchio, F.; Potenziani, M. Delivering and using 3D models on the web: Are we ready? *Virtual Archaeol. Rev.* **2017**, *8*, 1–9. [CrossRef]
15. Tucci, G.; Bonora, V.; Conti, A.; Fiorini, L. High-quality 3d models and their use in a cultural heritage conservation project. *Int. Arch. Photogramm. Remote Sens. Spat. Inform. Sci.* **2017**, *42*, 687. [CrossRef]
16. Thwaites, H. Digital Heritage: What Happens When We Digitize Everything? In *Visual Heritage in the Digital Age*; Springer: London, UK, 2013; pp. 327–348.
17. Moher, D.; Cook, D.J.; Eastwood, S.; Olkin, I.; Rennie, D.; Stroup, D.F. Improving the quality of reports of meta-analyses of randomised controlled trials: The QUOROM statement. *Revi. Esp. Salud Púb.* **2000**, *74*, 179–195. [CrossRef] [PubMed]
18. Rahaman, H.; Das, R.; Zahir, S. Virtual Heritage: Exploring Photorealism. In *Progress in Cultural Heritage Preservation*; Loannides, M., Fritsch, D., Leissner, J., Davies, R., Remondino, F., Caffo, R., Eds.; Springer: Berlin, Germany; Heidelberg, Germany, 2012; Volume 7616, pp. 191–200.
19. Hardy, Q. Google Says It Will Run Entirely on Renewable Energy in 2017. The New York Times. Available online: https://www.nytimes.com/2016/12/06/technology/google-says-it-will-run-entirely-on-renewable-energy-in-2017.html (accessed on 4 April 2019).
20. Vaughan, A. How Viral Cat Videos Are Warming the Planet. The Guardian. Available online: https://www.theguardian.com/environment/2015/sep/25/server-data-centre-emissions-air-travel-web-google-facebook-greenhouse-gas (accessed on 4 April 2019).
21. Jones, N. How to stop data centres from gobbling up the world's electricity. *Nature* **2018**, *561*, 163. [CrossRef] [PubMed]
22. Byrne, P.; Edwards, K.H. Three ways Making a Smartphone Can Harm the Environment. The Conversation. Available online: http://theconversation.com/three-ways-making-a-smartphone-can-harm-the-environment-102148 (accessed on 4 April 2019).
23. Huggett, J. Lost in information? Ways of knowing and modes of representation in e-archaeology. *World Archaeol.* **2012**, *44*, 538–552. [CrossRef]
24. Neamţu, C.; Popescu, D.; Mateescu, R.; Suciu, L.; Hurgoiu, D. About the Quality and Properties of Digital Artifacts. *Mediterr. Archaeol. Archaeom.* **2014**, *14*, 55–64.
25. Kiourt, C.; Koutsoudis, A.; Markantonatou, S.; Pavlidis, G. The "Synthesis" Virtual Museum. *Mediterr. Archaeol. Archaeom.* **2016**, *16*, 1–9.
26. Clarke, M. The digital dilemma: Preservation and the digital archaeological record. *Adv. Archaeol. Pract.* **2015**, *3*, 313–330. [CrossRef]
27. Denard, H. A new introduction to the London Charter. In *Paradata and Transparency in Virtual Heritage*; Bentkowska-Kafel, A., Denard, H., Baker, D., Eds.; Routledge: Farnham, UK, 2016; pp. 83–98.
28. Di Benedetto, M.; Ponchio, F.; Malomo, L.; Callieri, M.; Dellepiane, M.; Cignoni, P.; Scopigno, R. Web and mobile visualization for cultural heritage. In *3D Research Challenges in Cultural Heritage*; Springer: Berlin/Heidelberg, Germany, 2014; pp. 18–35.

© 2019 by the authors. Licensee MDPI, Basel, Switzerland. This article is an open access article distributed under the terms and conditions of the Creative Commons Attribution (CC BY) license (http://creativecommons.org/licenses/by/4.0/).

Article

The White Marbles of the Tomb of Christ in Jerusalem: Characterization and Provenance

Antonia Moropoulou [1,*], Ekaterini T. Delegou [1], Maria Apostolopoulou [1], Aikaterini Kolaiti [1], Christos Papatrechas [2], George Economou [2] and Constantinos Mavrogonatos [3]

1 Laboratory of Materials Science and Engineering, School of Chemical Engineering, National Technical University of Athens, 15780 Athens, Greece; edelegou@central.ntua.gr (E.T.D.); mairi_apostol@hotmail.com (M.A.); kolaitik@gmail.com (A.K.)
2 Institute of Geology and Mineral Exploration (I.G.M.E.), 13677 Athens, Greece; papatrechas@windowslive.com (C.P.); georgeoik7@gmail.com (G.E.)
3 Department of Geology & Geoenvironment, National and Kapodistrian University of Athens, 15784 Athens, Greece; kmavrogon@geol.uoa.gr
* Correspondence: amoropul@central.ntua.gr; Tel.: +30-210-7723276

Received: 9 March 2019; Accepted: 23 April 2019; Published: 28 April 2019

Abstract: In this work, samples of the white marbles enclosing the Tomb of Christ, as well as samples from the interior marble facades of the Holy Aedicule structure surrounding the Tomb of Christ in the Church of Resurrection in Jerusalem, are investigated using petrographic and isotopic analysis. The aim is to characterize the marble samples and investigate their provenance. The results demonstrate that all examined marble samples originate from Proconnesos (Marmara island), and can be attributed to the so-called Proconnesos-1 variety. Published maximum grain size (MGS) and isotopic ($\delta^{18}O$ and $\delta^{13}C$) values of Proconessos quarries are compared with the respective values displayed by the marble samples of the Holy Aedicule, aiming to achieve—to a certain degree—intra-site discriminations. A number of ancient quarries are excluded through this double parameter criterion as sources for the examined Holy Aedicule marbles. The discussion of petrographic and isotopic results in relation to historical testimonies and previously published archaeometry results, regarding the mortars of the Holy Aedicule, reveal that Proconnesos marble was the material of choice used at different construction phases of the Holy Aedicule, from the time of Constantine the Great and throughout the centuries, both for the cladding of the Holy Tomb and the interior facings of the Tomb Chamber and the Chapel of the Angel.

Keywords: Holy Sepulchre; Church of Resurrection; petrography; isotopic analysis; Proconnesos; cultural heritage; monument; history; archaeometry

---

1. Introduction

Natural rocks have been used as a raw material for constructive and decorative purposes since antiquity. Many rock-types have been selected for these purposes, (e.g., chert, granite, gneiss, schists, serpentinite, etc.), but undoubtedly, marble has been the most prominent type [1], especially concerning monumental, religious and important buildings. Marbles are metamorphic rocks, consisting almost entirely of calcite, with minor admixture of a number of accessory minerals. In the broad Mediterranean region, marble mining activities have been documented since the Neolithic period, with constantly increasing rates of exploitation, which flourished during the Greco-Roman period [1].

White and colored marbles have been long known as a favorite building and decoration material of the ancient times. The provenance identification of the raw materials used during antiquity, is a matter of great significance to both historians and archaeologists [2], as well as to other scientists in the field of monument protection. This is not only because such results could lead towards understanding commercial and socio-economic relations (e.g., trade routes) among ancient populations, but because it could also provide the proper material for conservation and restoration interventions in damaged monuments and sculptures [3]. Especially for marbles, which were preferably used as raw material for sculpturing, building and decoration purposes, attempts to determine their provenance date back to the end of the 19th century. At that time, Lepsius (1890) [4], was the first who tried to use "scientific" criteria in order to define the provenance of marble-curved items, based mostly on their macroscopic characterization, an approach which later proved to be erroneous and/or inadequate. Since then, and especially during the last three decades, archaeometric studies have expanded significantly beyond macroscopic examination, by including new analytical techniques or combinations of more than one technique, in order to obtain more accurate results [2,3,5,6].

Petrographic investigation of rock samples is among the first truly scientific methods applied for provenance determination [7–10], because it is indicative of the environment of marble formation. This method comprises a number of parameters related to the mineralogy and the fabric of the rock. The mineralogical examination aims to identify the main and the accessory mineral phases that compose the rock. Beyond calcite and less commonly dolomite, which often exceed 95% of the rock volume, a number of accessory phases have been described for many marbles around the world [11]. In addition, the fabric of the studied sample gives significant information and may be critical in the identification of various rock types, as it reflects the degree of rock deformation, mostly expressed by the maximum grain size (MGS) and the grain boundary shape (GBS). These two parameters, coupled with the mineralogical content, provided encouraging results, but soon proved to be inadequate for a secure provenance determination. This is due to: (i) the great number of accessory minerals that were commonly identified; and (ii) the sometimes significant inhomogeneity of rock samples from the same quarry (e.g., differences in the dolomite content, grain size variations, etc.) [12].

Geochemical analyses, which focus on measuring minor (e.g., Na, Mn) and trace (e.g., REE) elements' concentrations, commonly conducted by neutron activation analysis (NAA) or inductively-coupled plasma mass spectrometry (ICP-MS) resulted in a number of significant geochemical data sets. Unfortunately, the inhomogeneous content of marbles in the aforementioned chemical impurities, as well as the overlapping of many data sets from different areas, again proved to be discouraging in relation to provenance determination.

Measuring the isotopic ratios of C and O in marble samples, was proposed by Craig and Craig, (1972) [13], and, since then, it is considered, along with petrographic analysis (and/or other analytical techniques), the most classic approach in archaeometry studies, especially for marble artifacts. This combination of methods commonly suffices and results in reliable data sets which allow scientists to make a relatively safe identification. Other analytical methods comprise cathodoluminescence, which is related to the impurities of Fe and Mn, electronic spin resonance and electronic paramagnetic resonance (ESR/EPR), as well as strontium isotopic measurements. Although the aforementioned methods usually provide additional information, they cannot be used alone, as they remain unreliable in attributing a marble artifact to its quarry of origin [12]. Moreover, the high cost of the Sr-isotopic measurements is often a reason for researchers to avoid this method. X-ray diffraction (XRD) and electron-probe micro-analysis (EPMA) are two methods commonly used to obtain mineral-chemical data, especially for accessory phases that could be used as pathfinders towards the provenance identification of the sample [11,14].

The last decades, the Association for the Study of Marbles and Other Stones in Antiquity (ASMOSIA) and its members have provided a very useful dataset and have proposed new methodological approaches for the provenance identification problem. Based mostly on isotopic and petrographic characterizations, a number of researchers have published detailed studies, especially for the most common marble types used in antiquity and have provided accurate datasets that compare artifact samples and geological material. These studies refer mainly to the Proconnesian (Marmara island) marble [15,16], the Thasian marble [17], the Carrara [18], the Parian marbles [19], the Aphrodisias marble [20] and the Pentelic marble [21–23].

Among the many marble-producing areas during the ancient times, some have proved to be major producers, due to the quality of their material and on account of their location on important trade routes or their vicinity to important ancient cities. In the case of white marbles, the most important quarrying locations around the Mediterranean are scattered mainly in what is now modern Greece, Turkey and Italy (Figure 1). The most important and frequently used white marbles since antiquity are the white marbles from Penteli, in Greece, from the Greek islands of Thasos and Paros, the Italian Carrara marble, as well as the Proconnesos marble from modern day Turkey.

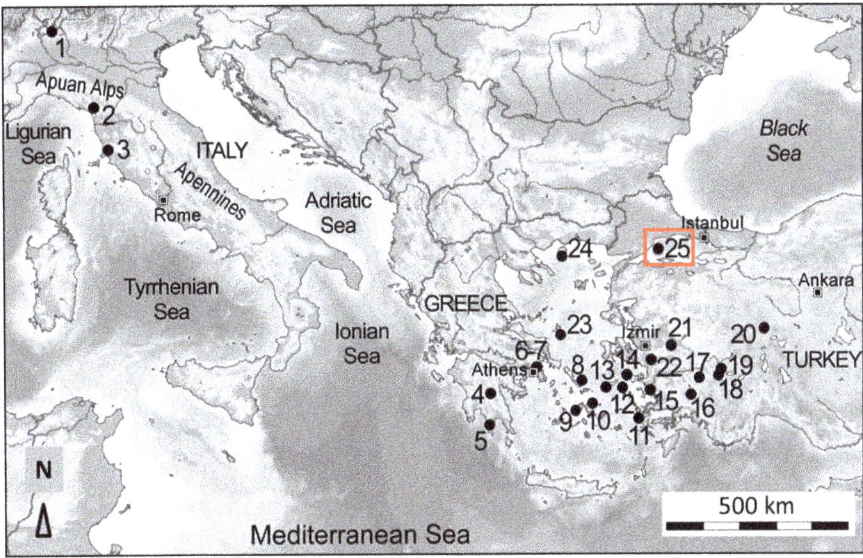

Figure 1. Map of the eastern Mediterranean region indicating the locations of ancient quarrying sites, modified after Antonelli and Lazzarini 2015 [12]: (1) Candoglia; (2) Carrara; (3) Campiglia; (4) Doliana; (5) Mani; (6–7) Mts Pentelic and Hymettus; (8) Tinos; (9) Paros; (10) Naxos; (11) Kos; (12) Fourni; (13) Ikaria; (14) Samos; (15) Miletus; (16) Göktepe; (17) Aphrodisia; (18) Denizli; (19) Thiounta; (20) Docimium; (21) Sardis; (22) Ephesos; (23) Skyros; (24) Thasos; (25) Proconnesos (Marmara), indicated by red rectangular.

Pentelic marble is perhaps the most famous marble used in ancient monuments, due to its excellent quality; in classical Athens it was the material of choice, used for the construction of the Parthenon in Athens and the Hephaestus Temple in the Ancient Athenian Agora [13,24,25]. Its use was continuous throughout the ages, as it also adorns important neoclassical buildings of the 19th and 20th century in Athens [26].

Parian marble was used in various monuments in the ancient Greek and Roman world and was highly appreciated [27]. Owing to its high purity and texture, it was used in decorative elements, for example the decorative reliefs of the ancient Temple of Hephaestus in Athens, as well as important sculptures [28] and anthropomorphic sarcophagi [29]. Thasian marble was also widely used, in monuments (such as the burial monument of Amphipolis in Macedonia, Greece), as well as in Greek and Roman statues and reliefs [30]. Carrara marble has been used in monumental roman structures, such as the Pantheon and Trajan's column, roman sculptures, and sculptures of the Renaissance, such as Michalangelo's David, as well as in more recent monumental, religious and important buildings throughout the world [31–33].

The northern side of Proconnesos Island has been well known as a major marble producing area since the archaic period [34]. Its production spiked during the Roman times, around the second half of the first century AD, while the local quarries were made dependent on the imperial treasury by Hadrian. In the fourth century, the quarries produced ready-to-sculpt blocks and building elements, tied to the large imperial buildings of Constantine the Great [35]. Proconnesos marble, due to its appearance and quality, was a greatly appreciated marble. In addition, the insular nature of its quarries from which marble could be directly transported by sea, greatly reduced transportation costs, thus increasing its popularity [35]. In addition, the vicinity of the quarries to Constantinople, the capital city of the Byzantine empire, made the Proconnesian marble one of the most popular raw materials during the Byzantine times [16,36]. Proconnesian marble has been used in numerous monuments, sculptures and sarcophagus throughout the ages [37–41], as well as in important modern-day buildings.

A relatively recent study regarding the provenance of the marbles comprising the Sanctuary of the Great Gods in Samothrace presented a complex pattern of provenance including Thasian, Pentelic, Parian and Proconnesian marble, sometimes in combination [42]. This is true for many monuments, as reuse of older members from other monumental buildings is usual throughout all historical periods, an issue which complicates provenance studies.

The great importance of the above mentioned marble monuments, as well as the complexity of their construction and multiple construction/restoration phases, implies that accurate provenance identification is a very critical field of research; not only for restoration purposes, but also because it can broaden conceptions and interpretations in relation to historical and archaeological aspects regarding the selection, transportation and use—or even re-use—of this valuable material. In this framework, the provenance of marbles used in monumental buildings, such as the Holy Aedicule, is especially interesting and is herein addressed.

The Holy Aedicule is an indoor structure located in the center of the Rotunda area in the Church of Resurrection in Jerusalem (Figure 2a). It is the structure surrounding and containing what is believed to be the Tomb of Christ and is thus a religious site of great importance for the Christian World. The Holy Aedicule today consists of two chambers; the Chapel of the Angel to the east and the Holy Tomb Chamber to the west (Figure 2b). The exterior of the structure is covered with stone facings, consisting of locally quarried compact stones (mizzy and slayeb stones) (Figure 3) [43–45]. The interior of both chambers is adorned with marble facings, mostly of white coloration (Figure 4a, b). The Tomb of Christ, located at the northern side of the Tomb Chamber, is also enclosed within marble facing (Figure 4b).

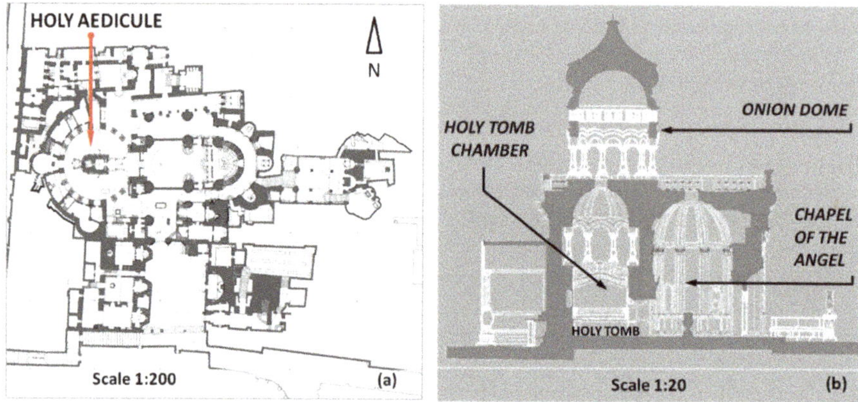

Figure 2. (**a**) Ground plan of the Church of Resurrection in Jerusalem, indicating the location of the Holy Aedicule within the Rotunda area, modified after Lavvas 2009 [46]; (**b**) Section of the Holy Aedicule North view, indicating the basic parts of the structure, modified after Balodimos et al 2003 [47], Lavvas 2009 [46].

Figure 3. (**a**) View of the South façade of the Holy Aedicule as it is today; (**b**) The main entrance of the Holy Aedicule from the East, where the pilgrims enter.

Figure 4. (**a**) Interior west façade of the Chapel of the Angel, showing the low entrance to the Tomb Chamber, with the marble pedestal containing part of the stone of the Angel; (**b**) Tomb Chamber adorned with marble facings; the Holy Tomb is indicated on the right-hand side of the image.

In addition to its religious significance, it is also a site of high archaeological and architectural interest, as its evolution during the last 17 centuries reflects and embeds the tumultuous history of the city of Jerusalem. Since the fourth century AD, when the cave-hewn Tomb was uncovered by Saint Helena and Constantine the Great and enclosed within a structure (Aedicule), it has evolved into the complex structure that stands today, visited by millions of pilgrims annually [48]. The major historical events, constructions, destructions and subsequent reconstructions which mark the evolution of the Holy Aedicule structure are presented in Table 1.

Table 1. Major historical phases of the Holy Aedicule—Construction, Destructions and Restorations.

Chronology	Major Historical Phases of the Holy Aedicule—Construction, Destructions and Restorations
136–137 AD:	At the site of the rock-cut Tomb of Christ, the Roman Emperor Hadrian builds a Capitol (Capitolium), as a statement of Roman domination and power. Eusebius (260/265–339/340 AD) mentions: *"Indeed with a great expenditure of effort they brought earth from somewhere outside and covered up the whole place, then leveled it, paved it, and so hid the divine cave somewhere down beneath a great quantity of soil"* [49].
325–326 AD:	Constantine the Great and Saint Helena discover the Tomb of Christ, which they uncover and enclose in a structure, known as the Holy Aedicule. Constantine requests that: *"the building should surpass all the Churches in the world, in the beauty of its walls, its columns and marbles"* [50]. Regardless of controversy as to the form of the Constantinean Aedicule—no entirely realistic depiction of this Aedicule is available—it is widely accepted that the Holy Tomb was at this point in the form of a polygonal ciborium-type structure (Aedicule) consisting of the burial chamber [49,51].
614 AD:	The Persians conquer the city of Jerusalem and infer severe damages on the Church of Resurrection [52]
626–630 AD:	In 626, Modestos, who was at the time standing in as Patriarch of Jerusalem, begins the reconstruction of the Church of Resurrection, which was concluded after four years, allowing Heraclius, *to make a victorious ceremonial entry into Jerusalem on 21 March 630, carrying the precious relic of the True Cross, as the trophy of victory, and reinstall it in the church of the Holy Sepulchre* [53–57]. There is no certain knowledge whether the Aedicule itself sustained any damage or whether any repairs were required [51]. In any case, the Aedicule remains in the form of a polygonal ciborium-type structure.
1009 AD:	The Fatimid Caliph, al-Ḥākim bi-Amr Allāh, conquers Jerusalem and nearly destroys the Holy Pilgrimage Sites [57]. Hakim ordered Yaruk, governor of Ramla, to *"demolish the church of the Resurrection and to remove its (Christian) symbols, and to get rid of all traces and remembrance of it." Yaruk's son and two associates "seized all the furnishings that were there, and knocked the church down to its foundations, except for what was impossible to destroy and difficult to grub up to take away"* (according to [51]).
~1012–1023 AD:	Al Hakim's Christian mother, Maria, begins to *"rebuild with well-dressed squared stones the Temple of Christ destroyed by her son's order"* [51].
~1037–1040 AD:	Major reconstruction takes place, following a strong earthquake in 1034, either during the time of the Byzantine Emperor Michael IV the Paphlagonian (1034–41 CE), or during the reign of Emperor Constantine IX Monomachos (1042–1048 CE) or both [51,58]; the exact dates of the 11th century Byzantine interventions are still under debate. Up to this point the Aedicule remains in a polygonal form.
1099–1187 AD:	The Crusaders expand the Aedicule to the east through the addition of an antechamber, the Chapel of the Angel, at the spot where the Angel appeared after the burial of Jesus Christ, according to the Gospels, thus leading to the characteristic, since then, horse-shoe shape of the Aedicule [51,57].
1244 AD:	Khwarismian horsemen sweep down from the north-west, and enter Jerusalem. According to testimonies, *"the Khwarismians entered the Church of the Holy Sepulchre, broke open the tombs of the kings, slaughtered the Christians who had taken refuge in front of the Edicule, decapitated the priests who were celebrating at the altars, and laying sacrilegious hands on the tomb of the Lord's resurrection, they defiled it in many ways, overturning from its base the marble cladding placed around it . . . The carved columns placed in front of the tomb of the Lord for decoration, they removed, sending them as a sign of victory to the tomb of the evil Mohammed, to the disgrace of the Christians"* [51].

Table 1. *Cont.*

Chronology	Major Historical Phases of the Holy Aedicule—Construction, Destructions and Restorations
1555 AD:	Major restoration of the Holy Aedicule by Fra Bonifacio da Ragusa, the Custos of the Franciscan Order [59]. Restoration was crucial, not only owing to the fact that the Aedicule structure had not been restored in 500 years and was thus in a deteriorated state, but was also necessitated due to damages inferred by strong earthquakes (1453 AD, 1545 AD) [51,57]. It is possible that, although Bonifacio states that the reconstruction was from the foundation, in fact, only the Dome was reconstructed, in addition to the repositioning of several members of the structure.
1808 AD:	A large fire breaks out in the Church of Resurrection, causing the Rotunda dome to collapse, affecting the Aedicule as well [51,57].
1809–1810 AD:	The architect Kalfa Komnenos restores the Holy Aedicule, giving it its present form. Although he states that it was rebuilt from the foundations, there is probability that the fire did not affect the interior of the Holy Aedicule [51,57].
1947 AD:	An external iron frame is installed by the British Mandate in 1947, as a temporary solution, to address the severe deformations of the Holy Aedicule structure, observed already as early as 1927, and to prevent the collapse of the Aedicule [51,60].
2015 AD:	The National Technical University of Athens (NTUA) was invited by His Beatitude, Patriarch of Jerusalem Theophilos III, to implement an "Integrated Diagnostic Research Project and Strategic Planning for Materials, Interventions Conservation and Rehabilitation of the Holy Aedicule of the Church of the Holy Sepulchre in Jerusalem" [61] in order to ascertain as to the causes for the severe buckling and serious deformations of the structure and design rehabilitation proposals.
2016–2017 AD:	A historical Common Agreement of the three Christian Communities responsible for the Holy Sepulchre is signed on 22 March 2016 to rehabilitate the Holy Aedicule. Based on the results of the study and the rehabilitation interventions proposed, a rehabilitation project was initiated, supervised by NTUA. The rehabilitation project was successfully implemented and completed in March 2017 [62–67].

The rehabilitation project implemented under the scientific supervision of the National Technical University of Athens (NTUA), provided the framework to examine the building and decorative materials of the Holy Aedicule. It should be highlighted that on the 26th of October, 2016, the NTUA team was granted permission by the three Christian Communities—Guardians of the Holy Tomb, to open the Tomb of Christ during the grouting of the structure, in order to protect the burial surface. The top marble slab was thus shifted and the interior of the burial monument came to light; a lower marble plate, fragmented in a direction from west to east and in position to the north of the Tomb, was revealed; on the southern part of the Tomb, the burial rock, the original burial surface of the Tomb, was revealed [62,68].

The current study will try to identify the provenance of marble samples collected from the most holy site of Christianity, the Tomb of Christ, as well as from various locations of the Holy Aedicule enclosing it, by using mineralogical, petrographic and C-O isotopic analyses. The combination of these methods enables the marbles to be identified with an acceptable degree of reliability [3]. The obtained results are compared to relevant literature data in order to establish their provenance and an attempt for an intra-site topographical discrimination is made. These results are discussed in relation to historical sources and testimonies, as well as to previously published archaeometric results regarding the historical mortars of the Holy Aedicule [69]. Thus, new data regarding the evolution of the Aedicule, as well as information regarding the selection of marbles throughout the monument's history, emerges.

2. Materials and Methods

2.1. Sampling

Marble samples were collected from the interior facings of the Holy Aedicule (Chapel of the Angel and Holy Tomb Chamber), as well as from the marble slabs enclosing the Tomb of Christ. Six marble samples were studied: four from the Holy Tomb Chamber and two from the Chapel of the Angel. In particular, in the Holy Tomb Chamber, one sample was taken from the marble facing of the interior south wall opposite the Tomb, while the other three were collected from the Holy Tomb cladding: one from the upper marble plate, worshipped today, one from the interior marble plate, which was revealed during the NTUA rehabilitation project, and one from a marble fragment found within the Tomb. Two marble samples were collected from the western wall of the Chapel of the Angel, northern and southern to the low entrance of the Holy Tomb Chamber, respectively.

A summarized description of the marble samples under investigation and their sampling location is presented in Table 2. Sampling locations on the ground plan of the Holy Aedicule are presented in Figure 5, while in Figures 6–9, views of the sampling areas are displayed.

Table 2. Codes, short description of the sampling locations and macroscopic features of the studied samples.

Sample Code	Brief Description of the Sampling Location	Macroscopic Features on Fragmented (Natural) Surface
OM-10	Tomb Chamber, North side, Holy Tomb interior, grey marble plate above the burial bed rock and under the amber hued* marble plate. See Point 1 at Figures 5 and 6.	White, medium to coarse grained, gray in parallel bands
OM-11	Tomb Chamber, North side, Holy Tomb exterior, amber hued marble plate. See Point 3 at Figures 5 and 6.	White, medium to coarse grained
OM-13	Tomb Chamber, South side, smooth surface of interior marble facing, opposite the Holy Tomb. See Point 4 at Figures 5 and 8.	White, medium to coarse grained
OM-49	Tomb Chamber, North side, Holy Tomb interior, marble fragment, found in the filling material of the Holy Tomb. See Point 2 at Figures 5 and 6.	White, medium to coarse grained, gray in parallel bands
OM-50	Chapel of the Angel, North-West side, northern to the Entrance of the Tomb Chamber, relief of interior marble facing. See Point 5 at Figures 5 and 9.	White, medium to coarse grained
OM-51	Chapel of the Angel, South-West side, southern to the Entrance of the Tomb Chamber, smooth surface of interior marble pilaster. See Point 6 at Figures 5 and 9.	White, coarse grained

* The amber hued coloration is noticed on the external surface of the marble plate and it is attributed to the application of myrrh for centuries as part of the liturgical functions.

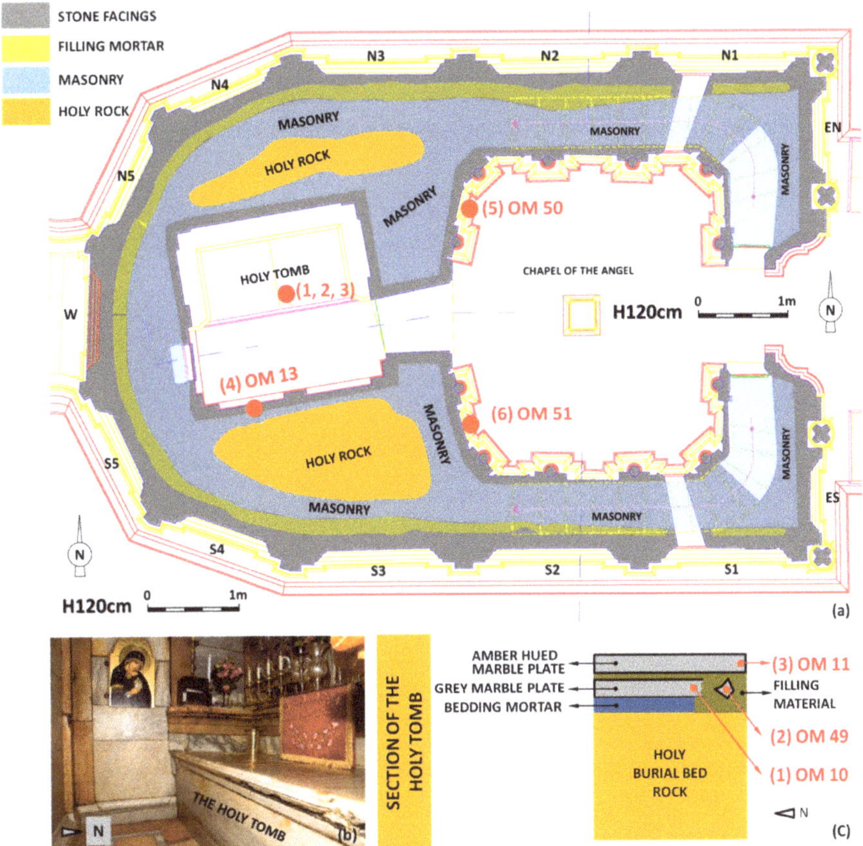

Figure 5. (**a**) Depiction of sampling locations on ground plan of the Holy Aedicule (ground plan modified from Lampropoulos et al 2017 [70]); (**b**) View of the Holy Tomb from the entrance on the east; (**c**) Schematic representation of the Holy Tomb: section depicting the layering of materials and the sampling areas of OM11, OM49, OM10.

Figure 6. Description of sampling areas of marble samples (1) OM10 and (3) OM11: (**a**) the Holy Tomb with the top amber hued marble plate, that it is worshiped today, in place; (**b**) View of the Holy Tomb during the shifting out of position of the top amber hued marble plate on October 26, 2016; (**c**) The open Holy Tomb where it was visible: the Holy Burial Bed Rock; the fragmented gray marble plate (sampling area of (1) OM10); the, shifted out of position, top amber hued marble plate (sampling area of (3) OM11).

Figure 7. Sample (2) OM49: the marble fragment, found within the Holy Tomb; (**a**) front view, where characteristic grey stripes can be macroscopically observed; (**b**) top view, where mortar residues are evident; (**c**) side view, where mortar residues and a special curved ending are evident; (**d**) the other side view, where, besides the special curved ending, characteristic grey stripes can be macroscopically observed.

Figure 8. Description of sampling area of marble sample (4) OM13: smooth interior marble facing, opposite the Holy Tomb in the Holy Tomb Chamber.

Figure 9. Description of sampling areas of marble samples (5) OM50 and (6) OM51 in the Chapel of the Angel; At the right hand side-northern to the Tomb Chamber entrance, the sampling area of sample (5) OM50, is displayed; At the left hand side-southern to the Tomb Chamber entrance, the sampling area of sample (6) OM51, is displayed.

2.2. Analytical Techniques

From the six (6) marble samples that were collected, thin-and-polished sections were created and underwent detailed petrographic investigation. The technique of optical microscopy was performed using a Leica DM2500P optical microscope mounted with a Nikon camera. MGS values were measured on images collected through a ProgRes-C14PLUS video-camera, using the ProgRes CapturePRO2.1 software.

Stable isotope (C, O) analyses of the studied marble samples were performed; marble chips, hand-picked for every sample under the stereoscope in order to avoid any contaminations of the samples (e.g., from mortar rests), were pulverized in an agate mill for this purpose. The marble powders were reacted with 100% phosphoric acid at 70°C using a Gasbench II connected to a ThermoFisher Delta V Plus Mass spectrometer. All values are reported in per mil relative to V-PDB (Vienna Pee Dee Belemnite). Reproducibility and accuracy were monitored by replicate analysis of laboratory standards calibrated by assigning $\delta^{13}C$ values of +1.95‰ to NBS19 and −47.3‰ to IAEA-CO9 and $\delta^{18}O$ values of −2.20‰ to NBS19 and −23.2‰ to NBS18. Reproducibility for $\delta^{13}C$ and $\delta^{18}O$ was ±0.01 and ±0.08, respectively.

3. Results

3.1. Petrographic Characterization of the Studied Samples

The studied samples were taken from an archaeological monument of great significance, thus each one will be described separately in the present section. Panoramic microphotographs, taken under crossed-polarized light are presented in Figure 10. Detailed textural and petrographic characteristics for each sample are depicted in Figures 11–16.

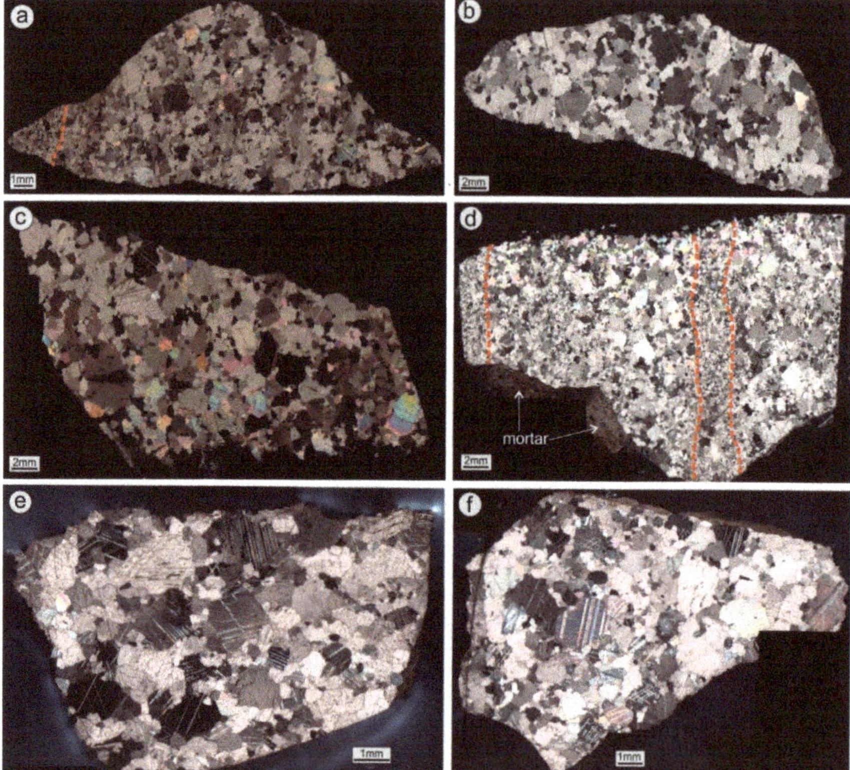

Figure 10. Transmitted light, panoramic microphotographs of the studied samples, taken under crossed-polarized light: (**a**) Sample OM10; (**b**) sample OM11, (**c**) sample OM13; (**d**) sample OM49, note the mortar rest on the right corner of the sample; (**e**) sample OM 50; (**f**) sample OM51. Note the characteristic heteroblastic fabric on all samples (mortar-type). Red dotted lines mark the areas that belong to grey-colored bands in samples OM10 and OM49.

A common characteristic of all six samples is that they emit an intense smell of sulfur upon scraping or hitting, fact that implies an organic component (bituminous compounds). Coloration of the studied samples is relatively homogeneous, around 180–200. Two of the samples (OM10 and OM49) are characterized by a relatively small number of parallel and mm-sized grayish bands that can be observed even in macroscopic scale. In microscopic scale, these bands are composed of fine-grained calcite, indication that the rock has suffered an extent of tectonic deformation. What is also common in all the studied samples is their characteristic heteroblastic fabric, known as "mortar-type". This fabric is characterized by two groups of crystal sizes: the first largest group comprises subhedral calcite porphyroblasts, which maximum grain size (MGS) ranging from 1.6 up to ~3 mm (most common values are >2 mm). This fact is helpful to characterize the studied samples as medium/coarse grained marbles, as the value of 2 mm is commonly used in archaeometry as a cutoff between the fine- and medium/coarse-grained marble varieties [3]. The second group is characterized by much smaller MGS (less than 1 mm). These grain sizes are detected in the Holy Aedicule samples under examination, in the grey-colored bands of samples OM10 and OM49, as well as in calcite crystals that form around the porphyroblasts, due to strain deformation and subsequent neocrystallization. Deformation of the rocks is also evidenced by bent twinning lines in calcite crystals. Accessory minerals in all samples include mainly mica (phlogopite), while minor apatite and pyrite were also observed. Dolomite was also identified, but its volumetric participation in the rocks is rather insignificant.

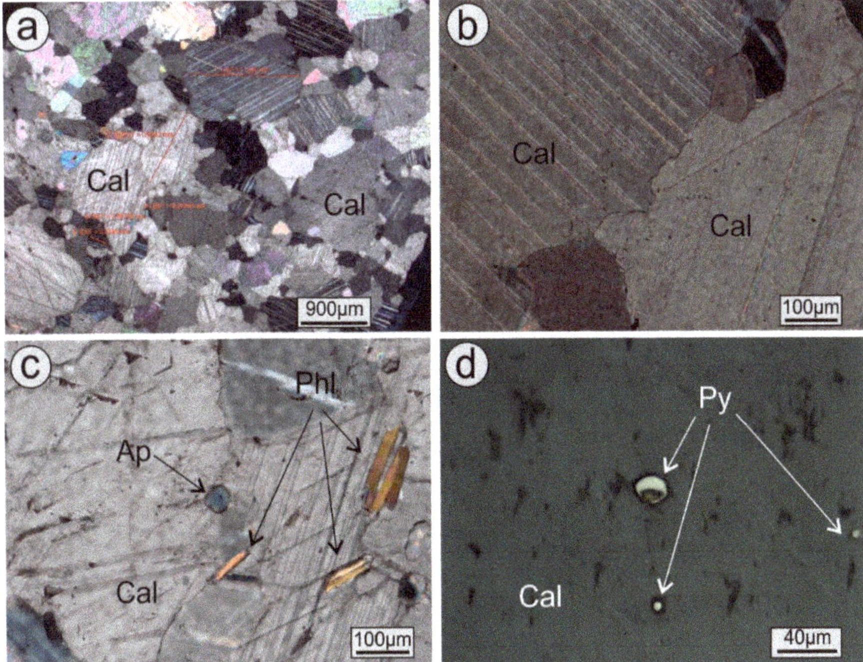

Figure 11. Transmitted (**a**–**c**) and reflected (**d**) light microphotographs of sample OM10: (**a**) Heteroblastic fabric ("mortar"), crossed-polarized light; (**b**) sutured and partly embayed grain boundaries, crossed-polarized light; (**c**) anhedral apatite (Ap) crystal included, along with phlogopite (Phl) in calcite (Cal), plane-polarized light; (**d**) round-shaped pyrite (Py) crystals included in calcite (Cal).

3.1.1. Sample OM10

Calcite is the dominant mineral component of the sample. Dolomite is present in minor amounts. The fabric is strongly heteroblastic, ("mortar-type", Figure 11a). Tectonic deformation is expressed by the presence of mm-sized bands, which are macroscopically observed as grey-colored stripes. They are composed of fine-grained calcite, along with parallel-oriented, mica-group minerals (phlogopite/muscovite). MGS values reach up to 2.0 mm for the coarse grain parts and are significantly lower, up to 0.4 mm, in the fine-grained bands. Xenomorphic crystals of calcite in the groundmass often exhibit polysynthetic twinning, as a result of deformation. Grain boundary shapes are typically sutured, in some cases embayments were also identified (Figure 11b). Accessory minerals are commonly phlogopite, rarely apatite and pyrite (Figure 11c,d).

3.1.2. Sample OM11

Calcite is the dominant mineral component of the sample and forms subhedral porphyroblastic crystals (MGS 1.7 mm) surrounded by smaller neoblasts (Figure 12a). The fabric is strongly heteroblastic, (mortar-type, Figure 12a) and many crystals display intense polysynthetic twinning. In some cases, twinning lines are bent, a fact that indicates tectonic deformation. Calcite grains usually display sutured boundaries, but in places embayed boundaries were also observed (Figure 12b,c) Phlogopite, apatite and pyrite, which forms round-shaped grains, were identified as accessory phases (Figure 12b–d).

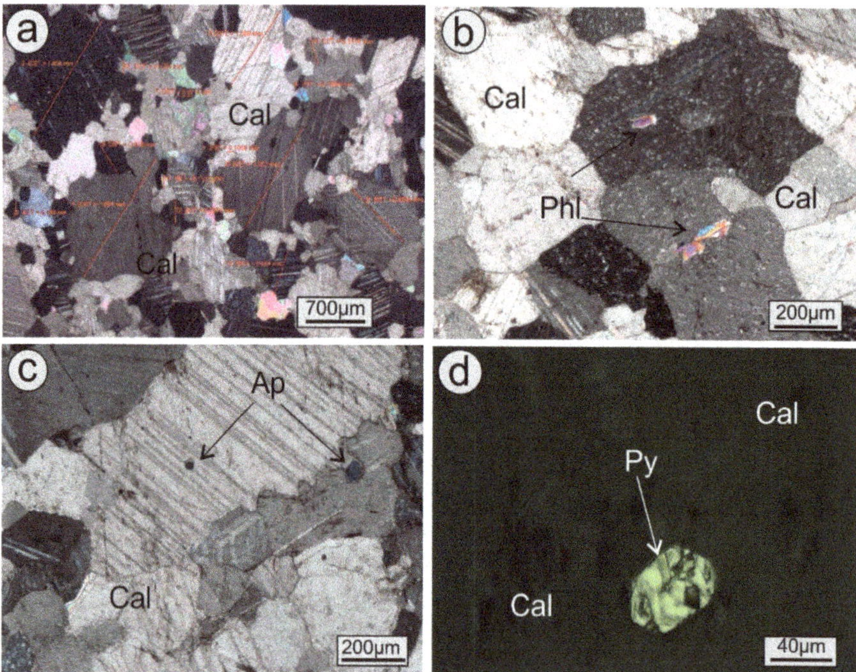

Figure 12. Transmitted (**a**–**c**) and reflected (**d**) light microphotographs of sample OM11: (**a**) Heteroblastic fabric ("mortar"), crossed-polarized light; (**b**) sutured calcite (Cal) grain boundaries. At places calcite includes phlogopite (Phl), crossed-polarized light; (**c**) anhedral apatite (Ap) crystal included in calcite (Cal). Note the sutured and partly embayed grain boundaries, crossed-polarized light; (**d**) round-shaped pyrite (Py) crystal included in calcite (Cal).

3.1.3. Sample OM13

Calcite is the dominant mineral component of the sample. Dolomite is present in minor amounts. The sample exhibits a heteroblastic fabric ("mortar-type", Figure 13a,c), with maximum grain sizes being slightly smaller compared to the other samples, around 1.6mm. Deformation is also remarked in this sample, as xenomorphic crystals of calcite in the groundmass often exhibit polysynthetic twinning.

Rarely, triple point junctions of calcite grains (Figure 13b) were observed. Micas (phlogopite muscovite), anhedral apatite and round-shaped pyrite were identified as main accessory phases (Figure 13c,d).

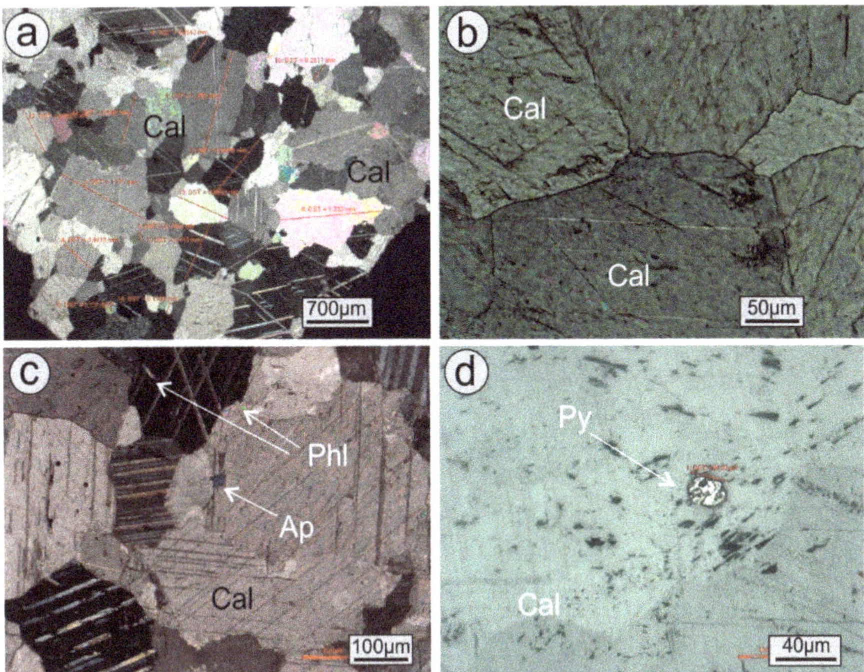

Figure 13. Transmitted (**a–c**) and reflected (**d**) light microphotographs of sample OM13: (**a**) Heteroblastic fabric ("mortar"), crossed polarized light; (**b**) slightly sutured calcite (Cal) grains forming triple-point junctions, plane-polarized light; (**c**) anhedral apatite (Ap) crystal included in calcite (Cal). At places calcite includes phlogopite (Phl). Note the sutured and partly embayed grain boudaries; (**d**) round-shaped pyrite (Py) crystal included in calcite (Cal).

3.1.4. Sample OM49

This sample is characterized by heteroblastic fabric (Figure 14a) and the presence of a fine-grained band, with a maximum thickness of about 500 μm (Figure 14b). This band, which is also remarked as a grey-colored stripe in macroscopic scale, is composed of anhedral calcite with maximum MGS up to 0.6 mm. These MGS values are much smaller compared to the MGS values of calcite grains in the rest of the rock, where they reach up to almost 2 mm. Bent twinning lines are common, along the band and the rest of the rock as well. Grain boundaries inside the band are commonly embayed, while in the rest of the sample, the most common grain boundary shape is sutured, although embayments were observed here as well. Minor mineralogical components like apatite and micas (phlogopite and/or muscovite) were identified (Figure 14c), with the latter being commonly found in the band, oriented parallel to its major dimension. Rarely, round-shaped pyrite was observed included in calcite (Figure 14d).

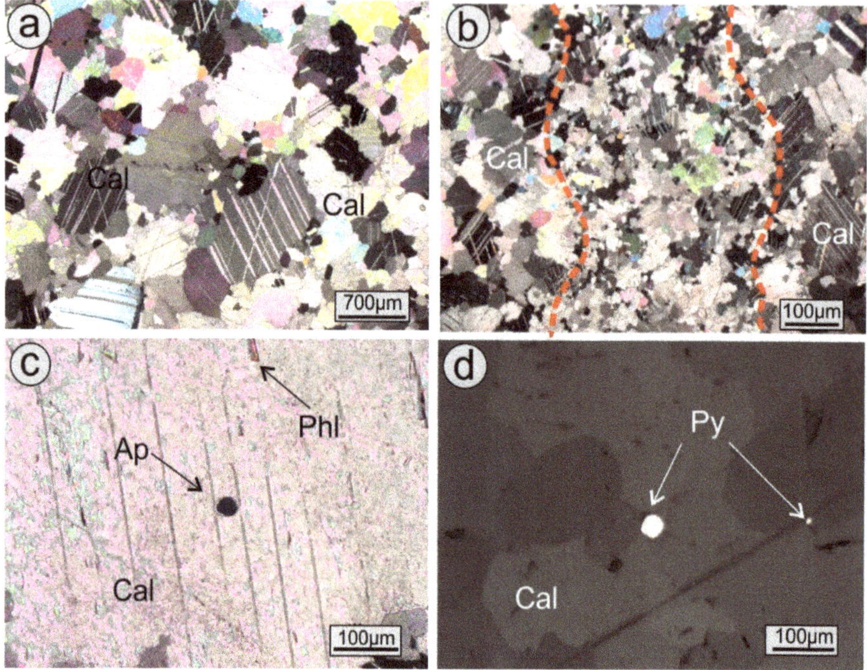

Figure 14. Transmitted (**a**–**c**) and reflected (**d**) light microphotographs of sample OM49: (**a**) heteroblastic fabric ("mortar") in calcite grains (Cal) most of which exhibit polysynthetic twinning, crossed-polarized light; (**b**) fine-grained calcite (Cal) crystals composing a band, which is grey-colored in macroscopic scale, crossed-polarized light; (**c**) anhedral apatite (Ap) crystal included in calcite (Cal), crossed-polarized light; (**d**) round-shaped pyrite (Py) crystals included in calcite (Cal).

3.1.5. Sample OM50

This sample exhibits a characteristic heteroblastic fabric ("mortar-type", Figure 15a). Calcite is the predominant mineralogical constituent. Grain boundaries are usually sutured, but in rare cases, straight-sided grain boundaries, leading to the formation of triple point junctions (120°) were observed (Figure 15b), perhaps suggesting local equilibrium conditions. Deformation is also traced in this sample, as xenomorphic crystals of calcite in the groundmass often exhibit polysynthetic twinning. Maximum grain sizes in this sample reach up to 1.7 mm. Anhedral apatite and round-shaped pyrite were identified as main accessory phases, while mica group minerals are quite rare. Apatite forms anhedral to subhedral rains, and its size does not exceed 100 μm (Figure 15c). Pyrite forms, fine-grained (up to 20 μm) rounded crystals, scattered in calcite.

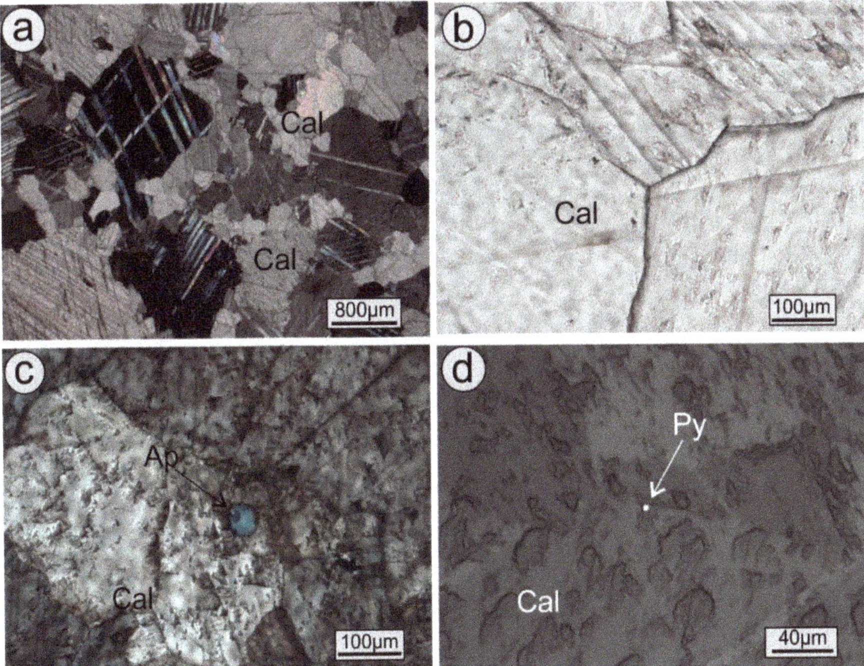

Figure 15. Transmitted (**a**–**c**) and reflected (**d**) light microphotographs of sample OM50: (**a**) heteroblastic fabric ("mortar") in calcite grains (Cal) most of which exhibit polysynthetic twinning, crossed-polarized light; (**b**) calcite (Cal) crystals forming a triple-junction and exhibiting slightly sutured grain boundaries, crossed-polarized light; (**c**) anhedral apatite (Ap) crystal included in calcite (Cal), plane-polarized light; (**d**) round-shaped pyrite (Py) crystal included in calcite (Cal).

3.1.6. Sample OM51

The sample is characterized by anhedral calcite grains that compose a heteroblastic fabric ("mortar" type, Figure 16a). The majority of these crystals are characterized by sutured and minor embayed grain boundaries (Figure 16b). Maximum grain size reaches up to 2.3 mm. Bent twinning lines in calcite crystals are common, indicating tectonic deformation. Accessory phases include minor dolomite, mica group minerals (phlogopite/muscovite), apatite and pyrite (Figure 16c,d) which forms round-shaped grains included in calcite.

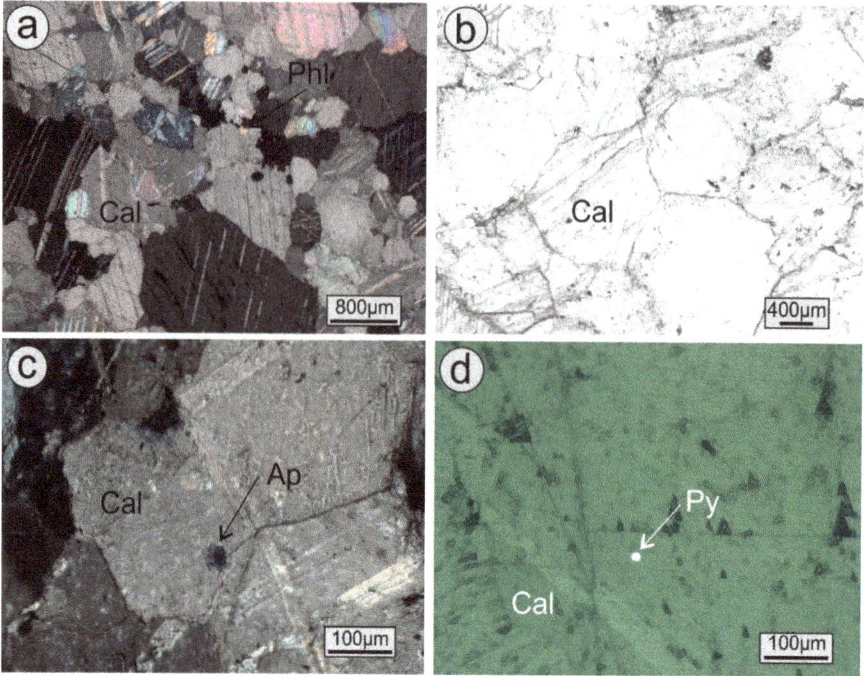

Figure 16. Transmitted (**a**–**c**) and reflected (**d**) light microphotographs of sample OM51: (**a**) heteroblastic fabric ("mortar") in calcite grains (Cal) most of which exhibit polysynthetic twinning. Minor phlogopite (Phl) is also present, crossed-polarized light; (**b**) calcite (Cal) crystals with partly sutured and embayed grain boundaries, plane-polarized light; (c) anhedral apatite (Ap) crystal included in calcite (Cal), crossed-polarized light; (**d**) round-shaped pyrite (Py) crystal included in calcite (Cal).

The above described microscopy results are summarized in Table 3, where the mineralogical content, the fabric, as well as the MGS and GBS parameters are stated.

Table 3. Petrographic features and mineralogical content of the studied samples.

Sample	Mineralogy	Fabric	MGS	GBS
OM-10	Cal (± Dol) ± Phl ± Ap ± Py	HE	2.0 0.4 *	Sutured, embayed
OM-11	Cal (± Dol) ± Phl ± Ap ± Py	HE	1.7	Sutured, embayed
OM-13	Cal (± Dol) ± Phl ± Ap ± Py	HE	1.6	Sutured, embayed
OM-49	Cal (± Dol) ± Phl ± Ap ± Py	HE	1.9 0.6 *	Sutured, embayed
OM-50	Cal (± Dol) ± Phl ± Ap ± Py	HE	1.7	Sutured, embayed
OM-51	Cal (± Dol) ± Phl ± Ap ± Py	HE	2.3	Sutured, embayed

Abbreviations: Cal = calcite; Dol = dolomite; Phl = phlogopite; Ap = apatite; Py = pyrite, HE = heteroblastic; MGS = maximum grain size in mm; GBS = grain boundary shape. Maximum grain size (MGS). Values with (*) refer to calcite grains from grey-colored bands.

3.2. *Isotopic Signature of the Studied Samples*

Isotopic values of the six samples can be generally characterized as homogeneous, despite the observed variation in their $\delta^{13}C$ and $\delta^{18}O$ val ues and are summarized in Table 4.

$\delta^{18}O$ values are negative in all samples and range from −2.49 to −1.13. The majority of the samples (OM10, OM13, OM49 and OM50) display very homogeneous $\delta^{18}O$ values (ranging from −1.89 to −1.13 ‰ V-PDB), while the rest two samples (OM11 and OM51) are isotopically lighter, with values −2.49 and −2.06 ‰ V-PDB respectively. Regarding the $\delta^{13}C$ values, they vary slightly from 2.43 to 3.37 ‰ V-PDB.

Values of the studied samples are plotted in the marble isotopic reference diagram (Figure 17, after Gorgoni et al. 2002 [18]). In this diagram, isotopic fields of white marbles that were commonly used in antiquity have been marked, based on published isotopic data for marbles from Carrara (Italy), Penteli, Naxos, Paros, Thassos (Greece) and Afyon, Aydin, Proconnesos (Turkey). Samples from this study plot in the central part of the Proconnesos-1 marble field. Overlapping in this area is noticed with parts of other fields, which represent marbles from other localities. In particular, some samples plot along or inside the subfields for Thasos (subfields T1,2,3), Paros (subfields Pa2,3) and Carrara (field C).

Table 4. $\delta^{18}O$ and $\delta^{13}C$ values (‰ V-PDB) of the studied samples.

Sample	$\delta^{18}O$	$\delta^{13}C$
OM-10	−1.89	3.08
OM-11	−2.49	2.59
OM-13	−1.58	3.37
OM-49	−1.58	3.31
OM-50	−1.13	2.43
OM-51	−2.06	2.71

V-PDB: Vienna Pee Dee Belemnite.

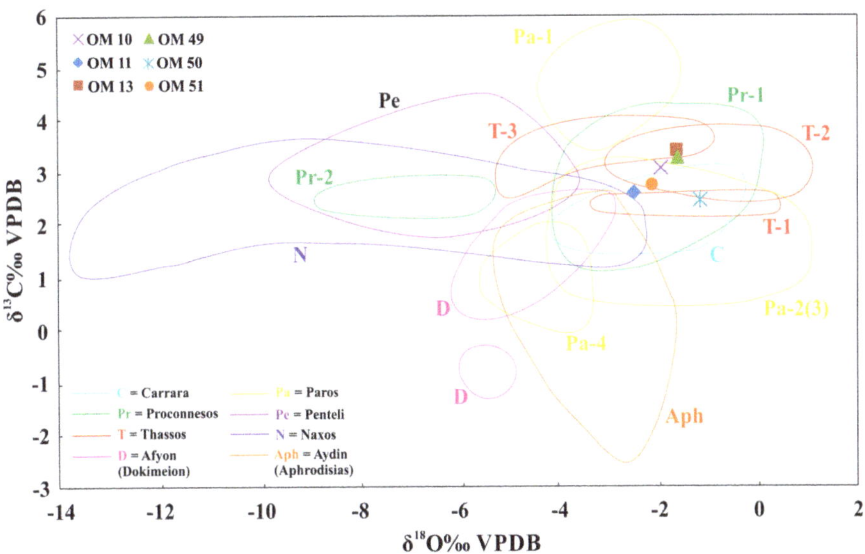

Figure 17. $\delta^{18}O$ versus $\delta^{13}C$ plot, displaying the isotopic values of the studied samples. Fields of ancient quarrying samples are from Gorgoni et al. 2002 [18]. Pr1,2 = Proconnesos, variety 1 and 2; T1,2 = Thasos-Alyki; T3 = Thasos-Vathy; C = Afyon (Dokimeion); Pa1 = Paros-Lychnites, Pa2,3 = Paros-Chorodaki, Pa4 = Paros-Lakkoi; Pe = Penteli; N = Naxos; Aph = Aydin.

4. Discussion

Provenance of the numerous marble varieties that have been used in archaeological sites worldwide still remains one of the most debated topics [11]. This matter is of critical importance due to the need of determining a suitable source that can supply the optimum restoration material, in terms of physicochemical, mechanical and historical compatibility [2,3,12]. Furthermore, it can also reveal crucial information regarding the history of a monument and shed light on transportation routes and commercial practices. In this section, the results of the petrographic and isotopic characterization of the examined samples will be used to establish the provenance of the Holy Aedicule marbles and a discussion will be made in relation to possible intra-site discriminations and in relation to historical testimonies and evidence.

4.1. Establishing the Provenance for the Holy Aedicule Marbles

Based on the petrographic study results, all six samples display critical similarities—textural and mineralogical—that indicate a common origin.

The studied marble samples are characterized by explicit textural similarities: all six samples display a strong heteroblastic fabric, which is typical for marbles that have suffered a certain degree of deformation during (or after) peak of metamorphism. This fabric comprises xenomorphic calcite crystals that commonly form two distinct generations: the first refers to calcite porphyroblasts, which are scattered in a matrix of fine-grained calcite neoblasts, forming the second generation. This deformation-inferred fabric is not observed in marbles that have been formed under equilibrium conditions with no subsequent deformation, thus exhibiting isodiametric grains comprising the so-called homeoblastic fabric (e.g., the marbles from Carrara).

The studied samples exhibit a strong heteroblastic fabric, commonly described as "mortar-type", which is characterizing marbles originating from Proconnesos and is a strong vector towards provenance determination (Antonelli and Lazzarini 2015 [12] and references therein). Beyond the "mortar-type" fabric, all the studied samples carry more evidence of deformation, like the polysynthetic twinning, often with bent twinning lines, in the calcite grains. In addition, two of the samples, namely OM10 and OM49, exhibit another feature: they are characterized by grey-colored bands that can be observed with naked-eye as well. This feature gives the rock a relatively foliated texture, which is common to many marbles (e.g., those from Pentelikon mountain, Proconnesos etc. [12]). In the samples under investigation, these gray-colored bands are composed by preferentially oriented calcite grains and minor presence of accessory minerals, mainly micas (phlogopite and/or muscovite). Despite the strong deformation features that characterize the studied marble samples, local equilibrium conditions were also identified, as evidenced by the rare presence of straight-sided grains that conjunct into triple points.

The maximum grain size is a parameter that could also be used to identify the provenance of a marble sample, because along with textural features, it is related to the tectonometamorphic conditions of the rock. In terms of archaeometry, marbles are distinguished in three categories: fine-grained (MGS up to 2 mm), average-medium grained (MGS 2–5 mm) and coarse-grained (MGS >5 mm). By using solely this parameter, and according to the coarse categorization proposed by Antonelli and Lazzarini 2015 [12], the studied samples, can be characterized as medium-grained, since they exhibit MGS that range mostly around 2 mm. This category consists of marbles from Proconnesos, Aphrodisias (Aydin) and some Parian and Thassian varieties (Pa2,3,4 and T3 respectively). This parameter allows the exclusion of marble quarries that produce both fine- and coarse-grained marbles. Based on this ascertainment, marble producing areas that should not be considered as raw material sources for the Holy Aedicule marbles are: Penteli and Hymmetos, Paros 1/Lychnites, Carrara and Afyon (Dokimeion) which produce fine-grained varieties and Thasos 1-2 and Naxos, which produce very coarse-grained marble varieties. MGS identified in the present study, range from 1.6 to 2.3 mm, gathered mostly around 2 mm, and fit well in the published MGS range for all the medium-grained marble varieties (Proconnesos, Aphrodisias, Paros 2,3,4 and Thasos 3), thus suggesting that by defining this parameter only, no safe conclusions regarding provenance can be made.

In parallel, accessory minerals have been also identified as possible key indicators useful in provenance determination [11]. The six studied samples contain the same accessory minerals paragenesis, which consist of (in decreasing volumetrically order) dolomite, mica (phlogopite, muscovite), apatite and pyrite. Mica group minerals, namely phlogopite and muscovite-paragonite-margarite (white micas, which are not easily distinguishable by optical microscopy only) are common accessory phases in many marbles (e.g., Proconnesos, Thassos, Paros, Naxos etc. [11]). The presence of apatite is also common in many marbles used in antiquity (e.g., Proconnesos, Carrara, Naxos, Thassos). Dolomite is present in marbles from a great number of different localities and so is pyrite. This fact precludes their use as provenance indicators.

By composing the above-mentioned data with the isotopic signature of the studied samples, their provenance should be searched between the Proconnesos, Thasos and Paros-3 marble quarries (see Figure 17), suggesting that a multi-parametric comparison of the studied samples to literature data for the above-mentioned localities is necessary.

Regarding the Thassian marble, despite its textural similarity to the studied samples, its heteroblastic texture is characterized as "mosaic-type" and not "mortar-type", as evident in the studied samples. Furthermore, the predominant grain boundary shape in Thassos marbles is the curved type, but in the samples from the Holy Aedicule, sutured and rarely embayed grain boundaries were the only identified types. The isotopic signature of the studied samples only partly (two out of six samples) overlaps the field of Thassos-3 (Vathy), indicating low isotopic relations. Finally, serpentine, which is common accessory mineral in the Thassian marbles (especially the dolomite-rich varieties), was not observed in any of the studied samples, and along with the previously-mentioned differences, Thassos should also be excluded as a possible source for the studied samples.

The Parian variety 3 (and 2), originating from the Chorodaki quarries, displays significant isotopic similarities to the studied samples, as four out of six, plot inside the relative subfield. Its MGS is also comparable to the observed MGS values. Textural and mineralogical variations though, are substantial to exclude this variety as well: the Parian marbles are characterized by mixed homeoblastic and heteroblastic, mosaic/lineated fabric, composed of curved and embayed crystals, and comprises accessory amphiboles, rutile, zircon ± serpentine [11,12]. Beyond scarce grey-colored bands in the studied samples, which could be considered as a common feature of the studied samples to the Parian marble, the absence of accessory phases like amphibole, zircon and serpentine also enhance another sourcing locality rather than Paros.

Although a partial overlapping of three out of the six studied samples is noticed with the Carrara subfield in the isotopic diagram (Figure 17), this provenance has to be excluded. Carrara marbles present significantly lower MGS values and they are characterized as a fine-grained variety in contrast to the Holy Aedicule samples, which are classified as medium to coarse-grained marbles. Furthermore, no evidence of plagioclase crystals was found in the Holy Aedicule samples, which is however, a characteristic accessory phase for the Carrara locality. In addition, their fabric is not homeoblastic-polygonal, as described for the Carrara marbles [12,18].

The final variety to be examined is the Proconnesian marble. In the isotopic diagram (Figure 17), it is clear that all the studied samples plot in a very narrow field in the center of the Proconnesos-1 subfield, indicating a critically similar isotopic signature between the Holy Aedicule samples and the published data on Proconnesos marbles ([15,16]). The observed "mortar-type" heteroblastic fabric is another significant similarity that should not be neglected and points toward a Proconnesian provenance for the studied samples. This is also enhanced by the remarked accessory mineralogical components, (phlogopite ± other micas, e.g., muscovite+apatite+minor dolomite+pyrite). Furthermore, in terms of MGS values, which range from 1.6–2.3 mm (mostly around 2 mm), the observed values fit well in the field of the published data.

Taking into consideration all aforementioned data, and having excluded other marble-producing localities with similar isotopic signature, it is strongly suggested that the examined marbles of the Holy Aedicule originate from the Proconnesos island quarries.

4.2. Proconnesos Quarries: Intra-Site Discriminations

The Proconnesian marble is one of the most popular white marbles used in antiquity [16], attracting the interest of the scientific community in relation to provenance studies in archaeometry.

Its use flourished during the Roman era, with numerous buildings and art materials (sarcophagi, sculptures etc.) made out of Proconnesos marble. During the second and third centuries AD, the use of Proconnesian marble has been documented throughout the whole Imperial territory [16,71–73]. This continued to the Byzantine times, since the close proximity of the quarries to the capital city of Constantinople was an additional advantage.

More than twenty-three ancient and contemporary quarrying locations have been identified so far in the northern part of Proconnesos (Marmara Island) [16]. In many of them, remnants of sculpted and or half-worked items are present (e.g. columns). The abundance of quarries and the widespread use of their marbles lead to the publication of detailed works, defining mineralogical and isotopic signatures of these quarries, in an attempt to define any possible intra-site discriminating features [15,16,18].

Based on isotopic characterization, Asgari and Mathews (1995) [15] already distinguished two possible Proconnesos marble varieties: The Proconnesos-1 and the Proconnesos-2, with the later exhibiting highly negative $\delta^{18}O$ values. This discrimination was later confirmed by Gornoni et al. 2002 [18] and Attanasio et al. 2008 [16], who extended knowledge on the Proconnesian marble properties. Attanasio et al. 2008 [16] proposed the use of an isotopic threshold ($\delta^{18}O = -5.00$), as a discriminating factor of the two varieties. As shown in Figure 18, the Holy Aedicule samples are characterized by relatively low-negative $\delta^{18}O$ values, ranging from −2.49 to −1.13, thus revealing an isotopic affinity to the Proconnesos-1 variety.

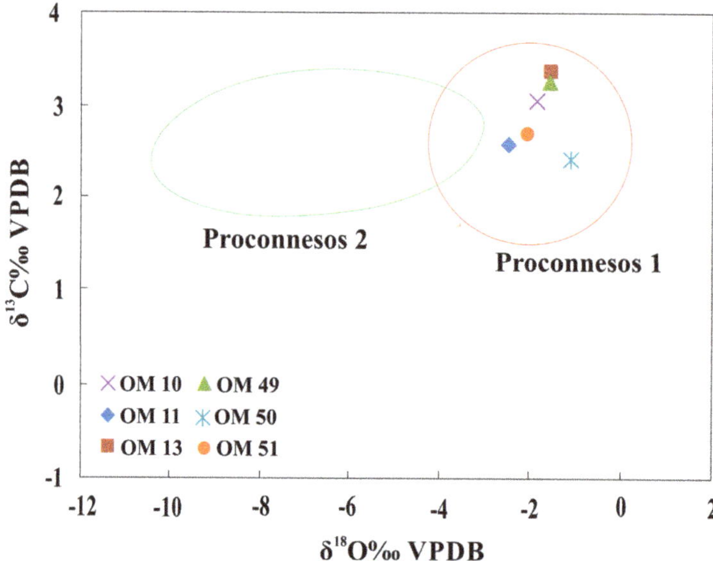

Figure 18. Isotopic discrimination plot between the so-called varieties Proconnesos-1 and Proconnesos-2. Fields for Proconnesos-1 and Proconnesos-2 varieties are adopted from Attanasio et al., 2008 [16].

Even though the isotopic values for the majority of the quarries display significant overlaps, a fact that allows limited chances for safe intra-site topographical discriminations, an effort was made to couple selected values of Proconnesos quarries with the respective values displayed by the marble samples of the Holy Aedicule.

In particular, for each sample of the Holy Aedicule, comparison was made with the published MGS values and isotopic values ($\delta^{18}O$ and $\delta^{13}C$), as given by Attanasio et al. 2008 [16], for the quarries that the later examined, excluding the modern-only quarries and the open-air museum artifacts.

In Table 5, the comparison of isotopic and MGS values between the Holy Aedicule samples and the literature data for quarrying locations on Proconnesos island [16] is presented. Highlighted boxes indicate pairing of the isotopic values of the Holy Aedicule samples (both $\delta^{18}O$ and $\delta^{13}C$) with the published range of values for a certain quarry, while (√) stands for pairing of the MGS values as well. Obviously, when both aforementioned criteria are satisfied, the probability of a sample deriving from that specific quarry is increased.

Table 5. Comparison of isotopic and MGS values between the Holy Aedicule samples and literature data for quarrying locations on Proconnesos island (Attanasio et al. 2008 [16]). Highlighted boxes indicate pairing of the isotopic values of a Holy Aedicule sample (both $\delta^{18}O$ and $\delta^{13}C$) to the published range of values for a certain quarry; (√) stands for respective pairing regarding the MGS values.

Quarry	Samples from the Holy Aedicule					
	OM10	OM 11	OM13	OM49	OM50	OM51
Altintaş	√				√	√
Harmantaş					√	
Filiz						
OC13		√				
Mandira		√			√	√
Saraylar						
C1		√		√		
C2	√	√				
C3	√					
C4		√				
C5						
C5b	√		√	√		
C5t		√				
C6		√			√	
C6b						
C7						
C7i		√				
C8						
C9						
C10						
C11		√				
C12	√					
C12s						
C13		√				
C14	√	√				
C15		√				
C16						

Based on this comparison, seven quarries, namely the Filiz, Saraylar, C5, C6b, C7, C8 and C16, demonstrate values which exclude them as sources of raw material for the examined samples of the Holy Aedicule.

According to the set criteria, sample OM13, which corresponds to the internal facing opposite the Holy Tomb, could have derived only for the C5b quarry. Sample OM49, which corresponds to the marble fragment found inside the Holy Tomb, could have originated only from quarries C1 and C5b. Although intra-site discrimination can be considered successful for the aforementioned samples, this is not the case for samples OM10 and OM11, which correspond to the lower and the upper plate of the Holy Tomb respectively. OM 10 could have originated from quarries Altintas, C2, C3, C12, C14 and C5b; the latter being the only common quarry with samples OM13 and OM49. OM11 is correlated with quarries OC13, Mandira, C1, C2, C4, C5t, C6, C7i, C11, C13, C14, C15, making any comparison with other samples difficult.

Samples OM50 and OM51 were taken from the internal facings of the Chapel of the Angel, right and left of the entrance to the Holy Tomb Chamber, respectively. OM50 could have been quarried from Altintas, Harmantas, Mandira and C6, while OM51 could have originated only from Altintas and Mandira quarries.

4.3. Discussing the Petrographic and Isotopic Analysis Results in Relation to Historical Sources, Data and Testimonies

Mineralogical, petrographical and isotopic investigations indicate that the Holy Aedicule marble samples were quarried at the island of Proconnesos. Intra-site discrimination demonstrated that the samples are of the Proconnesos-1 variety, while further topographical investigation suggested possible quarries from which each sample could have originated from.

These results will herein be discussed in relation to existing historical data and a previous archaeometry study of mortar samples from the Holy Aedicule [69], in order to discuss the use of Proconnesian marble to enclose the Holy Tomb, as well as to decorate the interior of the Holy Aedicule structure.

It must be noted that, although historical testimonies of pilgrims visiting the Tomb of Christ throughout the centuries are of great significance, their interpretation is difficult and may sometimes be misleading or even contradictory. Usually, their intention is not to document the Holy Tomb and the structure surrounding it, but to describe this most holy site glorifying the resurrection of Christ (it is frequently addressed as "holy of holies"), while the variations in metric systems throughout the ages does not facilitate comparisons. Thus, their accounts are useful to the interpretation of archaeometric findings, however, cannot always be taken into account in a literal and conclusive manner. In Table 6, the most important historical sources and testimonies related to the presence of marble facings surrounding the Tomb of Christ in the Holy Aedicule structure, are presented in chronological order.

Table 6. References to the Tomb of Christ and the interior marble facings of the Holy Aedicule.

Era	Basic References to Marble Facings of the Tomb and the Holy Aedicule
326/327 AD:	Constantine, in a letter to Macarius, declares: *"As to the columns or marble, you should after a survey yourself write promptly to us about what you may consider to be of most value and use, so that whatever quantity and kind of materials we may learn from your letter to be needful may be competently supplied from all sources" in order for the "world's most miraculous place" to "be worthily embellished"* [49]. Eusebius (Eusebius of Caesarea, counselor of Constantine the Great) states that: *"First of all, then, he (Constantine) adorned the sacred cave itself, as the chief part of the whole work, and the hallowed monument at which the Angel radiant with light had once declared to all that regeneration which was first manifested in the Savior's person"* [50], the cave was *"decked out and decorated with superb columns and full ornamentation, brightening the solemn cave with all kinds of artwork"* [49].
~380 or 450 AD:	Egeria offers a detailed account of her pilgrimage to the Holy Sites and describes: *"And what shall I say of the decoration of the fabric itself, which Constantine, at his mother's instigation, decorated with gold, mosaic, and costly marbles, as far as the resources of his kingdom allowed him, that is, the greater church, as well as the Anastasis, at the Cross, and the other holy places in Jerusalem"* [74].
1105–1107 AD:	Abbot Daniel (Daniel of Kiev) describes: *"it is a small cave hewn in the rock, having an entrance so low that a man can scarcely get through by going on bended knees; its height is inconsiderable, and its dimensions, equal in length and breadth, do not amount to more than 4 cubits. When one has entered the grotto by the little entrance, one sees on the right hand a sort of bench, cut in the rock of the cavern, upon which the body of our Lord Jesus Christ was laid; it is now covered by marble slabs. This sacred rock, which all Christians kiss, can be seen through three small round openings on one side"* [75].

Table 6. *Cont.*

Era	Basic References to Marble Facings of the Tomb and the Holy Aedicule
1322 AD:	John Maundeville in his pilgrimage recount in 1322, states that *"it is not long since the Sepulchre was all open, that men might kiss it and touch it"* [76].
1345 AD:	An anonymous English traveler describes the marble plate covering the Tomb of Christ: The Tomb is *"decorated with a porphyry slab, that had lips on the sides, and in the middle of the slab there was cut a streak"* [51]. This description is close to the upper marble plate in place today. However, it must be noted that either the use of the word "porphyry" inaccurately describes the distinctive amber hue of this marble plate or the marble plate in place today is a later addition presenting rounded edges and a streak in the middle as well.
1347 AD:	Niccolò da Poggibonsi describes three port-holes in a vertical slab encasing the Tomb, from which the rock burial surface is visible [51]. This is the last pilgrim account of the three port-holes in the vertical slab.
1570 AD:	Fra Bonifacio da Ragusa, describing his reconstruction: *"It seemed necessary to level the structure to the ground, so that that which would be constructed might be stronger and last longer. When the existing one was destroyed, the tomb of the Lord appeared clearly to our eyes, carved in the rock. In it could be seen depicted two angels, of whom one had an inscription which said: 'He is risen and is not here', whereas the other, indicating the tomb, proclaimed: 'Behold the place where he has been laid'. As soon as the figures of these two angels came into contact with the air, they almost completely disappeared. When it was necessary to remove one of the slabs of alabaster that covered the tomb on which the holy mystery of the Mass is celebrated, there appeared to us that ineffable place in which for three days the Son of Man reposed...."* [51,77].
1724–1744 AD:	Elzear Horn claims that Bonifacio covered the exterior of the Holy Aedicule with marble slabs and columns in order to beautify it. Also, when describing his own designs of the Holy Aedicule he says that: *"No. 20"* [this is in reference to the exterior facades of the Aedicule] *"indicates old marble slabs of white colour, which are here and there striped with grey; with these (slabs) the zealous piety of the Most Rev. Fr. Boniface, Superior of the Holy Land, caused also the interior parts of the Tomb to be covered in the year 1555 AD"* [78]. This description of white marbles with grey stripes could refer to marbles of Proconnesos. *"The said slabs* [referring to the exterior white marble facings with grey stripes] *had first been attached elsewhere, and then were placed here, as also the 10 small columns ... "* [78]. This is a direct and absolute reference to re-use of marble members. Describing the Chapel of the Angel, the western part of it *"is in the form of a half circle"* and *"the upper part of which is adorned with multicoloured tapestries. The lower is covered with marble slabs"*, while the eastern part of the Chapel of Angel is *"square in form"*. While describing the Tomb Chamber *" ... .between the marble slabs which cover the walls, there is a distance of 6 feet and 2 palms.."* [78]. Thus, both the Tomb Chamber and the Chapel of Angel where adorned with marble facings at the time of Horn's recount.
1809 AD:	Maximos Simaios in his description of the works conducted during the reconstruction by Kalfa Komnenos at the Holy Tomb area states that: *"Komnenos opened the west end and encountered an indescribably sweet Odour rising up [...] There was a blocking on the south side consisting of two 'marbles'. Above 'this' there were two more 'marbles', one above another, each one the same"* (quoted from Biddle 1999 [51]). He also states that: *"the Aedicule, in need of total repair, was demolished ... All the marbles present in the cave of the Tomb were collected and stored ... Thus, the holy burial cave was revealed ... "* and *"At its north and south part, the burial chamber is hewn in the natural rock while the eastern and western part, as well as the roof, are built"*

Historical sources and the surviving accounts of pilgrims visiting the Holy Aedicule, already from the fourth century, describe a glorious monument surrounding the Tomb of Christ adorned with marbles and other decorative features. Constantine declares his intention of bringing materials from *"all sources"*. From Eusebius's account it seems that the Tomb of Christ *"the sacred cave"* was the chief part of the whole work during the construction of the Constantinean complex and was lavishly adorned and *"brightened up"*. Historical sources reveal that Constantine and Helena brought large quantities of marble to Jerusalem [54]. Thus, according to the historical data, it is likely that the Holy

Aedicule structure was adorned with marble facades from the Constantinean era, and the Tomb of Christ was cladded with marble in order to protect it.

Optical Stimulated Luminescence (OSL) dating conducted in a previous study [69] supports this theory. Indeed, the bedding mortar connecting the fragmented lower marble plate (corresponding to sample OM10) with the original burial rock surface, was found to be of an age of 1670 ± 230 years, thus the calendar centered age of 345 AD, corresponding to the era of the Constantinean constructions. The fact that a number of pilgrims speak of the Tomb as *"cut out of the rock"* is perhaps connected to the three port-holes of the vertical marble slab, which allowed visibility to the original burial rock surface itself (e.g., Abbot Daniel, in the early 11th century states *"This sacred rock, which all Christians kiss, can be seen through three small round openings on one side ... "*).

The cladding of the Holy Tomb with marble from as early as the Constantinean era, was obviously installed as, without protection, the pilgrims would have all cut out a piece of the rock surface of the Holy Tomb material as a keepsake. The use of white marble from Proconnesos is an excellent selection for the Tomb of Christ, as its color and texture reflect light, symbolizing regeneration, with the proper amount of required simplicity. Furthermore, as already mentioned in the Introduction section, Proconnesos marble was extremely popular and widely used in imperial buildings during the Constantinean era. Thus, the lower plate, originating from Proconnesos, was installed at the time of Constantine; it was preserved throughout the centuries, even after being fragmented, and is still present, although not visible, within the Tomb of Christ under the upper plate worshiped today.

The marble fragment (sample OM49), found within the Tomb of Christ when it was opened by the NTUA interdisciplinary team, was also determined as Proconnesian marble in this study. This fragment is of the same thickness as the lower Constantinean marble plate (~3.3 cm). The location where it was found (inside the Tomb, indicating its importance) and its identical dimension of thickness with the lower marble plate (OM10), suggests that it is perhaps a fragment of the decorative edge from the missing part of the lower plate. If this is the case, the original marble member was extracted from the quarry C5b (Table 5).

The upper plate (sample OM11), still worshiped today, is also Proconnesian marble and, as already mentioned, its surface discoloration, giving the stone an amber-hue, is related to the daily application of myrrh on the surface, as a religious ritual, continuing for centuries up to today. The upper plate, due to its position, was definitely placed at a later date than the lower Constantinean plate, however it is very difficult to pinpoint the exact date it was placed; it was probably in place by 1345 AD, as it is at that time mentioned by an anonymous English pilgrim. The presence of more than one slabs is mentioned by Bonifacio da Ragusa, as stated in Table 6, however it cannot be certain which slabs he is referring to and whether he installed the upper plate worshiped today or if it preexisted. Sample OM11 could have been quarried from a number of quarrying locations of the Proconnesos island (Table 5), thus any comparison with other samples could not lead to safe conclusions.

In any case, a marvelous continuity of material both in relation to its memory and its use, is highlighted by the fact that the two marble plates covering the Tomb of Christ, although placed centuries apart, are both marbles that originate from the island of Proconnesos.

The marble sample taken from the marble facings of the interior of the Tomb Chamber, opposite the Tomb of Christ, and in particular from a slab of the south interior wall (sample OM13), is also proven to be Proconnesian marble in the current study. In a previous study, OSL dating was conducted on a mortar behind this particular slab and found to correspond to an age of 450 ± 68 years, thus to the central calendar age of 1570, corresponding to the Bonifacio da Ragusa restoration [69]. In parallel, a mortar selected from behind a marble slab above the one examined in this study (OM13), was dated to 335AD (calendar centered age) ±235years, that is the Constantinean Aedicule.

Thus, it is highly likely that the interior of the Tomb Chamber was adorned with Proconnesian marbles from the time of Constantine. Therefore, Bonifacio da Ragusa, either substituted certain marble slabs of the Tomb Chamber facings, using Proconnesos marble in order to be consistent with the material already adorning the Tomb Chamber, or reinstalled certain marble slabs already present, which had shifted out of position or unfastened. The later is more possible, especially taking into account Horne's reference related to the Bonifacio interventions *"The said slabs had first been attached elsewhere, and then were placed here, as also the 10 small columns ... "*. Horne, of course, is referring to the exterior facings, however, his statement is indicative of marble members re-use as a common practice throughout the ages, which could have been applied to the interior facings as well. This complicates the interpretation and interconnection of the historical and analytical evidence related to the architectural and structural evolution of the Holy Aedicule.

OM13 is the only sample which could have originated only from one of the examined quarries and in particular from C5b. Thus, if this marble member was first installed in the Constantinean era and Bonfacio da Ragusa reinstalled it, and taking into account that the lower tomb plate (OM10) was also installed in the Constantinean era, perhaps both members were quarried from C5b, which is the only common quarry of origin for both samples. Furthermore, quarry C5b is also a common origin quarry with sample OM49, which is the marble fragment, perhaps the decorative edge of the lower marble plate, as already mentioned. However, it is certain that the marble facing slab examined (OM13) does not match with any of the possible origin quarries either of the upper tomb plate (OM11) or of the facings of the Chapel of the Angel (OM50, OM51).

The two marble samples (OM50, OM51) from the west interior wall of the Chamber of the Angel are also Proconnesian marble. The Chamber corresponding to the Chapel of the Angel, as already mentioned, is a later addition, thus the marble facings were probably installed at least after 1099 AD in the era of the Crusaders. However, it should be noticed, that their presence from much earlier, as part of the exterior marble facings of the ciborium-type earlier structure, cannot be entirely excluded. In addition, they could also have been installed at later era, either during the Bonifacio renovation or during the major reconstruction of Kalfa Komnenos in 1810. OM 50, in accordance to Table 5, could have originated from the quarries Altintaş, Harmantaş, Mandira and C6, while OM51, could have originated from Altintaş and Mandira. They therefore present two common origin quarries, that is, Altintaş and Mandira. In any case, they do not present any common quarries with OM13 and with OM49, thus strengthening interpretation that they are a later addition.

It is extremely interesting and of great archaeological importance, that throughout the centuries, Proconnesian marble was the material of choice for the interior of the Holy Aedicule, both in the Tomb Chamber and the Chapel of the Angel. The continuity of the memory of the material and its grasping symbolism survived over 15 centuries, from its first use from Constantine's architects to the time of Komnenos, whom brought a large quantity of Proconnesos marbles to Jerusalem in order to use for architectural elements throughout the Church of Resurrection [57].

Thus, one more monument is added to the long and magnificent list of important monuments adorned with Proconnesian marble. In parallel, if the need arises for replacement of slabs in the interior of the Holy Aedicule, Proconnesian marble can be selected as the most compatible material and one that carries the light from the beginning of the Holy Aedicule.

5. Conclusions

Petrographic and isotopic analysis was implemented to study the white marbles of the Holy Aedicule and the Tomb of Christ. The examined marble samples display a characteristic heteroblastic fabric, which is characterized as "mortar-type". They comprise mostly of calcite with minor presence of dolomite, micas (phlogopite, muscovite), apatite and pyrite. Their MGS values range from 1.6 to 2.3 mm, gathered mostly around 2 mm. Their isotopic signature is characterized by $\delta^{18}O$ values ranging from −2.49 to −1.13 (‰ V-PDB) and by $\delta^{13}C$ values ranging from 2.43 to 3.37 (‰ V-PDB). The techniques employed proved to be adequate for the identification of their provenance and the aforementioned data suggest that the Holy Aedicule white marbles examined herein, originate from the island of Proconnesos and in particular they belong to the variety type Proconnesos-1.

Furthermore, an intra-site discrimination was attempted by comparing published MGS values and isotopic analysis data of Proconnesos quarries, with the respective data of the Holy Aedicule samples as measured herein. These results in correlation with historical data, allows for further interpretations. The fragmented lower marble plate (corresponding to sample OM10), is in fact the initial cladding of the original burial rock surface attributed to the Constantinean era. The marble fragment (sample OM49), found within the Tomb of Christ, which presented the same thickness as the lower Constantinean marble plate, is perhaps a fragment of the decorative edge from the missing part of the lower plate; if this is the case, the original marble member was extracted from the quarry C5b, which is the only common quarry between these two samples. The upper plate of the Tomb (sample OM11), was added several centuries later, however, no definite conclusion regarding the era that it was placed can be drawn; furthermore, intra-site discrimination could not provide any safe conclusions regarding the exact Proconnesos quarry it originated from.

The marble sample taken from the marble facings of the interior of the Tomb Chamber, opposite the Tomb of Christ (sample OM13) was most probably placed into its current position at the time of the Bonifacio da Ragusa restoration, without excluding the possibility that this member was present from the Constantinean era, since marble members re-use was a common practice throughout the centuries. The particular sample could have originated only from the ancient quarry C5b. The two marble samples (OM50, OM51) from the west interior wall of the Chamber of the Angel, were collected from marble members, which were placed probably during the Crusaders construction phase, without, however, excluding the Bonifacio renovation or even the major reconstruction of Kalfa Komnenos in 1810. Altintaş and Mandira are the only common origin quarries for these two samples and it is definite that they could not have originated from the same quarry as OM13 and OM49 (C5b).

Hopefully, the discussion made in the current research, interconnecting the results with aspects and events of the Holy Aedicule's construction history, will assist in revealing its evolution. It is extremely interesting and of great archaeological importance, that throughout the centuries, from the Constantinean era up to the Komnenos restoration, Proconnesian marble, and in particular the Proconnesos-1 variety, was the material of choice both for the cladding of the Holy Tomb, as well as for the interior facings of the Holy Aedicule.

Author Contributions: A.M. scientifically supervised and coordinated all aspects of the presented research; E.T.D. and M.A. conducted the in-situ investigation and sampling, conducted the correlation of the research results with the historical sources and data; A.K. assisted in the preparation of the original draft regarding the historical data; C.P. and G.E. conducted and evaluated the mineralogical and petrographical investigations; C.M. evaluated the isotopic analyses and correlated mineralogical, petrographical and isotopic data; E.T.D., M.A. and C.M. conducted the original draft preparation, as well as the writing-review and editing of the manuscript.

Funding: This research received no external funding. It is a pro bono post-program research, following the rehabilitation of the Holy Aedicule, aiming to shed light on the construction history and highlight the values of the Holy Aedicule.

Acknowledgments: The study and the rehabilitation project of the Holy Aedicule became possible and were executed under the governance of His Beatitude Patriarch of Jerusalem, Theophilos III. The Common Agreement of the Status Quo Christian Communities provided the statutory framework for the execution of the project; His Paternity the Custos of the Holy Land, Archbishop Pierbattista Pizzaballa (until May 2016—now the Apostolic Administrator of the Latin Patriarchate of Jerusalem), Fr. Francesco Patton (from June 2016), and His Beatitude the Armenian Patriarch of Jerusalem, Nourhan Manougian, authorized His Beatitude the Patriarch of Jerusalem, Theophilos III, and NTUA to perform the project. Contributions from all over the world secured the project's funding. Worth noting Mica Ertegun's and Jack Shear's donations through WMF, Aegean Airlines et al. The interdisciplinary NTUA team for the Protection of Monuments, Em. Korres, A. Georgopoulos, A. Moropoulou, C. Spyrakos, Ch. Mouzakis, were responsible for the rehabilitation project and A. Moropoulou, as Chief Scientific Supervisor, was responsible for its scientific supervision.

Conflicts of Interest: The authors declare no conflict of interest. The funders had no role in the design of the study; in the collection, analyses, or interpretation of data; in the writing of the manuscript, or in the decision to publish the results.

References

1. Mrozek-Wysocka, M. Ancient marbles: Provenance determination by archaeometric study. In *Geoscience in Archaeometry. Methods and Case Studies*; Michalska, D., Szczepaniak, M., Eds.; Wydawnictwo Naukowe Bogucki: Poznań, Poland, 2014; pp. 99–122.
2. Gorgoni, C.; Lazzarini, L.; Pallante, P.; Turi, B. An updated and detailed mineropetrographic and C-O stable isotopicreference database for the main Mediterranean marbles used in antiquity. In Proceedings of the 5th ASMOSIA Conference, Boston, MA, USA, 12–15 June 1998; pp. 1–25.
3. Lazzarini, L. Archaeometric aspects of white and coloured marbles used in antiquity: The state of the art. Periodico di Mineralogia. *Per. Miner.* **2004**, *73*, 113–125.
4. Lepsius, R. *Griechische Marmorstudien*; Philosophisch-historische Klasse; Abhandlungen der Königlichen Akadamie der Wissenschaften zu Berlin: Berlin, Germany, 1890; pp. 1–135.
5. Attanasio, D.; Brilli, M.; Ogle, N. *The isotopic Signature of Classical Marbles*; Studia Archaeologica; L'Erma di Bretschneider: Roma, Italy, 2006; p. 145.
6. Attanasio, D.; Bruno, M.; Yavuz, A.B. Quarries in the region of Aphrodias: The black and white marbles of Göktepe (Mugla, Turkey). *J. Roman Archaeol.* **2009**, *22*, 312–348. [CrossRef]
7. Herz, N.; Prichett, W.K. Marble in Attic Epigraphy. *Am. J. Archaeol.* **1953**, *57*, 71–83. [CrossRef]
8. Herz, N. Petrofabrics and classical archaeology. *Am. J. Sci.* **1955**, *253*, 299–305. [CrossRef]
9. Renfrew, C.; Springer, P.J. Aegean marble: A petrological study. *Annu. Br. Sch. Athens* **1968**, *63*, 45–66. [CrossRef]
10. Young, W.J.; Ashmole, B. The Boston relief and the Ludovisi Throne. *Boston Bull. Mus.* **1968**, *66*, 124–166.
11. Capedri, S.; Venturelli, G. Accessory minerals as tracers in the provenancing of archaeological marbles used in combination with isotopic and petrographic data. *Archaeometry* **2004**, *46*, 517–536. [CrossRef]
12. Antonelli, F.; Lazzarini, L. An updated petrographic and isotopic reference database for white marbles used in Antiquity. *Rend. Lincei* **2015**, *26*, 399–413. [CrossRef]
13. Craig, H.; Craig, V. Greek marble determination of provenance by isotopic analysis. *Science* **1972**, *176*, 401–403. [CrossRef]
14. Babcock, L.N. Accessory Minerals, $\delta^{13}C$ and $\delta^{18}O$, and Petrographic Structures of Pentelic and Proconnesian Quarry Marbles: Analysis with Nonmetric Multidimensional Scaling for Artifact Provenance. Ph.D. Thesis, The University of Georgia, Athens, GA, USA, 2012; 158p.
15. Asgari, N.; Matthews, K.J. The stable isotope analysis of marble from Proconnesos. In *The Study of Marble and Other Stones Used in the Antiquity*; Maniatis, Y., Herz, N., Basiakos, Y., Eds.; Archetype Publication: London, UK, 1995; pp. 123–129.
16. Attanasio, D.; Brili, M.; Bruno, M. The properties and identification of marble from Proconnesos (Marmara Island, Turkey): A new database including isotopic, EPR and petrographic data. *Archaeometry* **2008**, *50*, 747–774. [CrossRef]
17. Bruno, M.; Conti, L.; Lazzarini, L.; Pensabene, P.; Turi, B. The marble quarries of Thasos: An archeometric study. In *Interdisciplinary Studies in Ancient Stone*; ASMOSIA VI; Lazzarini, L., Ed.; Padova: Bottega dErasmo, Italy, 2002; pp. 157–162.
18. Gorgoni, C.; Lazzarini, L.; Pallante, P.; Turi, B. An updated and detailed mineropetrographic and C-O stable isotopic reference database for the main Mediterranean marbles used in antiquity. In *Interdisciplinary Studies on Ancient Stone*; Herrmann, J.J., Jr., Herz, N., Newman, R., Eds.; Archetype: London, UK, 2002; pp. 115–131.
19. Bruno, M.; Lazzarini, L.; Soligo, M.; Turi, B.; Varti-Matarangas, M. A recently discovered ancient quarry at Karavos (Paros, Greece) and the characterization of its marble. In *Parian Quarries, Marble and Workshops of Sculpture*; Schilardi, D.U., Ed.; The Paros and Cyclades Institute of Archaeology: Athens, Greece, 2000.
20. Lazzarini, L.; Ponti, G.; Preite Martinez, M.; Rockwell, P.; Turi, B. Historical, technical, petrographic, and isotopic features of Aphrodisian marbles. In *Interdisciplinary Studies in Ancient Stone*; ASMOSIA 5; Herrmann, J.J., Herz, N., Newton, R., Eds.; Bottega dErasmo: Padova, Italy, 2002.
21. Matthews, K.J.; Moens, L.; Walker, S.; Waelkens, M.; De Paepe, P. The re-evaluation of stable isotopedata for Pentelic marble. In *Ancient Stones: Quarrying, Trade and Provenance*; Waelkens, M., Herz, N., Moens, L., Eds.; University Press: Leuven, Belgium, 1992; pp. 203–212.

22. Pike, S. Preliminary results of asystematic characterization study of Mount Pentelikon, Attica, Greece. In *Archéomatériaux: Marbles et Autres Roches*; Schvoerer, M., Ed.; Presses Universitaires de Bordeaux: Pessac, France, 1999; pp. 165–170.
23. Goette, H.R.; Polikreti, K.; Vakoulis, T.; Maniatis, Y. Investigation of theblue-gray marble of Pentelikon andthe equivalent Hymettian: Possibleuses in antiquity. In *Archéomatériaux: Marbles Etautres Roches*; Schvoerer, M., Ed.; Presses Universitaires de Bordeaux: Pessac, France, 1999; pp. 83–90.
24. Maravelaki-Kalaitzaki, P. Black crusts and patinas on Pentelic marble from the Parthenon and Erechtheum (Acropolis, Athens): Characterization and origin. *Anal. Chim. Acta* **2005**, *532*, 187–198. [CrossRef]
25. Korres, E. *From Pentelicon to the Parthenon*; Melissa Publications: Athens, Greece, 2001.
26. Delegou, E.T.; Ntoutsi, I.; Kiranoudis, C.T.; Sayas, J.; Moropoulou, A. Advanced and Novel Methodology for Scientific Support on Decision-Making for Stone Cleaning. In *Advanced Materials for the Conservation of Stone*; Hosseini, M., Karapanagiotis, I., Eds.; Springer: Cham, Switzerland, 2018.
27. Palagia, O. Parian Marble and the Athenians. In *Paria Lithos: Parian Quarries, Marble and Workshops of Sculpture, Proceedings of the First International Conference on the Archaeology of Paros and the Cyclades, Paros 2-5 October 1997*, 1st ed.; Andromeda Books: Athens, Greece, 2000; pp. 347–354.
28. Bradley, M. The importance of colour on ancient marble sculpture. *Art hist.* **2009**, *32*, 427–457. [CrossRef]
29. Bashar, M.; Abbas, N. New marble sarcophagus from Syrian coast. *Sci. Cult.* **2015**, *1*, 17–26.
30. Tykot, R.H.; Herrmann, J.J.; van der Merwe, N.J.; Newman, R.; Allegretto, K.O. Thasian marble sculptures in European and American collections: Isotopic and other analyses. *Asmosia* **2002**, *5*, 188–195.
31. Larson, J. The conservation of marble monuments in churches. *Conservator* **1978**, *2*, 20–25. [CrossRef]
32. Bonazza, A.; Sabbioni, C.; Messina, P.; Guaraldi, C.; De Nuntiis, P. Climate change impact: Mapping thermal stress on Carrara marble in Europe. *Sci. Total Environ.* **2009**, *407*, 4506–4512. [CrossRef]
33. Herz, N.; Waelkens, M. (Eds.) *Classical marble: Geochemistry, Technology, Trade*; Kluwer Academic Publishers: Dordrecht, The Netherlands, 1988; Volume 153.
34. Monna, D.; Pensabene, P. *Marmi dell' Asia Minore*; Consiglio Nazionale delle Ricerche, Scienze Sussidiarie dell' Archeologia: Roma, Italy, 1977.
35. Marano, Y.A. The circulation of marble in the Adriatic Sea at the time of Justinian. In *Ravenna: Its Role in Early Medieval Change and Exchange*; Institute of Historical Research: London, UK, 2016; pp. 111–132.
36. Asgari, N. Roman and early Byzantine marble quarries of Proconnesos. In Proceedings of the X International Congress of Classical Archaeology, Ankara-tIzmir, Turkey, 23–30 April 1973; Akurgal, E., Ed.; Türk Tarih Kurumu Basimevi: Ankara, Turkey, 1978; Volumes 1–3.
37. Walker, S. The Marble Quarries of Proconnesos: Isotopic Evidence for the Age of the Quarries and for Lenos-Sarcophagi carved at Rome. In *Marmi Antichi: Problemi d'impiego, di Restauro e d'identificazione*; Pensabene, P., Ed.; L'Erma di Bretschneider: Rome, Italy, 1985; pp. 57–65.
38. Walker, S.; Matthews, K.J. The marbles of the Mausoleum. In *Sculptors and Sculpture of Caria and me Dodecsnese*; Jenkins, I., Waywell, G.B., Eds.; British Museum Press: London, UK, 1997; pp. 49–59.
39. Cramer, T. Die Marmore des Telephosfrieses am Pergamonaltar. *Berliner Beirrage zur Archiiomeuie* **1998**, *15*, 95–198.
40. Ward-Perkins, J.B. *Roman Imperial Architecture*; The Yale University Press: New Haven, CT, USA, 1981; p. 139.
41. Naddaf, M.; Al-Bashaireh, K.; Al Wacked, F. Characterization and provenance of marble chancel screens, Northern Jordan. *Mediter. Archaeol. Archaeom.* **2010**, *10*, 75–83.
42. Maniatis, Y.; Tambakopoulos, D.; Dotsika, E.; Wescoat, B.D.; Matsas, D. The sanctuary of the Great Gods on Samothrace, Greece: An extended marble provenance study. In *Interdisciplinary Studies on Ancient stone, Proceedins of the IX ASMOSIA Conference (Tarragona, 2009)*; Istituto Català d'Arqueologia Clàssica: Tarragona, Spain, 2012; pp. 263–278.
43. Alexakis, E.; Delegou, E.T.; Lampropoulos, K.C.; Apostolopoulou, M.; Ntoutsi, I.; Moropoulou, A. NDT as a monitoring tool of the works progress and the assessment of materials and rehabilitation interventions at the Holy Aedicule of the Holy Sepulchre. *Constr. Build. Mater.* **2018**, *189*, 512–526. [CrossRef]

44. Moropoulou, A.; Korres, E.; Georgopoulos, A.; Spyrakos, C.; Mouzakis, C.; Lambrou, E.; Pantazis, G.; Kavvadas, M.; Marinos, P.; Moropoulos, N.; Zafeiris, V.; Lampropoulos, K.; et al. Faithful Rehabilitation. History, culture, religion, and engineering all intertwined in a re-cent project in Jerusalem to rehabilitate and strengthen the site believed to be the tomb of Jesus of Nazareth. Modern methods, including 3-D laser scanning, Non Destructive Techniques and ground-penetrating radar, combined with careful documentation and design, revealed the secrets of how to help preserve and sustain the ancient site for future generations. *J. Am. Soc. Civ. Eng.* **2017**, *78*, 54–61.
45. Moropoulou, A.; Georgopoulos, A.; Korres, M.; Bakolas, A.; Labropoulos, K.; Agrafiotis, P.; Delegou, E.T.; Moundoulas, P.; Apostolopoulou, M.; Lambrou, E.; et al. Five-Dimensional (5D) Modelling of the Holy Aedicule of the Church of the Holy Sepulchre Through an Innovative and Interdisciplinary Approach. In *Mixed Reality and Gamification for Cultural Heritage*; Ioannides, M., Magnenat-Thalmann, N., Papagiannakis, G., Eds.; Springer: Cham, Switzerland, 2017.
46. Lavvas, G. *The Holy Church of the Resurrection in Jerusalem*; The Academy of Athens: Athens, Greece, 2009.
47. Balodimos, D.; Lavvas, G.; Georgopoulos, A. Wholly Documenting Holy Monuments. CIPA XIX. *Int. Arch. Photogramm. Remote Sens.* **2003**, *34*, 5.
48. Lampropoulos, K.C.; Korres, M.; Moropoulou, A. A transdisciplinary approach to reveal the structural evolution of the Holy Aedicule in the Church of the Holy Sepulchre. In *Nondestructive Evaluation and Monitoring Technologies, Documentation, Diagnosis and Preservation of Cultural Heritage*; Osman, A., Moropoulou, A., Eds.; Springer Nature: Basingstoke, UK, in press.
49. Cameron, A.; Hall, S. (Eds.) *Eusebius' Life of Constantine*; Clarendon Press: Oxford, UK, 1999.
50. Pamphilus, S. *Eusebius Book III*; Chapter XXXI The Sacred Writings of Eusebius Pamphilus (Annotated Edition); Jazzybee Verlag: Altenmunster, Germany, 2012.
51. Biddle, M. *The Tomb of Christ*; Sutton: Gloucestershire, UK, 1999.
52. Marek, M. How to reverse the decline of an empire? Two Byzantine case studies: Herakleios and Alexios Komnenos. *Graeco-Lat. Brun.* **2016**, *21*, 119–134.
53. Howard-Johnston, J. Heraclius' Persian Campaigns and the Revival of the East Roman Empire, 622–630. *War Hist.* **1999**, *6*, 1–44. [CrossRef]
54. Michael, G.; Past, M. *Monumental Present: Building with Antiquities in the Mediaeval*; Brill: Boston, MA, USA, 2009.
55. Varvounis, M.G. *The Holy Tomb and the Church of Resurrection*; Chelandion Publications: Athens, Greece, 2009; pp. 37–46. (In Greek)
56. Lavvas, G. The All-Holy Church of Resurrection in Jerusalem. *Corpus* **2004**, *60*, 30–47.
57. Mitropoulos, T. *The Church of Holy Sepulchre—The Work of Kalfas Komnenos*; European Centre of Byzantine and Post-Byzantine Monuments: Thessaloniki, Greece, 2009.
58. Patrich, J. The Edicule in the Church of the Holy Sepulchre—Biddle Martin, the Tomb of Christ (Sutton Publishing Ltd., Stroud, Glos. 1999). *J. Roman Archaeol.* **2002**, *15*, 688–690. Available online: https://www.cambridge.org/core/journals/journal-of-roman-archaeology/article/edicule-in-the-church-of-the-holy-sepulchre-martin-biddle-the-tomb-of-christ-sutton-publishing-ltd-stroud-glos-1999-pp-xii-172-103-figs-including-colour-isbn-0750919264-25/7E5FC29153F52B752E6E63B2B313C8BC (accessed on 25 April 2019). [CrossRef]
59. Pringle, D. *The Churches of the Crusader Kingdom of Jerusalem: A Corpus*; Cambridge University Press: New York, NY, USA, 2007; Volume 3.
60. Freeman, Fox, and Partners, Church of the Holy Sepulchre, Jerusalem. Report. 1947.
61. Moropoulou, A.; Korres, E.; Georgopoulos, A.; Spyrakos, C. *Materials and Conservation, Reinforcement and Rehabilitation Interventions in the Holy Edicule of the Holy Sepulchre*; NTUA: Athens, Greece, 2016; ISBN 978-618-82612-0-4.
62. Moropoulou, A.; Korres, E.; Georgopoulos, A.; Spyrakos, C.; Mouzakis, C.; Lampropoulos, K.C.; Apostolopoulou, M.; Delegou, E.T.; Alexakis, E. The rehabilitation of the Holy Aedicule. In *Proceedings of the XXXIII Convegno Internazionale Scienza e Beni Culturali*; Trasferimenti, Contaminazioni, Ibridazioni; Le Nuove Frontiere del Restauro: Bressanone, Italy, 2017; pp. 27–30.
63. Moropoulou, A.; Korres, E.; Georgopoulos, A.; Spyrakos, C.; Mouzakis, C. *Presentation upon completion of the Holy Sepulchre's Holy Aedicule Rehabilitation*; NTUA: Athens, Greece, 2017; p. 12. ISBN 978-618-82196-4-9.

64. Moropoulou, A.; Korres, M.; Georgopoulos, A.; Spyrakos, C.; Mouzakis, C.; Lampropoulos, K.C. and Apostolopoulou, M. The Project of the Rehabilitation of Holy Sepulchre's Holy Aedicule as a Pilot Multispectral, Multidimensional, Novel Approach Through Transdisciplinarity and Cooperation in the Protection of Monuments. In *International Conference on Transdisciplinary Multispectral Modeling and Cooperation for the Preservation of Cultural Heritage*; Springer: Berlin/Heidelberg, Germany, 2018; pp. 3–25.
65. Apostolopoulou, M.; Delegou, E.T.; Alexakis, E.; Kalofonou, M.; Lampropoulos, K.C.; Aggelakopoulou, E.; Bakolas, A.; Moropoulou, A. Study of the historical mortars of the Holy Aedicule as a basis for the design, application and assessment of repair mortars: A multispectral approach applied on the Holy Aedicule. *Constr. Build. Mater.* **2018**, *181*, 618–637. [CrossRef]
66. Spyrakos, C.C.; Maniatakis, C.A.; Moropoulou, A. Preliminary Assessment of the Structural Response of the Holy Tomb of Christ Under Static and Seismic Loading. In *International Conference on Transdisciplinary Multispectral Modeling and Cooperation for the Preservation of Cultural Heritage*; Springer: Berlin/Heidelberg, Germany, 2018; pp. 44–57.
67. Alexakis, E.; Kapassa, E.; Touloupou, M.; Kyriazis, D.; Georgopoulos, A.; Moropoulou, A. Innovative Methodology for Personalized 3D Representation and Big Data Management in Cultural Heritage. In *International Conference on Transdisciplinary Multispectral Modeling and Cooperation for the Preservation of Cultural Heritage*; Springer: Berlin/Heidelberg, Germany, 2018.
68. Moropoulou, A.; Farmakidi, C.M.; Lampropoulos, K.; Apostolopoulou, M. Interdisciplinary planning and scientific support to rehabilitate and preserve the values of the Holy Aedicule of the Holy Sepulchre in interrelation with social accessibility. *Sociol. Anthropo.* **2018**, *6*, 534–546. [CrossRef]
69. Moropoulou, A.; Zacharias, N.; Delegou, E.T.; Apostolopoulou, M.; Palamara, E.; Kolaiti, A. OSL mortar dating to elucidate the construction history of the tomb chamber of the Holy Aedicule of the Holy Sepulchre in Jerusalem. *J. Archaeol. Sci.: Rep.* **2018**, *19*, 80–91. [CrossRef]
70. Lampropoulos, K.C.; Moropoulou, A.; Korres, M. Ground penetrating radar prospection of the construction phases of the holy Aedicula of the holy sepulchre in correlation with architectural analysis. *Constr. Build. Mater.* **2017**, *155*, 307–322. [CrossRef]
71. Asgari, N. Die halbfabrikate kleinasiatischen girlanden-sarkophage und ihre herkunft. *Archäologischer Anzeiger* **1977**, 329–380.
72. Ward-Perkins, J.B. *The trade in Sarcophagi, in Marble in antiquity, collected papers of J.B. Ward-Perkins. Archaeological Monographs of the British School at Rome*; Dodge, H., Ward-Perkins, B., Eds.; British School at Rome: London, UK, 1992; pp. 31–39.
73. Pensabene, P. Il fenomeno del marmo nel mondo Romano. In *I Marmi Colorati Della Roma Imperiale*; De Nuccio, M., Ungaro, L., Eds.; Marsilio: Roma, Italy, 2002; pp. 3–69.
74. McClure, M.M.H.; Feltoe, C.L. *The pilgrimage of Etheria*; Society for Promoting Christian Knowledge: London, UK; The Macmillan Company: New York, NY, USA, 1919.
75. Bogdanović, J. The rhetoric of architecture in the Byzantine context: The case study of the Holy Sepulchre/Реторика архитектуре у византијском контексту: пример црквеСветог гроба. *Zograf/ЗОГРАФ* **2014**, *38*, 1. Available online: https://lib.dr.iastate.edu/cgi/viewcontent.cgi?referer=https://www.google.com.hk/&httpsredir=1&article=1041&context=arch_pubs (accessed on 20 April 2019).
76. Wright, T. *Early Travels in Palestine*; H.G. Bohn: London, UK, 1848.
77. Piccirillo, M. The Role of the Franciscans in the Translation of the Sacred Spaces from the Holy Land to Europe. In *New Jerusalems: Hierotopy and Iconography of Sacred Spaces*; Indrik: Moscow, Russia, 2009; pp. 363–394. Available online: http://hierotopy.ru/contents/NewJerusalems_16_Picirillo_RoleOfFranciscans_2009_EngRus.pdf (accessed on 20 April 2019).
78. Horn, E.; Golubovich, G. *Ichnographiae Locorum et Monumentorum Veterum Terrae Sanctae: Accurate Delineatae et Descriptae. E codice Vaticano Lat. No. 9233 excerpsit, adnotavit et edidit (cum 75 figuris et appendice historica ex eodem codice) Hieronymus Golubovich*; Typis Sallustianis, 1902. Available online: https://opac.bibliothecaterraesanctae.org/cgi-bin/koha/opac-search.pl?q=au:Horn,%20Elzearius (accessed on 20 April 2019).

© 2019 by the authors. Licensee MDPI, Basel, Switzerland. This article is an open access article distributed under the terms and conditions of the Creative Commons Attribution (CC BY) license (http://creativecommons.org/licenses/by/4.0/).

Review

A Review on the Archaeological Chemistry of Shellfish Purple

Ioannis Karapanagiotis

Department of Management and Conservation of Ecclesiastical Cultural Heritage Objects, University Ecclesiastical Academy of Thessaloniki, 54250 Thessaloniki, Greece; y.karapanagiotis@aeath.gr; Tel.: +30-2310-301784

Received: 24 May 2019; Accepted: 17 June 2019; Published: 29 June 2019

Abstract: Shellfish purple, also known as Tyrian purple and royal purple, has a long history, which has been revealed and documented in recent years through valid physicochemical studies using sophisticated techniques. The aim of the work was to summarize the conclusions of these studies and to describe the results of two unpublished investigations regarding the (i) identification of shellfish purple in a textile (4th century BCE) from ancient Macedonia and (ii) dramatic effect of the dyeing conditions on the composition of the purple dye. Moreover, a critical discussion is included about the discovery of the shellfish pigment and dye based on the available scientific evidence. Previously published reports describing the identification of the shellfish colorant in objects of the cultural heritage were carefully summarized. Shellfish purple was not used only as colorant, but it served other purposes as emphasized in this review. In particular, examples for the use of shellfish purple in medicine, grave goods and fillers and plasters in walls, were described. Examples of materials and methods that were used in the past to produce "fake" purple, imitating the aesthetic result of the valuable royal marine material were summarized. Finally, the solubility of indigoids was discussed using modern approaches of physical chemistry.

Keywords: purple; shellfish; mollusk; Tyrian purple; indigo; pigment; dye; cultural heritage; HPLC

1. Introduction

Identification of the materials of cultural heritage objects is important to develop and apply effective conservation strategies, which ensure the preservation of the valuable historical and archaeological objects. Conservation and preservation are probably the most important conditions towards a sustainable market for cultural heritage, along with appropriate policies aiming at promoting the human civilization through the multidisciplinary area of archaeometry and other disciplines [1]. Moreover, the characterization and identification of cultural heritage materials leads to useful insights regarding the technological skills of civilizations and reveals economic, social and historic aspects [2]. For these reasons the application of physicochemical methods and instruments to archaeological research has increased substantially over the last half-century and today represents a major methodological subfield within archaeological science [3].

Ever since the beginning of human existence, humans made use of materials to color virtually everything they used or had: Bodies, caves, pottery, sculpture, stone structures, parchment, paper, clothing and other textiles [4]. The important role of color in human civilization becomes dominant for painting art and textiles. The former is essential for human nature, as evidenced for example by the Lauscax cave paintings, dated from about 17,000 BCE and the hundreds of other paleolithic cave paintings e.g., in Altamira, Chauvet and Cáceres. The astonishing findings of abalone shells that were used to hold ochre mixtures 100,000 years ago suggest that pigments were utilised from the very early days [5]. The vital, timeless and universal power of painting and, more general, pictorial art is evidenced in many different ways. For example, icons have been used for centuries by the Orthodox Church as a window to communicate with the unseen divine reality whereas the Pioneer plaques are pictorial messages for potential extraterrestrial life. Organic materials such as textiles and their colorants (dyes) are sensitive to degradation effects developed through ageing. Consequently, early known-identified use of dyes dates back to at around 6000 years ago [6], which falls short compared to the iron and manganese pigments found in the aforementioned prehistoric caves.

Among the various colorants, which have been used since the prehistoric times, shellfish purple originating from marine mollusks holds a prominent position [7–9]. Shellfish purple (in Greek, *porphyra* = πορφύρα) is known also as Tyrian purple, because of the significant deposits of shells found in the area of Tyre and the important role of the purple dye in the Phoenician history. The material is also known as royal purple and imperial purple, as the expensive purple-dyed textiles were status symbols. Shellfish purple is connected with the incredible story of tekhelet, the elusive sky-blue color mentioned throughout the Bible and the clothing of emperors and high priests in civilizations, which had scattered all around the Mediterranean basin.

The goal of this review is to summarize the current state of the art provided by the Natural Sciences for the shellfish purple in archaeology. Examples of historical written sources are briefly presented; biological mollusk sources and the chemical composition of shellfish purple are described. Critical discussions are included associated with the discovery of the shellfish pigment and dye based on the available scientific evidence. Previously published reports describing the identification of shellfish purple in objects of cultural heritage are carefully summarized, revealing the pros and cons of the physicochemical instruments employed for analytical purposes. Emphasis is placed on the results of shellfish purple in the prehistoric Aegean. Two unpublished studies are included (i) describing the identification of shellfish purple in a textile from ancient Macedonia and (ii) demonstrating the dramatic effect of the dyeing conditions on the composition of the purple dye. Shellfish purple was not used only as a paint and dye, but it served other purposes as described in the present review. Materials and methods that were used in the past to produce "fake" purple, imitating the aesthetic result of the valuable imperial marine material are discussed. Finally, the solubility of indigoids is discussed using modern approaches of physical chemistry.

2. Historical Written Sources and Legends for Shellfish Purple

The first inscription referring to the royal color is revealed in Amarna letters, dated to the 14th century BCE, in which blue-purple wool objects are described [10]. Another important written source comes from the Mycenaean Linear B clay tablets of Knossos dated to the 13th century BCE [11]. The masculine (po-pu-re-jo) and the feminine (po-pu-re-ja) adjectives are cited in tablets KN X976 and KN L474, respectively [11]. According to the Acts of the Apostles (Chapter 16), Lydia of Thyatira, who was baptized by St. Paul and is the first documented convert to Christianity in Europe, was in the purple-dye trade. Purple produced from mollusks is mentioned several times in the Bible as well as by Homer, Aristotle, Herodotus, Aeschylus and several others [12]. An impressive written example for the use of purple in Byzantine times is the inscription of the mosaic of the Byzantine Emperor, John II Komnenos (*Ἰωάννηςβ' Κομνηνός*) in Hagia Sophia (= Holy Wisdom), which was the cathedral and the spiritual center of the Byzantine empire, as it was the Parthenon for Athens, which was dedicated to the Goddess of Wisdom, Athena. The mosaic in Hagia Sophia is dated to the 12th century CE and the inscription reads: "John faithful in Christ God, emperor of the Romans the Komnenos and king born in the purple".

The history of shellfish purple is full of legends with the most famous being the one describing the discovery of Murex shellfish by Hercules' hound. The scene is represented in the exergue of coins with the ambrosial rocks [13,14], as well as in paintings, such as the oil on panel artwork by the 17th century Flemish master Peter Paul Rubens [8]. Murex shells have been represented in various objects since antiquity, as shown for instance in the Geek krater and coin of Figure 1 [15,16].

(a) (b)

Figure 1. (**a**) Pottery krater painted with red bands and row of parallel murex shells (13th century BCE). (**b**) Silver coin with the Head of Demeter (observe) and barley–ear with murex (reverse; 4th century BCE). Both objects were found in Greece [15,16]. © The Trustees of the British Museum. Shared under a Creative Commons Attribution–NonCommercial–ShareAlike 4.0 International (CC BY–NC–SA 4.0) license.

3. Biological Sources and Chemical Composition

Shellfish purple originates from marine mollusks of the Muricidae family, such as *Hexaplex trunculus* L. (*Murex trunculus*), *Bolinus brandaris* L. (*Murex brandaris*) and *Stramonita haemastoma* (*Thais haemastoma*). These three species live in the Mediterranean basin where the use of the purple pigment/paint in prestigious archaeological findings has been revealed. In the Aegean Sea, *M. trunculus* is the most abundant followed by *M. brandaris* and finally *T. haemastoma* species. However, it should be noted that the mollusk populations have been affected by the recent effects of sea pollution and global warming [17]. Surprisingly, *T. haemastoma* mollusks once described in the taxonomic literature as common in the Levant basin were not found in a recent survey [17]. In the same investigation, a decrease in the population of *M. trunculus* was reported. These observations are schematically presented in Figure 2.

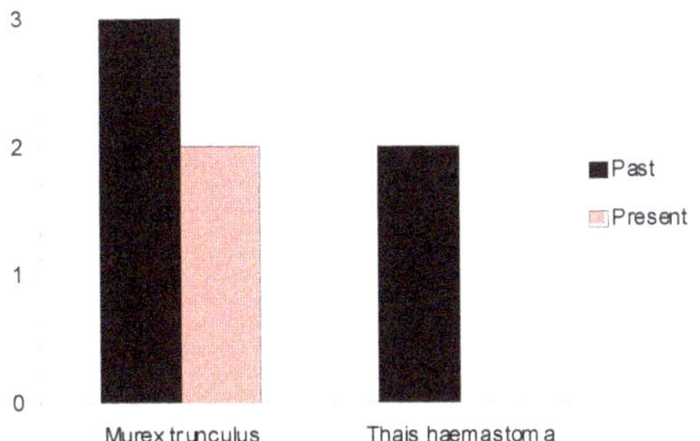

Figure 2. Past and present abundance categories for *Murex trunculus* and *Thais haemastoma* populations in the Levant basin [17]: 0 = not found; 1 = rare; 2 = common; 3 = very common.

Photographs of *M. trunculus* mollusks, which were collected in the Thermaic (Thessaloniki) gulf in the Aegean Sea, are shown in Figure 3a. The coloring compounds do not exist in the living mollusks, but they are produced by precursors that are called chromogens. The latter are present in hypobranchial glands of the mollusks and undergo spontaneous chemical reactions under exposure to light and air thus yielding to the coloring components of shellfish purple [9,18]. For example, the reaction series for the conversion of tyrindoxyl sulfate (precursor) to 6,6′–dibromoindigotin, DBI (coloring compound) is schematically illustrated in Figure 3b [9,18]. The conversion is fast and it takes about 5–10 minutes after exposure to air and sunlight. By analogous reaction series, the other coloring compounds of shellfish purple are produced: Indigotin (IND), 6–bromoindigotin (MBI), indirubin (INR), 6′–bromoindirubin (6′MBIR) 6–bromoindirubin (6MBIR) and 6,6′–dibromoindirubin (DBIR).

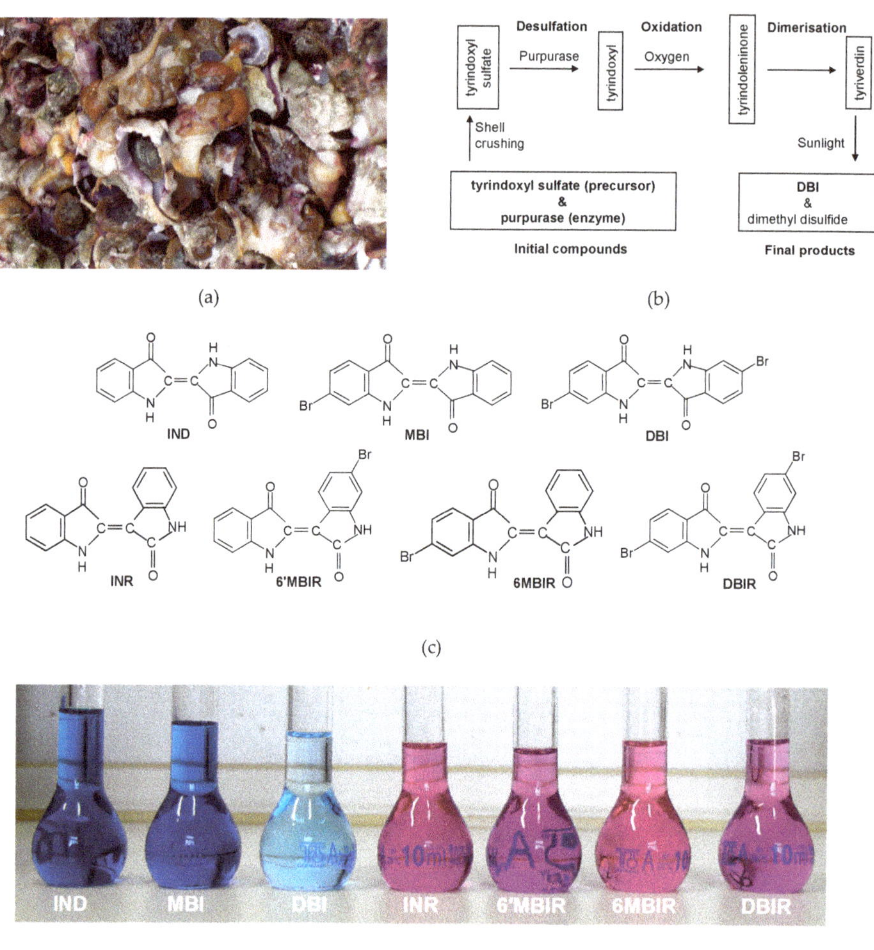

Figure 3. (**a**) *M. trunculus* mollusks. (**b**) Conversion of tyrindoxyl sulphate to DBI. (**c**) Structures of the coloring compounds of shellfish purple: Indigotin (IND), 6–bromoindigotin (MBI), 6,6′–dibromoindigotin (DBI), indirubin (INR), 6′–bromoindirubin (6′MBIR) 6–bromoindirubin (6MBIR) and 6,6′–dibromoindirubin (DBIR). (**d**) Solutions of the compounds in dimethyl sulfoxide (DMSO), 20 μg mL^{-1}.

The compounds of Figure 3c were detected in extracts of *M. trunculus* [19–21] and *M. brandaris* [22]. However, only MBI, DBI and DBIR were detected so far in extracts of *T. haemastoma*. Four more analogues of brominated and unbrominated indirubins, which were detected in *M. trunculus* extracts, were not included in Figure 3c [23]. These isomers have been scarcely reported, as they were found in extremely small amounts [23]. In early reports, DBI was considered to be the marker for the identification of shellfish purple, as this was the first compound isolated from the raw purple pulp [24]. However, none of the brominated molecules of Figure 3c is found in any colorant other than the shellfish purple and therefore any of these molecules can be considered as markers for the presence of shellfish purple. Moreover, DBI is not necessarily the most abundant component of the purple extract. Chromatographic results have shown that in some *M. trunculus* extracts, MBI is found in larger amounts than DBI [19–21].

Today, the compounds of Figure 3c can be produced in pure forms and in relatively large amounts following synthetic routes [20,25–27]. Photographs of standard solutions of synthesized shellfish components are provided in Figure 3d. Solutions of indigoids (IND, MBI, DBI) are bluish whereas solutions of indirubinoids (INR, 6′MBIR, 6MBIR, DBIR) are reddish. The latter are usually found in shellfish extracts in small amounts. Focusing now on the indigoids it is observed that the DBI solution has a bluish-purple color whereas the MBI and particularly the IND solutions are blue. Consequently, the color of a painted/dyed object can vary, depending on the relative composition of the fixed purple pigment/dye. For example, it was shown that wool dyed with (i) IND is blue, (ii) DBI is purple and (iii) MBI is in between blue-purple [28,29]. Microscopic and chromatographic studies in a 1500-year-old textile dyed with shellfish purple, showed that blue-purple and red-purple fibers were rich in IND and DBI, respectively [30]. In other studies on archaeological textiles, it was suggested that shellfish purple in antiquity was sometimes combined with red anthraquinones, when high proportions of IND were attached on dyed textiles [31–34]. High proportions of IND may have resulted because of the composition of the source mollusk or the dyeing conditions, as both affect the resulting color [28,29]. High proportions of IND may have resulted in purple dyes that looked too blue and therefore mixing with a red colorant was probably necessary to adjust the color and to make it look more purple [31].

4. The Discovery of the Purple Colorant

The origins of the purple dyeing or pigment production industry could have been in the Persian Gulf [9]. P. G. Moatsos wrote in 1932 that probably the Sumerians discovered the use of shellfish purple as a dye for textiles [35]. However, to date, there is no archaeological or scientific evidence to support this hypothesis. The earliest scientific evidence suggesting purple dyeing in the Gulf area goes back to only 3rd–5th century CE [36]. Fragments of pottery vessels used as dyed baths were analyzed revealing the presence of shellfish purple [36].

The first confirmed use of shellfish purple in wall paintings comes from the analyses of samples found in the islands of Santorini and Rhodes in the Aegean Sea [37–43]. The samples are dated to the 18th–17th century BCE. Among the findings in which the presence of shellfish purple was revealed by chemical analysis was a lump of pigment (Santorini). Chromatographic studies showed that the relative chemical composition of the pigment was similar with the compositions of purple paints removed from local wall paintings, proving indisputably the use of the pigment in the local painting art and suggesting pigment production in the Aegean in the Late Bronze Age [42,43]. The pigment was rich in chalk-aragonite and was presumably prepared by immersion of the chalk in the vat serving for the dyeing of textiles [43]. The procedure for the preparation of the pigment purpurissum, documented by Pliny the Elder [44], is reported to occur in the same dye bath, which also served for the dyeing of purple textiles [45]. It has been therefore suggested that the production of organic pigments in ancient times was probably carried out within the existing dyeing workshops using white inorganic substrates, which gave body to the purple pigment [43,45]. This argument is supported by the chemical results reported for samples that were extracted from a painted burial klinai and a textile fragment, both found within the same tomb, in Daskyleion (5th century BCE) [46]. The relative compositions of the shellfish purple pigment and dye used in the two objects were similar according to chromatographic results [46], thus offering support to the scenario that pigment production and dyeing were probably carried out within the same batch reactor-vessel. If this is true, then it serves as indirect evidence for the presence of a shellfish dyeing textile industry in the prehistoric Aegean where the purple pigment production has been confirmed.

The oldest textile fragments dyed with shellfish purple were found in archaeological sites in Syria and particularly in Chagar Bazar (18th–16th century BCE) [47] and in a tomb complex in Tell Mishrife (Qatna), which had remained sealed since the 14th century BCE, at the time of the Hittite conquest of Syria [48]. The earliest scientific result supporting the development of a shellfish dyeing industry was achieved by analyzing samples of pottery vessels from a Minoan facility (ca. 1800/1700 BCE) in Alatsomouri–Pefka, in the island of Crete in the Aegean Sea [49]. These Cretan findings are earlier than the discovered Phoenician pottery sherds, suggesting that the *Murex* purple of the Minoans became the Tyrian purple of the Phoenicians and later the imperial purple of Byzantium [50]. Consequently, the findings in Chagar Bazar and Alatsomouri–Pefka are the earliest textile fragments and dyeing facilities associated with shellfish purple.

The discussion now focuses on the shell deposits found in the Mediterranean. In the eastern shores of the basin, significant deposits of shells have been discovered in various sites in the Aegean islands, which are dated to the 18th century BCE [51,52]. The earliest archaeological evidence comes from Crete and corresponds to the Old Palace Period and particularly in 19th–18th century BCE [53–55]. There is no evidence suggesting that the famous heaps of mollusk fragments found in Tyre are associated with a dyeing industry, which is dated earlier than the 17th century BCE [56]. Consequently, the available scientific and archaeological evidence suggest that the Minoans in the Aegean and probably the Hurrians in Syria have pioneered the extraction of shellfish purple centuries before the Tyrians.

On the other hand, there is no doubt that the purple dye industry of the Phoenicians became famous and dominated the market from the Homeric times (9th–7th century BCE) onwards [54]. The Phoenicians spread their trade and founded cities, ports and bases up to the Black Sea and across the Mediterranean basin and beyond the Pillars of Hercules. For example, the Phoenicians founded Gadir (Cádiz, Spain) and probably had visited the Canary Islands. In fact, the word "Phoenician" comes from the Greek word "Phoenix" (φοῖνιξ) meaning "red-purple". Consequently, the Greeks had associated the purple pigment/dye with the Phoenicians. However, based on the available archaeological and scientific data, the shellfish purple industry was developed by the Minoans and probably the Hurrians before the Phoenicians.

Finally, it should be stressed that thousands of *M. trunculus* shells were found in Coppa Nevigata, Italy [56–59]. The accumulations of shells are present in the Early Proto-Apennine layers [58,59]. In particular, the collection of Murex began in the 19th–18th century BCE and continued throughout the entire Bronze Age, reaching its highest peak in the 15th and 14th centuries BCE [59]. The large quantity and the condition of the shells in Coppa Nevigata suggest that they were likely to have been used for the extraction of purple [57–59]. This may be the earliest known for the industry of purple, predating that from the Aegean islands [56,58]. It is interesting to note that excavation data indicate the existence of trading activities between Coppa Nevigata and the islands of the Aegean Sea [57].

5. Identification of Shellfish Purple in Archaeological Objects

5.1. Paints, Pigments and Identification Methods

The goal of this section is twofold: (i) To summarize the scientific identifications of shellfish purple pigments applied as paints on wall paintings and some other objects and (ii) to briefly discuss the analytical techniques used for identification purposes. Both, previously reported identifications and analytical techniques are summarized in Table 1 [37–43,46,60–74]. The latter shows that shellfish purple was used in different objects for painting purposes, uninterrupted by all major pre-Roman civilizations of the Aegean, starting with the Minoan period of the Late Bronze Age up to the Hellenistic period. The use of shellfish purple was disrupted abruptly with the conquest of Constantinople by the Ottomans (1453 CE). At that time the capital of the Byzantine Empire was the major processing center of the valuable material. Around the same time in the West, a decree (1464 CE) of Pope Paul II, stipulated the replacement of shellfish purple by kermes to dye ecclesiastical robes. The two events combined, brought the historical use of shellfish purple to an end. According to Table 1, the most recent historical object painted with true purple is a wall painting in France dated to the 12th century CE [74]. It was suggested that probably the purple pigment was transported there from the eastern coast of the Mediterranean Sea during the time of the Crusades [74].

Table 1. Identifications of shellfish purple in pigments/paints corresponding to various historical periods and geographical locations. The techniques used for identification purposes are included.

Date	Provenance	Location	Object/Samples	Technique [1]	Reference
18th–17th c. BCE	Minoan	• Akrotiri, Santorini, Greece	Pigment and wall paintings	XRF, HPLC–DAD, LC– APCI–MS, Raman, FTIR	[37–43]
		• Raos, Santorini, Greece	Wall painting	HPLC–DAD, Raman, FTIR	[41,42]
		• Trianda, Rhodes, Greece	Pigment	HPLC–DAD, Raman, FTIR	[41,42]
13th c. BCE	Mycenaean	Nestor Palace, Pelopennese, Greece	Wall paintings	XRF, XRF–imaging	[60]
6th–2nd c. BCE	Archaic–Hellenistic	Cave of Koroneia, Greece	Painted astragalos	SEM–EDX	[61]
5th BCE	Persian	• Daskyleion, Turkey	Painted klinai	FTIR, FTIR– imaging, SEM–EDX, HPLC–DAD	[46]
486/485 BCE		• Bible Lands Museum, Israel	Painted jar	HPLC–DAD	[62]
4th c. BCE	Classical	• J. Paul Getty Museum, USA	Greek marble vessel	XRF	[63]
336 BCE		• Tomb of Philip II, Vergina, Macedonia, Greece	Mask (?)	FTIR	[64]
4th c. –3rd c. BCE	Daunian	Sansone Collection, Italy	Ceramics (vases, kraters)	IR, MCT	[65]
4th–2nd c. BCE	Hellenistic	• Macedonian Tomb III at Agios Athanasios, Greece	Wall painting	XRF, HPLC–DAD	[66–68]
		• Macedonian Tomb of the Palmettes, Greece	Wall painting	HPLC–DAD	[67–69]
		• Crete, Greece	Figurines	FTIR, XRF	[70]
		• Macedonian Tomb of Philosophers, Pella, Greece	Wall painting	SEM–EDX	[71]
		• Arch. Museum of Delos, Greece	Pigment	XRF	[72]
2nd half of the 4th c. BCE	Etruscan	Sarcophagus of the Amazons, Arch. Museum of Florence, Italy	Sarcophagus (from Tarquinia)	n.a.	[68]
1st CE	Roman	Pompeii, Italy	Pigment	DE–MS	[69,73]
12th c. CE	French	Church of Sainte Madeleine, France	Wall painting	HPLC–DAD, LC–APPI –MS	[74]

[1] DE–MS: direct exposure–mass spectrometry; FTIR: Fourier transform infrared spectroscopy; HPLC–DAD: high performance liquid chromatography–diode array detector; IR: infrared spectroscopy; LC–APCI–MS: liquid chromatography with atmospheric pressure chemical ionization–mass spectrometry; LC–APPI–MS: liquid chromatography with atmospheric pressure photoionization–mass spectrometry; MCT: Microchemical testing; Raman: Raman spectroscopy; SEM–EDX: scanning electron microscopy (SEM) coupled to energy dispersive x–ray (EDX) spectroscopy; XRF: X–ray fluorescence; n.a.: not available.

As described in Table 1 a large variety of techniques were used to identify paints and pigments of shellfish purple. The presence of bromine (Br) in the composition of the purple material (Figure 3c) offers the option to apply elemental analysis techniques, such as X–ray fluorescence (XRF) and scanning electron microscopy (SEM) coupled to energy dispersive x–ray (EDX) spectroscopy (SEM–EDX). The results should be carefully interpreted considering that lichens can contain significant amounts of Br. An example of a SEM–EDX study is provided in Figure 4, which shows the results for a purple painted sample extracted from a burial klinai [46]. The detection of Br is indicative for the use of true purple. Furthermore, the presence of an aluminosilicate compound and gypsum is supported by the results of Figure 4, as Al, Si, Ca and S are included in the detected elements of the EDX spectrum. FTIR measurements confirmed the presence of shellfish purple, kaolinite and gypsum in the archaeological sample [46]. Interestingly, the use of kaolinite in painting techniques of shellfish purple were previously reported in the analyses of the Darius I stone painted jar [62] and the pigment powder of Pompeii [73]. According to Table 1, vibrational spectroscopy, and particularly FTIR and Raman spectroscopy, is another set of techniques which was effective to identify purple pigments and paints in objects of the cultural heritage.

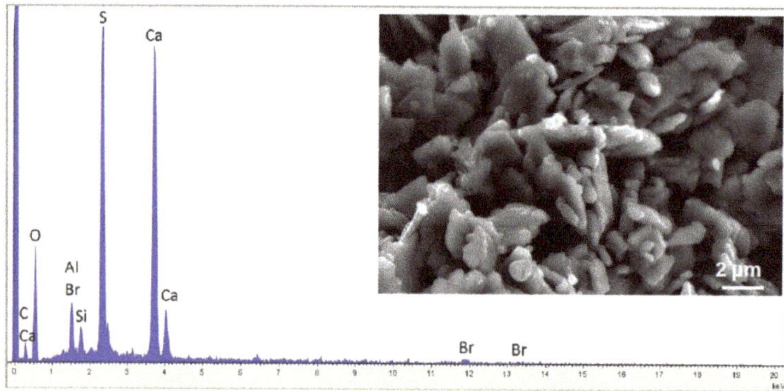

Figure 4. SEM–EDX analysis of a purple painted sample, which was extracted from a burial klinai, in Daskyleion, Turkey. Reprinted with permission form reference number [46].

High performance liquid chromatography–diode array detector (HPLC–DAD) is the most efficient among the techniques listed in Table 1 for the analysis of shellfish purple, as it provides: (i) Detection and identification of all seven components within a single measurement-chromatogram, (ii) semi-quantitative results relative to the composition of the detected purple pigment and (iii) complete quantitative results. Examples for these valuable sets of HPLC results are provided next.

The chromatogram for a purple pigment that was found in the Akrotiri settlement is shown in Figure 5. The sample was included in a previous study and labeled Ak3 [42]. The chromatogram is presented herein for the first time. INR was not detected in the sample extract. The other six components of shellfish purple are identified by HPLC, which therefore provides very rich results in terms of detected compounds.

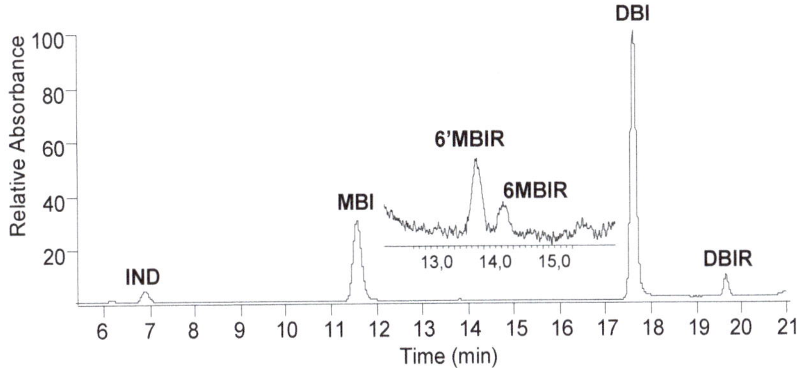

Figure 5. Chromatogram of a purple pigment from Akrotiri at 288 nm (Ak3 in [42]).

The HPLC peak areas (semi-quantitative results) are extremely useful to compare shellfish purple pigments from different samples. HPLC peak areas of purple components were reported in the literature for the samples of Minoan [42] and Persian origin [46,62] of Table 1 and these are summarized in Figure 6. In particular, relative integrated HPLC peak areas measured at 288 nm and normalized to the highest value (taken as 100%) are shown in Figure 6 for seven archaeological samples: Ak1 (pigment), Ak2 (wall painting), Ak3 (pigment) come from Akrotiri–Santorini, Ra (wall painting) from Raos–Santorini and Tri (pigment) from Trianda–Rhodes [42]. Samples Dar and Dask were extracted from the Darius I jar [62] and the burial klinai in Daskyleion [46], respectively. The HPLC peak areas for INR, 6′MBIR and 6MBIR were extremely small and therefore the sums of these areas are included in each sample column of the graphs of Figure 6.

The results of Figure 6a show that DBI is the dominant compound in the chromatograms of the five Minoan (Ak1, Ak2, Ak3, Tri and Ra) samples. The HPLC graphs gave moderate areas for MBI and small areas for IND and DBIR. The sum, INR + 6′MBIR + 6MBIR, is negligible or even zero (Ak2 and Ra) as revealed in Figure 6b. A similar HPLC profile was collected for the Dar sample [62], which, as shown in Figure 6, does not have any major difference compared to the Minoan samples. Trade routes between the southern Aegean islands, such as Santorini and Rhodes, and the Phoenician coast were developed since the prehistoric times and this may be depicted in the similar results of the Dar and the Minoan samples. However, the Dask sample is different: The HPLC peak area of DBI is small whereas the areas of IND, MBI, DBIR and the sum of INR + 6′MBIR + 6MBIR are large compared to the Minoan and the Dar samples. Consequently, a different processing method and/or molluskan sources should had been developed/used in Daskyleion, which is located close to the Marmara Sea, in the mainland of Turkey.

The relative peak areas reported in Figure 6 are not actual mass compositions of the samples. As mentioned previously, HPLC can provide complete quantitative results provided, however, that a HPLC calibration method is developed. The coloring components of interest should be available in pure form to be used as standards for quantitation purposes. A calibration procedure for shellfish purple pigment was previously developed and showed that the relative percentage of IND reported commonly using the HPLC peak areas at 288 nm (e.g., Figure 6) is in fact an overestimation of the actual mass content of IND [75]. On the other hand, the corresponding relative percentage of DBI, which is calculated using the HPLC peak areas at 288 nm, is an underestimation of the actual mass content of DBI in the shellfish pigment [75]. Moreover, the calibration process showed that low limits-of-detection (LODs) can be achieved with HPLC for the seven components of shellfish purple, ranging within 0.02 to 0.05 $\mu g\ mL^{-1}$ [75].

Finally, as described in Table 1, other methods employed to analyze and identify shellfish purple pigments/paints in objects of the cultural heritage are mass spectrometry (LC–MS, direct exposure–mass spectrometry—DE–MS) as well as imaging techniques including FTIR and XRF imaging. The latter are useful to reveal the distribution of shellfish purple on a painted surface, which, in turn, is sometimes useful to understand the applied painting technique.

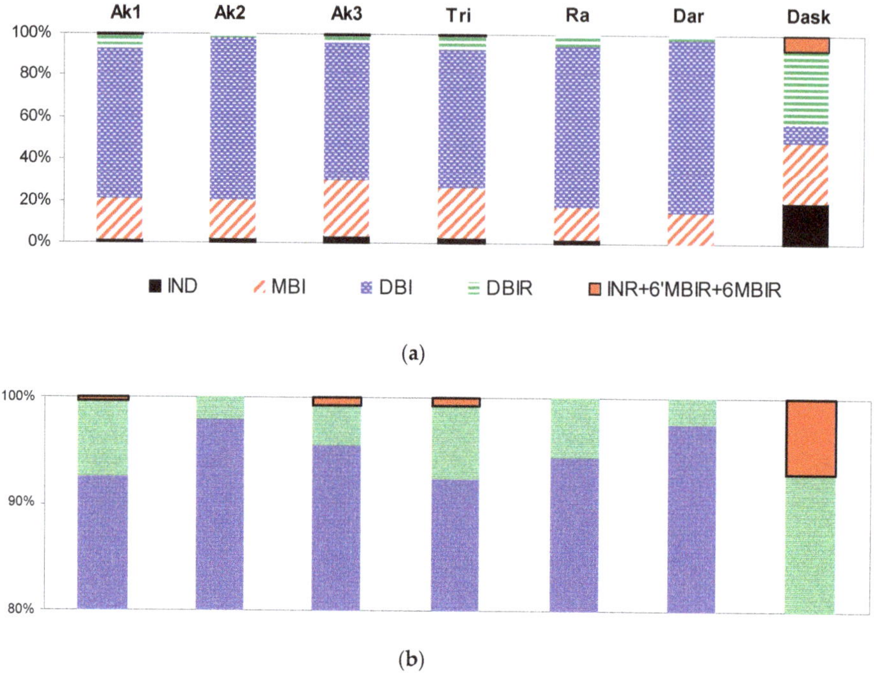

Figure 6. (**a**) Relative integrated HPLC peak areas measured at 288 nm and normalized to the highest value (taken as 100%) are shown for seven archaeological samples. The 18th-17th century BCE samples come from the Minoan sites in Akrotiri–Santorini (Ak1, Ak2, Ak3), Raos–Santorini (Ra) and Trianda–Rhodes (Tri) [42]. Samples Dar and Dask were extracted from the Darius I painted jar [62] and the burial klinai in Daskyleion [46], respectively, both dated to the 5th century BCE. The sums of the HPLC peak areas of INR, 6′MBIR and 6MBIR are included in the graph. (**b**) The upper (80%–100%) part of the (a) graph is enlarged.

5.2. Textiles

As mentioned previously, the earliest direct scientific evidence of the shellfish dyeing industry were reported for textile fragments unearthed in Chagar Bazar (18th–16th century BCE) [47] and Tell Mishrife [48] and pottery vessels from Crete (ca. 1800/1700 BCE) [49]. The most recent historical textile dyed with shellfish purple is a Byzantine epitaphios, which is dated to the 14th century CE [42] i.e., roughly about the time of the fall of Constantinople.

Shellfish purple has been identified in several textiles and objects related to the textile industry of antiquity e.g., post sherds [7,30–34,36,42,46–49,58,76–104]. As summarized in Table 2, it has been scientifically proved that the valuable dye was used by the Minoans, Hurrians, Phoenicians, Persians, Greeks (in various periods), Etruscans, Jews, Romans, Sarmatians, Copts-Egyptians, Arabs (pre-Islamic period) and Byzantines. These results show that the shellfish purple dye has a long established history of use that spans roughly 3.5 millennia, as it was used uninterrupted from the prehistoric up to the Byzantines times.

Table 2. Identifications of shellfish purple in textiles corresponding to various historical periods.

Provenance	Reference
Minoan (Crete)	[49]
Hurrian (Chagar Bazar, Qatna)	[47,48]
Phoenician	[76–81]
Phoenician–Punic	[82]
Submycenaean/Protogeometric (Greek)	[83]
Orientalising/early Archaic (Etruscan)	[84]
Archaic (Greek)	[85]
Persian	[46]
Classical (Greek)	[7,86–88]
Hellenistic (Greek & Etruscan)	[58,89–91]
Jewish	[92]
Roman	[31–33,93–96]
Sarmatian	[97,98]
Coptic–Egyptian	[34,97,99–104]
Arab (pre–Islamic period)	[36]
Byzantine	[7,30,42]

Case Study: Identification of Shellfish Purple in a Hellenistic Textile

In 1987 an excavation was carried out in a burial mount in the area of Lakkoma (Macedonia, Greece), which revealed a large cist tomb dating in the last quarter of the 4th c. BCE (Hellenistic period) [89]. Lakkoma is located around 60 km east of Aigai (Vergina), where the Royal Tomb of Philip II, the father of Alexander the Great, was found. The aforementioned dating of the burial is supported by a golden quarter stater of Philip II [89]. Textile purple residues and other organic materials were found within the tomb and investigated in the past using microscopic techniques [89].

A small piece of the textile residues was treated with hot dimethyl sulfoxide (DMSO) to extract the purple dye, which was analyzed using an established HPLC–DAD method [42,75]. This is the first reported HPLC analysis of a purple textile found in the mainland of Macedonia. The chromatogram is presented in Figure 7. The HPLC peaks corresponding to MBI, DBI and DBIR are dominant in the graph. IND, 6′MBIR and 6MBIR correspond to very small peaks, whereas INR was not detected.

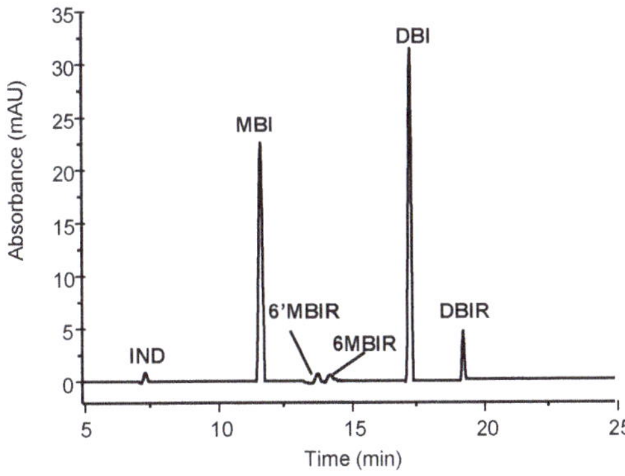

Figure 7. Chromatogram of a purple textile sample from the Lakkoma tomb at 288 nm.

It is interesting to note that the HPLC profile in Figure 7 is not very different from that of Figure 5, which is representative of the Minoan purple paints and pigments. Consequently, based on the results of the present study it is reported that no major difference is observed in the chemical compositions of the shellfish purple materials used by the Minoans (Figure 5) and the Macedonians (Figure 7).

Apparently this is not a general conclusion; it is an observation based on the chemical results of the tested samples.

5.3. Manuscripts

The scientific investigation of historical manuscripts is a challenging task. While samples can be removed from large-scale objects with relatively little impact on the integrity of the object, the removal of even tiny samples from manuscripts is practically prohibited. Consequently, the study of historical manuscripts is limited to non-sampling techniques, such as, for instance, XRF [105–108], fiber optics UV-Vis diffuse reflectance spectrophotometry with optical fibres (FORS) [106,108] and Raman spectroscopy [108,109]. Recent advances in the development of non-sampling methods to identify shellfish purple offer new perspectives for the investigation of historical manuscripts, artworks and other objects of the cultural heritage [110,111]. In any case, the limitation to use only non-sampling techniques increases the degree of difficulty in studying manuscripts.

The results reported in the literature do not suggest a widespread use of shellfish purple in historical manuscripts. In particular, the use of shellfish purple has been suggested in very few objects such as, for instance, in a Byzantine (6th century CE) [105], Italian (6th century CE) [106] and Anglo–Saxon (8th century CE) [107] manuscript. The use of shellfish purple was suggested according to XRF results, which revealed the presence of Br [105–107]. FORS results, however, did not confirm the presence of shellfish purple. Analyses yield the information that the chromatic features of manuscripts are dominated by the presence of inorganic and organic colorants other than shellfish purple [105–112].

5.4. Other Uses of Shellfish Purple

In a recently published report, shellfish purple was identified by XRF and HPLC in a mass of clay material, which was stored within a compartment of a small case [113]. The lidded box was found in the grave B of Derveni (Macedonia, Greece) dated to the 4th century BCE, together with other goods, including the famous Derveni krater and a gold coin of Philip II. Based on the current archaeological data, related with other metal cases that have been unearthed in Macedonian burials, it is estimated that the content of the Derveni metal case was probably used for medical purposes [113]. If true, then this is the first scientific evidence about the medical use of shellfish purple in antiquity. It is noteworthy, that recent studies revealed that mollusks from the family Muricidae produce biological active compounds with anti-inflammatory, anti-cancer, muscle-relaxing and pain relieving properties as well as steroidogenic activity [114,115]. Some of these properties are found in the Murex extracts and have been known since antiquity [116].

The finding in Derveni is not the only one revealing the use of shellfish purple as grave good. Shellfish purple was identified by DE–MS, Raman spectroscopy and HPLC in residues collected from a Gallo-Roman burial (Naintré, France) dated to the 3rd century CE [117]. Purple was widely spread after the deposition of the body for burial, suggesting that the valuable purple grave good was part of the funeral practice [117].

Murex snails and shells have been probably used in jewels, tools and foods [118], as well as in fillers and plasters as discussed next. It was previously described that the chalk base of the Akrotiri–Santorini pigment was rich in aragonite, indicating that this was probably obtained from crushed Murex shells [43]. Shell fragments of mollusks are sometimes visible in plaster surfaces, dated from the Middle Bronze Age onwards [118,119]. Murex shells in the perimeter of the 5th century BCE wall of Hermione, Greece are shown in the photograph of Figure 8. Moreover, results provided from experimental archaeology demonstrated that Murex shells can be used as a raw material for lime making, but that this is not necessarily visible in its end product if done thoroughly and properly [118]. Consequently, the use of Murex shells in the lime plaster production process can be older than the reported visible archaeological evidence. Murex species might have been used to add color in plaster, as suggested by XRD analysis in painted plaster samples from Gla, Greece [120]. XRD lead to the identification of aragonite in the investigated plaster samples, but the presence of Murex was not confirmed by SEM–EDX [120] or Raman [121] results.

Figure 8. Murex shells used as binder in the wall of Hermione, next to the Bisti area where the famous local dyeing workshops were located (photograph by V. Gatsos).

6. Purple but not True Purple

Shellfish purple was held in high esteem, as it is hard to be produced in large amounts from the molluskan raw source. In his pioneering work, Friedländer collected 12,000 *M. brandaris* extracting only 1.4 g of dry pigment [24]. It is estimated that 10,000 *M. trunculus* mollusks are needed to dye a kilogram and a half of wool [122]. Consequently, shellfish purple had always been an expensive material, which was sparingly used. Alternative procedures to mimic true purple had been developed and therefore "not all purples were equal, and not all purple was purple" [123].

For example, mixtures of Egyptian blue with red pigments were used in painting plasters to achieve a purple hue [124]. For the same reason, blue pigments and paints of Lapis lazuli [125] and indigo/woad [121] had been used in painting backgrounds since the Mycenaean period. Egyptian blue mixed physically with a red lake, which however is unidentifiable, was revealed in wall paintings fragments in Egypt [63]. Mixture of Egyptian blue and a red-pink lake was also found in 3rd century BCE oinochoe (British Museum) [45]. According to HPLC results, the lake was derived from purpurin-rich madder, cochineal (*Porphyrophora* spp.) and, unexpectedly, lac (*Kerria Lacca* Kerr) [45]. Mixtures of purpurin-rich madder and Armenian cochineal *Porphyrophora hamelii* Brandt were found in funeral figurines dated to 3rd–2nd century BCE [126,127].

A common practice for the Egyptian dyers was to mix madder and indigo/woad [8,34,93] achieving a remarkable purple hue. The purple mixture was sometimes enriched with a coccid dye, such as, kermes [128]. An analogous practice was revealed in Roman-Egyptian mummy portraits (2nd century CE) in which mixtures of madder lakes and indigo were identified and were probably applied as cheaper substitutes for shellfish purple [129]. Kermes in mixture with indigo/woad—but not madder—was identified in purple parts of a textile from the Topkapi Palace [130].

Folium and particularly orchil were widely used as alternatives to shellfish purple in the coloring of parchments in which the use of true purple was extremely rare [112,131,132], as discussed previously. Orchella weeds, when properly processed, provide a bright purple, which has been often detected in illuminated manuscripts [108,133].

7. Dyeing with Shellfish Purple: From Purple to Blue

Dyeing with shellfish is complicated. The colored water-insoluble extract must be chemically reduced in the dye vat to give the water-soluble and colorless leuco-forms of the compounds of Figure 3c. Reduction is induced by bacteria according to the mechanism that was elucidated for the reduction of woad [134]. The dyeing process can have an enormous effect on the composition and therefore the color of the attached dye considering that several parameters have to be adjusted in the vat (treatment time, temperature, pH, etc.). Moreover, sunlight can induce debromination of the leuco forms, which, moreover, do not have the same affinity for textile fibers [29].

The dramatic effect of the dyeing conditions on the composition of the attached dye is demonstrated in the following experiment. *M. trunculus* mollusks were collected by V. Gatsos from the sea of Hermione, Greece. The city of Hermione was famous for the fine dyeing with shellfish purple, an industry that had been flourished for more than 1000 years, from 6th century BCE to 6th century CE [12]. It was the Hermione purple that caused the admiration of Alexander the Great when he took Susa in 331 BCE and found purple garments in the palace of Darius III [7,12]. Wool and cotton were dyed by V. Gatsos following two recipes. Large mollusks (*M. trunculus*) were collected at a depth of 2–3 m, from the sea, just next to the ancient wall of Hermione.

Recipe I: The shells were crushed using a stainless steel tod at the third helix where the gland of the mollusk could be easily reached. The glands were quickly removed to preserve the secretions and placed in a flask containing 30 ml of water and 2 g of common salt. The flask was left open and was rigorously agitated 3–4 times per day for 15 days. After this process the pulp became purple. For the dyeing process, the purple pulp was transferred to a glass container. Ten (10) g of honey and 3 g of salt were added. The container was airtight sealed, agitated and remained into a water bath which was heated at 45 °C for two days. Then a 1 cm × 3 cm piece of fabric (wool or cotton) was immersed into the mixture and additional amounts of honey (2 g) and salt (1 g) were added. The container was airtight sealed, agitated and remained into a water bath which was heated at 45 °C for one day. The fabric was removed and washed with warm water (70 °C) and soap solution.

Recipe II: The shells were crushed and the glands were placed in a flask containing 80 mL of water and huge amount (20 g) of common salt. The flask was left in the sunlight for five days. Then, the dyeing procedure described in recipe I was followed.

Consequently, in recipe II, extreme conditions related to the quantity of salt and the duration of sunlight exposure were selected. The possible role of salt in the dyeing process is critically discussed in detail by C. Cooksey [18]. The four samples, two wools and two cottons, dyed with the two recipes, were analyzed using HPLC. A DMSO bath at 80 °C was used to extract the purple dyes [42]. The % relative integrated HPLC peak areas were measured and the results are provided in Table 3. The two recipes gave totally different results, as a dominant debromination process was developed in recipe II, resulting in reduced amounts of brominated indigoids (MBI and DBI) attached to the fibers. IND is clearly the major coloring compound in the samples, which were prepared using recipe II. On the contrary, large amounts of MBI were attached to fibers dyed with recipe I. The difference of the two recipes in the dyeing results was visible by naked eye, as samples prepared using recipe I were purple whereas samples that were dyed using recipe II were blue.

Table 3. Relative (%) integrated HPLC peak areas measured at 288 nm for wool and silk, which were dyed with two recipes, as described in the text.

Compound.	Recipe I		Recipe II	
	Wool	Cotton	Wool	Cotton
IND	11.9	8.5	70.3	94.4
MBI	56.5	44.7	16.0	3.3
DBI	29.1	4.5	8.8	1.7
INR	–	–	–	–
6′MBIR	0.2	0.2		–
6MBIR	0.2	0.1	1.2	–
DBIR	2.0	1.9	3.7	0.6

8. Solubility Issues

Solubility data is available in the open literature for indigo, which contains indigotin and indirubin but not their brominated derivatives. Table 4 shows the solubility of indigo in ten solvents [135]. The results of Table 4 were calculated using the COSMO–RS (conductor-like screening model for real solvents), which is a quantum-mechanical approach [135]. The best solvent found is sulfuric acid, in which COSMO–RS predicts complete miscibility with indigo. On the other hand, water practically does not dissolve indigo.

Table 4. Results of the conductor-like screening model for real solvents (COSMO–RS) solubility screening for indigo. The results were adapted from elsewhere [135].

Recipe I		Recipe II	
Sulfuric acid	100.0000	Pyrrole	4.1934
Dimethyl sulfoxide	11.9093	Chloroform	1.6337
Phenol	10.4013	Diethylether	1.4813
Propanone	6.5225	Ethanol	0.4694
Pyridine	4.3329	Water	0.0001

Efficient extraction of dyes from fiber samples is important for successful HPLC analysis of textiles of the cultural heritage. Among the solvents included in Table 4, DMSO [42,62] and pyridine [42,97] were suggested for the extraction of shellfish purple from archaeological samples. Moreover, N,N–dimethylformamide (DMF), which is not included in Table 4, was also used for the extraction of the purple material [42,78]. DMSO and DMF have similar solubility properties considering that they have comparable Hansen solubility parameters [136]. The latter are summarized in Table 5 for the three aforementioned solvents.

Table 5. Hansen solubility parameters: δ_d, δ_p and δ_{hb} are the dispersion, polar and hydrogen bonding parameters, respectively [136].

Solvent	δ_d	δ_p	δ_{hb}
Dimethylsulfoxide	18.4	16.4	10.2
N,N–dimethylformamide	17.4	13.7	11.3
Pyridine	19.0	8.8	5.9

The efficiencies of DMSO, DMF and pyridine to solubilize shellfish purple were compared using HPLC [42]. The experimental results showed that DMSO and pyridine result in very good and poor yields, respectively [42], which is in agreement with the prediction for indigo provided by the COSMO–RS (Table 4). Moreover, the experimental study showed that DMSO and DMF have comparable efficacy in solubilizing shellfish purple [42], which is in agreement with the similar Hansen solubility parameters of these two solvents (Table 5).

9. Conclusions

The conclusions of the review are summarized as follows:

- From the Minoan period to the cease of the Byzantine empire with the conquest of Constantinople by the Ottomans, shellfish purple had a significant role in painting art and textile industry, as suggested by chemical analyses (Tables 1 and 2; Figures 4–7) and depicted in various objects of cultural heritage (Figure 1).
- Not all purples were true purple, as procedures to mimic the purple hue had been developed by mixing blue and red pigments, lakes or dyes. Imitation of true purple was a common practice in purple manuscripts, as physicochemical analyses do not suggest a widespread use of the shellfish material on codices and manuscripts.
- Other past uses of shellfish purple in medicine, grave goods and wall construction (Figure 8) are supported by the literature.
- Dyeing conditions, such as sunlight and salt concentration, can have a dramatic effect on the composition of the purple dye and can therefore affect the color of the dyed textile, which can range from blue to purple (Figure 3; Table 3).
- The first HPLC analysis of a purple textile dated in the Hellenistic period was reported, herein. The textile fragment was found in the Macedonian tomb of Lakkoma (Northern Greece). The chromatogram was dominated by the peaks, which corresponded to 6–bromoindigotin, 6,6′–dibromoindigotin and 6,6′–dibromoindirubin (Figure 7).
- Dimethylsulfoxide and dimethylformamide are good solvents for indigoids dyes as suggested by experimental studies and theoretical predictions based on statistical mechanics (Tables 4 and 5).
- Marine species, including the mollusks of the Muricidae family are threatened by sea pollution and global warming (Figure 2). Action is needed to protect the shellfish, which for so many centuries have offered so much to us.

Funding: This research received no external funding.

Acknowledgments: The author thanks M.T.–A. and V.V. for providing the archaeological sample from the tomb at Lakkoma, A.V. for her assistance in the analysis of the sample, V.G. and S.P. for providing the dyed textiles and M.G. and S.S. for useful discussions.

Conflicts of Interest: The author declares no conflict of interest.

References

1. Liritzis, I.; Korka, E. Archaeometry's role in cultural heritage sustainability and development. *Sustainability* **2019**, *11*, 1972. [CrossRef]
2. Degano, I.; Ribechini, E.; Modugno, F.; Colombini, M.P. Analytical methods for the characterization of organic dyes in artworks and in historical textiles. *Appl. Spectrosc. Rev.* **2009**, *44*, 363–410. [CrossRef]
3. Nigra, B.T.; Faull, K.F.; Barnard, K. Analytical chemistry in archaeological research. *Anal. Chem.* **2015**, *87*, 3–18. [CrossRef] [PubMed]
4. Orna, M.V. Historic Mineral Pigments: Colorful Benchmarks of Ancient Civilizations. In *Chemical Technology in Antiquity*; Rasmussen, S.C., Ed.; ACS Symposium Series; American Chemical Society: Washington, DC, USA, 2015; Volume 1211, pp. 17–69.
5. Henshilwood, C.S.; D'Errico, F.; Van Niekerk, K.L.; Coquinot, Y.; Jacobs, Z.; Lauritzen, S.-E.; Menu, M.; García-Moreno, R. A 100,000-year-old ochre-processing workshop at Blombos Cave, South Africa. *Science* **2011**, *334*, 219–222. [CrossRef] [PubMed]
6. Splitstoser, J.C.; Dillehay, T.D.; Wouters, J.; Claro, A. Early pre-Hispanic use of indigo blue in Peru. *Sci. Adv.* **2016**, *2*, e1501623. [CrossRef] [PubMed]
7. Hofenk–de Graaff, J.H. *The Colourful Past: Origins, Chemistry and Identification of Natural Dyestuffs*; Archetype Publications Ltd: London, UK, 2004.
8. Cardon, D. *Natural Dyes–Sources, Tradition Technology and Science*; Archetype Publications Ltd: London, UK, 2007.

9. Cooksey, C. Tyrian purple: The first four thousand years. *Sci. Prog.* **2013**, *96*, 171–186. [CrossRef] [PubMed]
10. Moran, W.L. *The Amarna Letters;* Johns Hopkins University Press: Baltimore, MD, USA, 1992; pp. 53–79.
11. Ventris, M.G.F.; Chadwick, J. *Documents in Mycenaean Greece*, 2nd ed.; Cambridge University Press: Cambridge, UK, 1973; p. 573.
12. Kalaitzaki, A.; Vafiadou, A.; Frony, A.; Reese, D.S.; Drivaliari, A.; Liritzis, I. Po-pu-re: Workshops, use and archaeometric analysis in pre-Roman central eastern Mediterranean. *Mediterr. Archaeol. Archaeom.* **2017**, *17*, 103–130.
13. Graves, D.E. What is the Madder with Lydia's Purple? A re-examination of the Purpurarii in Thyatira and Philippi. *Near East Archaeol. Soc. Bull.* **2017**, *62*, 3–29.
14. British Museum Coins Phoenicia, Supplementary Plate XLIV, 7
15. Available online: https://www.britishmuseum.org/research/collection_online/collection_object_details.aspx?objectId=462267&partId=1&searchText=murex&page=1 (accessed on 22 May 2019).
16. Available online: https://www.britishmuseum.org/research/collection_online/collection_object_details.aspx?objectId=1298437&partId=1&searchText=murex&page=1 (accessed on 22 May 2019).
17. Rilov, G. Multi-species collapses at the warm edge of a warming sea. *Sci Rep.* **2016**, *6*, 36897. [CrossRef]
18. Cooksey, C. Pliny's first century AD recipe for a purple dye-vat–decoded. In *Purpureae Vestes V: Textiles, Basketry and Dyes in the Ancient Mediterranean World;* Ortiz, J., Alfaro, C., Turell, L., Martínez, J., Eds.; Universitat de València: València, Spain, 2016; pp. 217–220.
19. Wouters, J.; Verhecken, A. High-performance liquid chromatography of blue and purple indigoid natural dyes. *J. Soc. Dyers Colour.* **1991**, *107*, 266–269. [CrossRef]
20. Karapanagiotis, I.; de Villemereuil, D.; Magiatis, P.; Polychronopoulos, P.; Vougogiannopoulou, K.; Skaltsounis, A.-L. Identification of the coloring constituents of four natural indigoid dyes. *J. Liq. Chromatogr. Relat. Technol.* **2006**, *29*, 1491–1502. [CrossRef]
21. Koren, Z.C. HPLC-PDA analysis of brominated indirubinoid, indigoid, and isatinoid dyes. In *Indirubin, the Red Shade of Indigo;* Meijer, L., Guyard, N., Skaltsounis, A.-L., Eisenbrand, G., Eds.; Life in Progress Editions: Roscoff, France, 2006; pp. 45–53.
22. Mantzouris, D.; Karapanagiotis, I. Identification of indirubin and monobromoindirubins in *Murex Brandaris*. *Dyes Pigment.* **2014**, *104*, 194–196. [CrossRef]
23. Surowiec, I.; Nowik, W.; Moritz, T. Mass spectrometric identification of new minor indigoids in shellfish purple dye from *Hexaplex trunculus*. *Dyes Pigment.* **2012**, *94*, 363–369. [CrossRef]
24. Friedländer, P. Über den Farbstoff des antiken Purpurs aus *Murex brandaris*. *Ber. Dtsch. Chem. Ges.* **1909**, *42*, 765–770. [CrossRef]
25. Clark, R.J.H.; Cooksey, C.J. Monobromoindigos: A new general synthesis, the characterization of all four isomers and an investigation into the purple colour of 6,6′-dibromoindigo. *New J. Chem.* **1999**, *23*, 323–328. [CrossRef]
26. Imming, P.; Imhof, I.; Zentgraf, M. An improved synthetic procedure for 6,6′-dibromoindigo (Tyrian purple). *Synth. Commun.* **2001**, *31*, 3721–3727. [CrossRef]
27. Wolk, J.L.; Frimer, A.A. Preparation of Tyrian Purple (6,6′-dibromoindigo): Past and present. *Molecules* **2010**, *15*, 5473–5508. [CrossRef] [PubMed]
28. Ramig, K.; Lavinda, O.; Szalda, D.J.; Mironova, I.; Karimi, S.; Pozzi, F.; Shah, N.; Samson, J.; Ajiki, H.; Massa, L.; et al. The nature of thermochromic effects in dyeings with indigo, 6-bromoindigo, and 6,6′-dibromoindigo, components of Tyrian Purple. *Dyes Pigment.* **2015**, *117*, 37–48. [CrossRef]
29. Ramig, K.; Islamova, A.; Scalise, J.; Karimi, S.; Lavinda, O.; Cooksey, C.; Vasileiadou, A.; Karapanagiotis, I. The effect of light and dye composition on the color of dyeings with indigo, 6-bromoindigo, and 6,6′ dibromoindigo, components of Tyrian Purple. *Struct. Chem.* **2017**, *28*, 1553–1561. [CrossRef]
30. Koren, Z.C.; Verhecken-Lammens, C. Microscopic and chromatographic analyses of molluskan purple yarns in a late Roman period textile. *e-PS* **2013**, *10*, 27–34.
31. Cardon, D.; Nowik, W.; Granger-Taylor, H.; Marcinowska, R.; Kusyk, K.; Trojanowicz, M. Who could wear true purple in Roman Egypt? Technical and social considerations on some new identifications of purple from marine molluscs in archaeological textiles. In *Textiles y Tintes en la Ciudad antigua;* Alfaro, C., Brun, J.-P., Borgard, P., Pierobon Benoît, R., Eds.; Universitat de València: València, Spain, 2011; pp. 197–214.
32. Sukenik, N.; Iluz, D.; Shamir, O.; Varvak, A.; Amar, Z. Purple-dyed textiles from Wadi Murabba'at. *Archaeol. Text. Newslett.* **2013**, *55*, 46–54.

33. Sukenik, N.; Varvak, A.; Amar, Z.; Iluz, D. Chemical analysis of Murex-dyed textiles from wadi Murabba'at, Israel. *J. Archaeol. Sci. Rep.* **2015**, *3*, 565–570. [CrossRef]
34. Karapanagiotis, I.; Verhecken-Lammens, C.; Kamaterou, P. Identification of dyes in egyptian textiles of the first millennium AD from the collection Fill-Trevisiol. *Archaeol. Anthropol Sci.* **2019**, *11*, 2699–2710. [CrossRef]
35. Moatsos, P.-G. *Porphyra*; Kasimati & Iona: Alexandria, Egyt, 1932. (In Greek)
36. Ribechini, E.; Pérez-Arantegui, J.; Pavan, A.; Degano, I.; Zanaboni, M.; Colombini, M.P. First evidence of purple pigment production and dyeing in southern Arabia (Sumhuram, Sultanate of Oman) revealed by massspectrometric and chromatographic techniques. *J. Cult. Herit.* **2016**, *19*, 486–491. [CrossRef]
37. Aloupi, E.; Karydas, A.G.; Paradellis, T. Pigment analysis of wall paintings and ceramics from Greece and Cyprus. *X-ray Spectrom.* **2000**, *29*, 18–24. [CrossRef]
38. Sotiropoulou, S.; Karapanagiotis, I. Conchylian purple investigation in prehistoric wall paintings of the Aegean area. In *Indirubin, the Red Shade of Indigo*; Meijer, L., Guyard, N., Skaltsounis, A.-L., Eisenbrand, G., Eds.; Life in Progress Editions: Roscoff, France, 2006; pp. 71–78.
39. Karapanagiotis, I. Identification of indigoid natural dyestuffs used in art objects by HPLC coupled to APCI-MS. *Am. Lab.* **2006**, *38*, 36–40.
40. Van Elslande, E.; Lecomte, S.; Le Ho, A.S. Micro-Raman spectroscopy (MRS) and surface-enhanced Raman scattering (SERS) on organic colourants in archaeological pigments. *J. Raman Spectrosc.* **2008**, *39*, 1001–1006. [CrossRef]
41. Karapanagiotis, I.; Sotiropoulou, S.; Valianou, L. Identification of Tyrian purple in Aegean Bronze age pigments. In Proceedings of the 30th Dyes in History and Archaeology meeting, Derby, UK, 12–15 October 2011.
42. Karapanagiotis, I.; Mantzouris, D.; Cooksey, C.; Mubarak, M.S.; Tsiamyrtzis, P. An improved HPLC method coupled to PCA for the identification of Tyrian Purple in archaeological and historical samples. *Microchem. J.* **2013**, *110*, 70–80. [CrossRef]
43. Karapanagiotis, I.; Sotiropoulou, S.; Chryssikopoulou, E.; Magiatis, P.; Andrikopoulos, K.S.; Chryssoulakis, Y. Investigation of Tyrian purple occurring in prehistoric wall paintings of Thera. In *The Diversity of Dyes in History & Archaeology*; Kirby, J., Ed.; Archetype Publications: London, UK, 2017; pp. 82–89.
44. Pliny. Naturalis Historia, XXXV.xxvi.44
45. Dyer, J.; Tamburini, D.; Sotiropoulou, S. The identification of lac as a pigment in ancient Greek polychromy—The case of a Hellenistic oinochoe from Canosa di Puglia. *Dyes Pigment.* **2018**, *149*, 122–132. [CrossRef]
46. Papliaka, Z.E.; Konstanta, A.; Karapanagiotis, I.; Karadag, R.; Akyol, A.A.; Mantzouris, D.; Tsiamyrtzis, P. FTIR imaging and HPLC reveal ancient painting and dyeing techniques of molluskan purple. *Archaeol. Anthropol. Sci.* **2017**, *9*, 197–208. [CrossRef]
47. Breniquet, C.; Desrosiers, S.; Nowik, W.; Rast-Eicher, A. Les textiles découverts dans les tombes de l'age du Bronze moyen a Chagar Bazar (Syrie). In *Chagar Bazar (Syrie) VIII. Les Tombes Ordinaires d' L'āge du Bronze Ancien et Moyen des Chantiers D-F-H-I (1999–2011)*; Tunca, O., Baghdo, A., Eds.; Peeters Publishers: Lueven-Liège, Belgium, 2018; pp. 11–32.
48. James, M.A.; Reifarth, N.; Mukherjee, A.J.; Crump, M.P.; Gates, P.J.; Sandor, P.; Robertson, F.; Pfälzner, P.; Evershed, R.P. High prestige royal purple dyed textiles from the Bronze Age royal tomb at Qatna, Syria. *Antiquity* **2009**, *83*, 1109–1118. [CrossRef]
49. Koh, A.A.; Betancourt, P.P.; Pareja, M.N.; Brogan, T.M.; Apostolakou, V. Organic residue analysis of pottery from the dye workshop at Alatsomouri-Pefka, Crete. *J. Archaeol. Sci. Rep.* **2016**, *7*, 536–538. [CrossRef]
50. Edmonds, J. The trail of the purple snail. *J. Weavers Spinners Dyers* **2005**, *214*, 5.
51. Stieglitz, R.R. The Minoan origin of Tyrian purple. *Biblic. Archaeol.* **1994**, *57*, 46–54. [CrossRef]
52. Reese, D.S. Shells from Sarepta (Lebanon) and east Mediterranean purple-dye production. *Mediterr. Archaeol. Archaeom.* **2010**, *10*, 113–141.
53. Reese, D.S. Palaikastro shells and Bronze Age purple dye production in the Mediterranean basin. *Annu. Br. Sch. Athens* **1987**, *82*, 202–206. [CrossRef]
54. Veropoulidou, R. The Tyrian Purple, a "royal" dye. In *Greeks and Phoenicians at the Mediterranean Crossroads*; Adam-Veleni, P., Stefani, E., Eds.; Hellenic Ministry of Culture and Tourism, Archaeological Museum of Thessaloniki: Thessaloniki, Greece, 2012; pp. 103–105.

55. Reese, D.S. The invertebrates. In *Palaikastro, Building 1. Supplementary Volume I.*; Macgillivray, J.A., Sackett, L.H., Eds.; The British School of Athens: Athens, Greece, 2019; pp. 387–409.
56. Reese, D.S. Shell Purple-dye Production in the Mediterranean Basin and Environs. In press.
57. Minniti, C. Shells at the Bronze Age settlement of Coppa Nevigata (Apulia, Italy). In *Archaeomalacology. Molluscs in Former Environments of Human Behaviour*; Mayer, D.E.B.-Y., Ed.; Oxbow Books and the individual authors: Oxford, UK, 2005; pp. 71–81.
58. Gleba, M.; Vanden Berghe, I. Textiles from Strozzacapponi (Perugia/Corciano). New evidence of purple production in pre-Roman Italy. In *Purpureae Vestes IV: Production and Trade of Textiles and Dyes in the Roman Empire and Neighbouring Regions*; Alfaro, C., Tellenbach, M., Ortiz, J., Eds.; Universitat de València: València, Spain, 2014; pp. 167–173.
59. Minniti, C.; Recchia, G. New evidence on purple dye production from the Bronze Age settlement of Coppa Nevigata (Apulia, Italy). In *Purpureae Vestes VI: Textiles and Dyes in the Mediterranean Economy and Society*; Busana, M.S., Gleba, M., Meo, F., Tricomi, A.R., Eds.; Libros Pórtico: Zaragoza, España, 2018; pp. 87–97.
60. Kokiasmanou, E. Portable XRF Analysis and MA-XRF Imaging of Mycenaean Wall-Painting Pigments from the Palace of Nestor at Pylos. Master's Thesis, Aristotle University of Thessaloniki, Thessaloniki, Greece, 2018.
61. Colombini, M.P.; Carmignani, A.; Modugno, F.; Frezzato, F.; Olchini, A.; Brecoulaki, H.; Vassilopoulou, V.; Karkanas, P. Integrated analytical techniques for the study of ancient Greek polychromy. *Talanta* **2004**, *63*, 839–848. [CrossRef]
62. Koren, Z.C. Archaeo-chemical analysis of Royal Purple on a Darius I stone jar. *Microchim. Acta* **2008**, *162*, 381–392. [CrossRef]
63. Kakoulli, I. Late Classical and Hellenistic painting techniques and materials: A review of the technical literature. *Rev. Conserv.* **2002**, *3*, 56–67. [CrossRef]
64. Maniatis, Y.; Arvaniti, T.; Antikas, T.G.; Wynn-Antikas, L.; Orsini, S.; Ribechini, E.; Colombini, M.P. Investigation of an unusual composite material found in the larnax with cremated bones in royal tomb II at Vergina. In Proceedings of the 40th International Symposium on Archaeometry, Los Angeles, CA, USA, 19–23 May 2014.
65. Sanpaolo, E.A.; Follo, L.; Sfrecola, S. Alcuni aspetti tecnologici della ceramica policroma della Daunia nella collezione Sansone di Mattinata (FG). L'apporto dell'archeometria. In *Atti dell' 11 Convegno Nazionale sulla Preistoria-Protostoria, Storia della Daunia*; Gravina, A., Ed.; Gerni Editori: San Severo, Italy, 1990; pp. 139–170.
66. Karydas, A.G. In situ XRF analyses of wall-painting pigments on ancient funeral Macedonian monument. In *La Peinture Funeraire de Macedoine. Emplois et Fonctions de la Couleur IVe-IIe s. av. J. -C, Vol II: Planches & Tableaux, Appendix IV*; Brecoulaki, H., Ed.; National Hellenic Research Foundation: Athens, Greece, 2006.
67. Andreotti, A.; Carmignani, A.; Colombini, M.P.; Modugno, F. Characterization of paint organic materials in wall decorations of Macedonian tombs. In *La Peinture Funeraire de Macedoine. Emplois et Fonctions de la Couleur IVe-IIe s. av. J. -C, Vol II: Planches & Tableaux, Appendix IV*; Brecoulaki, H., Ed.; National Hellenic Research Foundation: Athens, Greece, 2006.
68. Brecoulaki. Precious colours in ancient Greek polychromy and painting: Material aspects and symbolic values. *Rev. Archéol.* **2014**, *1*, 3–35.
69. Andreotti, A.; Colombini, M.P.; Ribechini, E.; D'Alessio, A.; Frezzato, F. The characterisation of red-violet organic colours in ancient samples. In *The Diversity of Dyes in History & Archaeology*; Kirby, J., Ed.; Archetype Publications: London, UK, 2017; pp. 97–106.
70. Maravelaki-Kalaitzaki, P.; Kallithrakas-Kontos, N. Pigment and terracotta analyses of Hellenistic figurines in Crete. *Anal. Chim. Acta* **2003**, *497*, 209–225. [CrossRef]
71. Maniatis, G.; Sakelari, H.; Kavousanaki, D.; Minos, N. Physicochemical characterisation of pigments from the cist tomb of Pella. In *Cist Tomb with Painting Decoration from Pella*; Lilibaki-Akamati, M., Ed.; Ministry of Culture, 17th Ephorate of Prehistoric and Classical Antiquities: Athens, Greece, 2007; pp. 138–175. (In Greek)
72. Karydas, A.G.; Brecoulaki, H.; Bourgeois, B.; Jockey, P.H. In situ X-ray fluorescence analysis of raw pigments and traces of polychromy on Hellenistic sculpture and the Archaeological Museum of Delos. In Proceedings of the 7th International Conference of Association for the Study of Marble and Other Stones in Antiquity, Thassos, Greece, 15–20 September 2003; Maniatis, Y., Ed.; Suppl. 51. École française d'Athènes, BCH: Athens, Greece, 2009; pp. 811–829.

73. Clarke, M.; Fredrick, P.; Colombini, M.P.; Andreotti, A.; Wouters, J.; Van Bommel, M.; Eastaugh, N.; Walsh, V.; Chaplin, T.; Siddall, R. Pompei purpurissum pigment problems. In *Proceedings Art 2005: 8th International Conference on Non-Destructive Testing and Microanalysis for the Diagnostics and Conservation of the Cultural and Environmental Heritage*; Parisi, C., Buzzanca, G., Paradisi, A., Eds.; Italian Society of Non-Destructive testing Monitoring Diagnostics AIPnD: Lecce, Italy, 2005.
74. March, R.E.; Papanastasiou, M.; McMahon, A.W.; Allen, N.S. An investigation of paint from a mural in the church of Sainte Madeleine, Manas, France. *J. Mass Spectrom.* **2011**, *46*, 816–820. [CrossRef] [PubMed]
75. Vasileiadou, A.; Karapanagiotis, I.; Zotou, A. Determination of Tyrian purple by high performance liquid chromatography with diode array detection. *J. Chromatogr. A* **2016**, *1448*, 67–72. [CrossRef] [PubMed]
76. McGovern, P.E.; Michel, R.H. Boyal purple and the pre-Phoenician dye industry of Lebanon. *MASCA* **1984**, *3*, 67–70.
77. McGovern, P.E.; Michel, R.H. Royal purple dye: Tracing chemical origins of the industry. *Anal. Chem.* **1985**, *57*, 1514A–1522A.
78. Koren, Z.C. High-performance liquid chromatographic analysis of an ancient Tyrian purple dyeing vat from Israel. *Isr. J. Chem.* **1995**, *35*, 117–124. [CrossRef]
79. Karmon, N.; Spanier, E. Archaeological evidence of the purple dye industry from Israel. In *The Royal Purple and the Biblical Blue: Argaman and Tekhelet the Study of Chief Rabbi Dr Isaac Herzog on the Dye Industries in Ancient Israel and Recent Scientific Contributions*; Spanier, E., Ed.; Keter Publ House: Jerusalem, Israel, 1987; pp. 147–158.
80. Karmon, N.; Spanier, E. Remains of a purple dye industry at Tel Shiqmona. *Isr. Explor. J.* **1988**, *38*, 184–186.
81. Sukenik, N.; Iluz, D.; Amar, Z.; Varvak, A.; Bar, S. New evidence of the purple-dye industry at Tel Shiqmona, Israel. *Archaeometry* **2017**, *59*, 775–785. [CrossRef]
82. Domínguez-Bella, S.; March, R.J.; Gener, J.M.; Martínez, J. Analisis de restos organicos de la tumba punica de la Casa del Obispo, Cadiz. Reconstruyendo la memoria fenicia en el occidente del Mediterraneo. In *Gadir y el Círculo del Estrecho Revisados*; Domínguez-Pérez, J.C., Ed.; Servicio de Publicaciones de la Universidad de Cadiz: Cadiz, Spain, 2011; pp. 307–319.
83. Kolonas, L.; Sarri, K.; Margariti, C.; Vanden Berghe, I.; Skals, I.; Nosch, M.-L. Heirs from the loom? Funerary textiles from Stamna (Aitolia, Greece). A preliminary analysis. In *Hesperos. The Aegean Seen from the West, Proceedings of the 16th International Aegean Conference, Ioannina, Greece, 18–21 May 2016*; Fotiadis, M., Laffineur, R., Lolos, Y., Vlachopoulos, A., Eds.; Peeters Publishers: Lueven-Liège, Belgium, 2017; pp. 533–542.
84. Gleba, M.; Mandolesi, A.; Lucidi, M.R. New textile finds from *Tomba Dell' Aryballos Sospeso*, Tarquinia: Context, analysis and preliminary interpretation. *Origini* **2017**, *40*, 29–44.
85. Metallinou, G.; Moulhérat, C.; Spantidaki, G. Archaeological textiles from Kerkyra. *Arachne* **2009**, *3*, 30–51.
86. Margariti, C.; Protopapas, S.; Allen, N.; Vishnyakov, V. Identification of purple dye from molluscs on an excavated textile by non-destructive analytical techniques. *Dyes Pigment.* **2013**, *96*, 774–780. [CrossRef]
87. Moulhérat, C.; Spantidaki, Y. A study of textile remains from the 5th century BC discovered in Kalyvia, Attica. In *Ancient Textiles: Production, Craft and Society, Proceedings of the First International Conference on Ancient Textiles, Lund, Sweden, and Copenhagen, Denmark, 19–23 March 2003*; Gillis, C., Nosch, M.-L., Eds.; Oxbow Books: Oxford, UK, 2007; pp. 163–166.
88. Spantidaki, Y.; Moulhérat, C. Greece. In *Textiles and Textile Production in Europe from Prehistory to AD 400*; Gleba, M., Mannering, U., Eds.; Oxbow Books: Oxford, UK, 2012; pp. 185–200.
89. Τσιμπίδου-Αυλωνίτη, Μ. Ένα *Αδιατάρακτο Ταφικό Σύνολο του 4ου αι. π.Χ. στο Λάκκωμα Χαλκιδικής*; Στο Ήχάδιν ΙΙ; Τιμητικός Τόμος για τη Στέλλα Δρούγου; Κώτσου, Ε., επιμέλεια; Έκδοση του Ταμείου Αρχαιολογικών Πόρων και Απαλλοτριώσεων: Αθήνα, Ελλάδα, 2016; Σελ 756–784.
90. Daniels, V. Dye analysis on two fragments from Enkomi. *Dyes Hist Archaeol* **1985**, *4*, 15–18.
91. Gleba, M.; Vanden Berghe, I.; Cenciaioli, L. Purple for the masses? Shellfish purple-dyed textiles from the quarry workers' cemetery at Strozzacapponi (Perugia/Corciano), Italy. In *Treasures from the Sea*; Energen, H.L., Meo, F., Eds.; Oxbow Books: Oxford, UK, 2017; pp. 131–137.
92. Koren, Z.C. *The Unprecedented Discovery of the Royal Purple Dye on the Two Thousand Year-Old Royal Masada Textile*; The Textile Specialty Group Postprints; American Institute for Conservation of Historic and Artistic Works: Washington, DC, USA, 1997; Volume 7, pp. 23–34.

93. Wouters, J.; Vanden Berghe, I.; Richard, G.; Breniaux, R.; Cardon, D. Dye analysis of selected textiles from three Roman sites in the eastern desert of Egypt: A hypothesis on the dyeing technology in Roman and Coptic Egypt. *Dyes Hist. Archaeol.* **2008**, *21*, 1–16.
94. Bruni, S.; Guglielmi, V.; Pozzi, F. Surface-enhanced Raman spectroscopy (SERS) on silver colloids for the identification of ancient textile dyes: Tyrian purple and madder. *J. Raman Spectrosc.* **2010**, *41*, 175–180. [CrossRef]
95. Doherty, B.; Nowik, W.; Miliani, C.; Clementi, C. Tyrian purple in archaeological textiles: DMF extraction and recrystallization for the Raman identification of precursors and derivatives. *J. Raman Spectrosc.* **2017**, *48*, 744–749. [CrossRef]
96. Karapanagiotis, I.; Sotiropoulou, S.; Vasileiadou, S.; Karagiannidou, E.; Mantzouris, D.; Tsiamyrtzis, P. Shellfish purple and gold threads from a late antique tomb excavated in Thessaloniki. *Arachne* **2018**, *5*, 64–77.
97. Wouters, J. A new method for the analysis of blue and purple dyes in textiles. *Dyes Hist. Archaeol.* **1992**, *10*, 17–21.
98. Sato, M.; Sasaki, Y. Blue and purple dyestuffs used for ancient textiles. *Dyes Hist. Archaeol.* **2003**, *19*, 100–105.
99. Michel, R.H.; Lazar, J.; McGovern, P.E. Indigoid dyes in Peruvian and Coptic textiles of the university museum of archaeology and anthropology. *Archaeomaterials* **1992**, *6*, 69–83.
100. Hofmann-de Keijzer, R.; Van Bommel, M.R.; De Keijzer, M. Coptic textiles: Dyes, dyeing techniques and dyestuff analysis of two textile fragments of the MAK Vienna. In *Methods of Dating Ancient Textiles of the 1st Millennium AD from Egypt and Neighbouring Countries, Proceedings of the 4th Meeting of the Study Group 'Textiles from the Nile Valley', Antwerp, Belgium, 16–17 April 2005*; De Moor, A., Fluck, C., Eds.; Lannoo Publishers: Tielt, Belgium, 2007; pp. 214–228.
101. Wouters, J. Dye analysis in a broad perspective: A study of 3rd to 10th century Coptic textiles from Belgian private collections. *Dyes Hist. Archaeol.* **1995**, *13*, 38–45.
102. Gulmini, M.; Idone, A.; Davit, P.; Moi, M.; Carrillo, M.; Ricci, C.; Dal Bello, F.; Borla, M.; Oliva, C.; Greco, C.; et al. The Coptic textiles of the Museo Egizio in Torino (Italy): A focus on dyes through a multi-technique approach. *Archaeol. Anthropol. Sci.* **2017**, *9*, 485–497. [CrossRef]
103. Hofmann-de Keizer, R.; Van Bommel, M. Dyestuff analysis of two textile fragments from late antiquity. *Dyes Hist. Archaeol.* **2008**, *21*, 17–25.
104. Walton, P. Shellfish purple in a Coptic textile. *Dyes Hist. Archaeol.* **1985**, *4*, 33–34.
105. Aceto, M.; Agostino, A.; Fenoglio, G.; Baraldi, P.; Zannini, P.; Hofmann, C.; Gamillscheg, E. First analytical evidences of precious colourants on Mediterranean illuminated manuscripts. *Spectrochim. Acta A* **2012**, *95*, 235–245. [CrossRef] [PubMed]
106. Aceto, M.; Idone, A.; Agostino, A.; Fenoglio, G.; Gulmini, M.; Baraldi, P.; Crivello, F. Non-invasive investigation on a VI century purple codex from Brescia, Italy. *Spectrochim. Acta A* **2014**, *117*, 34–41. [CrossRef] [PubMed]
107. Porter, C.A.; Chiari, G.; Cavallo, A. The analysis of eight manuscripts and fragments from the fifth/sixth century to the twelfth century, with particular reference to the use of and identification of "real purple" in manuscripts. In *Proceedings Art 2002: 7th International Conference on Non-Destructive Testing and Microanalysis for the Diagnostics and Conservation of the Cultural and Environmental Heritage*; Van Grieken, R., Janssens, K., Van't dack, L., Meersman, G., Eds.; University of Antwerp: Antwerp, Belgium, 2002.
108. Bicchieri, M. The purple Codex Rossanensis: Spectroscopic characterisation and first evidence of the use of the elderberry lake in a sixth century manuscript. *Environ. Sci. Pollut. Res.* **2014**, *21*, 14146–14157. [CrossRef]
109. Andrikpolous, K.S.; Dannilia, S. Issues relating to the common origin of two Byzantine miniatures: In situ examination with Raman spectroscopy and optical microscopy. *J. Raman Spectrosc.* **2007**, *38*, 332–343.
110. Clementi, C.; Nowik, W.; Romani, A.; Cardon, D.; Trojanowicz, M.; Davantès, A.; Chaminade, P. Towards a semiquantitative non invasive characterisation of Tyrian purple dye composition: Convergence of UV-Visible reflectance spectroscopy and fast-high temperature-high performance liquid chromatography with photodiode array detection. *Anal. Chim. Acta* **2016**, *926*, 17–27. [CrossRef] [PubMed]
111. Verri, G.; de Fonjaudran, C.M.; Acocella, A.; Accorsi, G.; Comelli, D.; D'Andrea, C.; Nevin, A.; Zerbetto, F.; Saunders, D. An 'imperial radiation': Experimental and theoretical investigations of the photo-induced luminescence properties of 6,6'-dibromoindigo (Tyrian purple). *Dyes Pigment.* **2019**, *160*, 879–889. [CrossRef]

112. Aceto, M.; Calà, E.; Agostino, A.; Fenoglio, G.; Gulmini, M.; Idone, A.; Porter, C.; Hofmann, C.; Rabitsch, S.; Denoël, C.; et al. Mythic dyes or mythic colour? New insight into the use of purple dyes on codices. *Spectrochim. Acta A* **2019**, *215*, 133–141. [CrossRef] [PubMed]
113. Katsifas, C.S.; Ignatiadou, D.; Zacharopoulou, A.; Kantiranis, N.; Karapanagiotis, I.; Zachariadis, G.A. Non-destructive X-ray spectrometric and chromatographic analysis of metal containers and their contents from ancient Macedonia. *Separations* **2018**, *5*, 32. [CrossRef]
114. Benkendorff, K.; Rudd, R.; Nongmaithem, B.D.; Liu, L.; Edwards, V.; Avila, C.; Abbott, C.A. Are the traditional medical uses of Muricidae molluscs substantiated by their pharmacological properties and bioactive compounds? *Mar. Drugs* **2015**, *13*, 5237–5275. [CrossRef] [PubMed]
115. Nongmaithem, B.D.; Mouatt, P.; Smith, J.; Rudd, D.; Russell, M.; Sullivan, C.; Benkendorff, K. Volatile and bioactive compounds in opercula from Muricidae molluscs supports their use in ceremonial incense and traditional medicines. *Sci. Rep.* **2017**, *7*, 17404. [CrossRef] [PubMed]
116. Voultsiadou, E. Therapeutic properties and uses of marine invertebrates in the ancient Greek world and early Byzantium. *J. Ethnopharmacol.* **2010**, *130*, 237–247. [CrossRef] [PubMed]
117. Devièse, T.; Ribechini, E.; Baraldi, P.; Farago-Szekeres, B.; Duday, H.; Regert, M.; Colombini, M.P. First chemical evidence of royal purple as a material used, for funeral treatment discovered in a Gallo–Roman burial (Naintré, France, third century AD). *Anal. Bioanal. Chem.* **2011**, *401*, 1739–1748. [CrossRef] [PubMed]
118. Brysbaert, A. Murex uses in plaster features in the Aegean and eastern Mediterranean Bronze Age. *Mediterr. Archaeol. Archaeom.* **2007**, *7*, 29–51.
119. Brysbaert, A. Rotating angles in measuring the Aegean Bronze Age. The technology of bronze painted plaster from the Aegean and Eastern Mediterranean. In *Measuring the Aegean Bronze Age, Proceedings of the 9th International Aegean Conference, New Haven, CT, USA, 18–21 April 2002*; Foster, K.P., Laffineur, R., Eds.; Université de Liége and University of Texas at Austin: Liège, Belgium; Austin, TX, USA, 2003; pp. 167–178.
120. Brysbaert, A.; Perdikatsis, V. Bronze Age painted plaster from the Greek mainland: A comparative study of its technology by means of XRD analysis and optical microscopy techniques. In *Proceedings of the 4th Symposium of the Hellenic Society for Archaeometry*; Facorellis, Y., Zacharias, N., Polikreti, K., Eds.; National Hellenic Research Foundation: Athens, Greece, 2008; pp. 421–429.
121. Brysbaert, A.; Vandenabeele, P. Bronze Age painted plaster in Mycenaean Greece: A pilot study on the testing and application of micro-Raman spectroscopy. *J. Raman Spectrosc.* **2004**, *35*, 686–693. [CrossRef]
122. Koren, Z. New chemical insights into the ancient molluskan purple dyeing process. In *Archaeological Chemistry VIII.*; Armitage, R., Burton, J.H., Eds.; ACS Symposium Series; American Chemical Society: Washington, DC, USA, 2013; pp. 43–67.
123. Bogensperger, I. Purple and its various kinds in documentary papyri. In *Textile Terminologies from the Orient to the Mediterranean and Europe, 1000 BC to 1000 AD.*; Gaspa, S., Michel, C., Nosch, M.-L., Eds.; Univeristy of Nebraska and Zea Books: Lincoln, NE, USA, 2017; pp. 235–249.
124. Westlake, P.; Siozos, P.; Philippidis, A.; Apostolaki, C.; Derham, B.; Terlixi, A.; Perdikatsis, V.; Jones, R.; Anglos, R. Studying pigments on painted plaster in Minoan, Roman and Early Byzantine Crete. A multi-analytical technique approach. *Anal. Bioanal. Chem.* **2012**, *402*, 1413–1432. [CrossRef]
125. Brysbaert, A. Lapis lazuli in an enigmatic 'purple' pigment from a thirteenth-century BC Greek wall painting. *Stud. Conserv.* **2006**, *51*, 252–266. [CrossRef]
126. Mantzouris, D.; Karapanagiotis, I. Armenian cochineal (*Porphyrophora hamelii*) and purpurin-rich madder in ancient polychromy. *Color. Technol.* **2015**, *131*, 370–373. [CrossRef]
127. Fostiridou, A.; Karapanagiotis, I.; Vivdenko, S.; Lampakis, D.; Mantzouris, D.; Achilara, L.; Manoudis, P. Identification of pigments in Hellenistic and Roman funeral figurines. *Archaeometry* **2016**, *58*, 453–464. [CrossRef]
128. Sasaki, Y.; Matsubara, J.; Sasaki, K. Red dye analysis for historical textiles with multiple spectroscopic methods: An example for Coptic textiles in the Nara Silk Road Exchange Museum. In *The Diversity of Dyes in History & Archaeology*; Kirby, J., Ed.; Archetype Publications: London, UK, 2017; pp. 107–115.
129. Salvant, J.; Williams, J.; Ganio, M.; Casadio, F.; Daher, C.; Sutherland, K.; Monico, L.; Vanmeert, F.; De Meyer, S.; Janssens, K.; et al. A Roman Egyptian painting workshop: Technical investigation of the portraits from Tebtunis, Egypt. *Archaeometry* **2018**, *60*, 815–833. [CrossRef]
130. Karadag, R.; Torgan, E. Advantages and importance of natural dyes in the restoration of textile cultural heritage. *Int. J. Conserv. Sci.* **2016**, *7*, 357–366.

131. Aceto, M.; Agostino, A.; Fenoglio, G.; Idone, A.; Crivello, F.; Griesser, M.; Kirchweger, F.; Uhlir, K.; Puyo, P.R. Analytical investigations on the Coronation Gospels manuscript. *Spectrochim. Acta A* **2017**, *171*, 213–221. [CrossRef] [PubMed]
132. Denoël, C.; Puyo, P.R.; Brunet, A.-M.; Siloe, N.P. Illuminating the Carolingian era: New discoveries as a result of scientific analyses. *Herit. Sci.* **2018**, *6*, 28. [CrossRef]
133. Melo, M.J.; Nabais, P.; Guimarães, M.; Araújo, R.; Castro, R.; Oliveira, M.C.; Whitworth, I. Organic dyes in illuminated manuscripts: A unique cultural and historic record. *Philos. Trans. A Math. Phys. Eng. Sci.* **2016**, *374*, 20160050. [CrossRef] [PubMed]
134. Padden, A.N.; Dillon, V.M.; John, P.; Edmonds, J.; Collins, M.D.; Alvarez, N. Clostridium used in mediaeval dyeing. *Nature* **1998**, *396*, 225. [CrossRef]
135. Koudous, I.; Kunz, W.; Strube, J. Panorama of sustainable solvents for green extraction processes. In *Green Extraction of Natural Products*; Chemat, F., Strube, J., Eds.; Wiley-VCH: Weinheim, Germany, 2015; pp. 173–236.
136. Hansen, C.M. *Hansen Solubility Parameters: A User's Handbook*, 2nd ed.; CRC Press: Boca Raton, FL, USA, 2007.

© 2019 by the author. Licensee MDPI, Basel, Switzerland. This article is an open access article distributed under the terms and conditions of the Creative Commons Attribution (CC BY) license (http://creativecommons.org/licenses/by/4.0/).

Article

Investigation of the Optical, Physical, and Chemical Interactions between Diammonium Hydrogen Phosphate (DAP) and Pigments

Xiao Ma [1,2,*,†], Hélène Pasco [3], Magdalena Balonis [1] and Ioanna Kakoulli [1,2,4]

1 Department of Materials Science and Engineering, University of California at Los Angeles, 410 Westwood Plaza, Los Angeles, CA 90095-1595, USA
2 Molecular and Nano Archaeology Laboratory, Henry Samueli School of Engineering and Applied Science, Engineering V-1230B, 410 Westwood Plaza, Los Angeles, CA 90095-1595, USA
3 Sorbonne Université, Faculté des Sciences et Ingénierie, UFR 926, F-75005 Paris, France
4 UCLA/Getty Conservation Program, A210 Fowler Building, University of California at Los Angeles, Los Angeles, CA 90095-1510, USA
* Correspondence: xiaomapurdue@gmail.com
† Current address: Scientific Research Department, National Gallery of Art, 2000B South Club Drive, Landover, MD 20785, USA.

Received: 19 May 2019; Accepted: 8 July 2019; Published: 11 July 2019

Abstract: This research investigates and evaluates the optical, physical, and chemical interactions between diammonium hydrogen phosphate (DAP) and seven pigments commonly encountered in archaeological and historic fresco and secco wall paintings and polychrome monuments. The pigments include cinnabar, French ochre, chalk, lapis lazuli, raw sienna, burnt umber, and red lead. The raw pigments were analyzed before and after the interaction with DAP, and the reaction products resulting from the contact of the pigments with the DAP solution were evaluated to obtain a comprehensive understanding of the effects of diammonium hydrogen phosphate on the color, morphology, and chemical composition of the pigments. The results indicated no significant change of the color or of the chemistry of cinnabar, French ochre, and lapis lazuli. Carbonate-containing pigments, such as chalk and calcium carbonate, were transformed into calcium phosphate, though without a significant change in color. Phase and strong color changes occurred only for the red lead pigment, associated with the transformation of red lead into hydroxypyromorphite. These data established the parameters and identified the risks of the direct application of DAP solutions on pigments. Further research will be undertaken to assess the potential use of DAP as a consolidant of wall paintings and other polychrome surfaces through testing on wall painting/polychromy mockups and on-site archaeological/historic painted surfaces.

Keywords: hydroxyapatite; diammonium hydrogen phosphate; pigment alteration; wall painting consolidation; cultural heritage

---

1. Introduction

Cultural heritage materials including wall paintings and other forms of polychromy and painted architectural surfaces were central to the culture of ancient people. These complex, heterogeneous, and multilayer systems are usually composed of the paint layer (a binary system of pigment(s) and a binding medium) and the substrate (a rock surface or plaster(s) layers) [1,2]. In ancient and historical times, two different techniques were predominantly employed for painting on walls: the fresco (from the Italian, meaning 'fresh') and the secco (from the Italian, meaning 'dry') techniques. In the fresco technique, a small number of pigment powders—compatible with fresco application—are mixed with water and applied on a freshly laid or moist calcium hydroxide/lime ($Ca(OH)_2$)-rich plaster layer.

Setting (hardening) of the lime plaster involves a chemical reaction between $Ca(OH)_2$ and carbon dioxide (CO_2) present in the atmosphere to form a calcium carbonate lattice within which the pigments are 'fixed' and become an integral part of the wall. These chemical reactions help produce durable wall paintings. Owing to the high alkalinity of the lime and the exothermic reaction associated with the setting of the lime, only a small number of pigments are compatible with the fresco technique, and therefore ancient fresco paintings contain a limited palette of colors. The secco technique, on the other hand, involves no chemical reaction for the fixation of the pigments. In a secco application, the pigments are mixed with any film-forming binding medium such as egg, siccative oil, gum, and others [1] and applied on any type of finished plaster layer including gypsum, earth-based plasters, and lime plaster layers (fully carbonated).

As wall paintings constitute an integral part of the architectural ensemble where they are found, they are inevitably exposed to an open system of environmentally-linked events. As a result, the physical and chemical attributes of the system and individual constituent materials (i.e., plaster(s), pigment(s), binding media) can be impacted by fluctuations in the temperature, relative humidity, and presence of salts, microorganisms, and pollution in the surrounding environment. These conditions can compromise the stability of the system, resulting in the delamination of the plaster layers, staining, flaking, and losses of the paint and plaster layers, and powdering [1,3]. For archaeological wall paintings in particular, the risks for their preservation are even greater as the sudden change in the environmental conditions at the time of the excavation—mainly of the temperature, relative humidity, and light—can cause irreversible damage and degradation [1].

Over the course of the past decades, extensive studies have been carried out using a variety of consolidation treatments to improve the condition and re-establish the lost cohesion of decorated architectural surfaces and wall paintings [4–12]. These studies critically indicate that choosing a proper consolidating agent for these porous materials, especially those found in situ, is challenging. An appropriate consolidant for wall paintings needs to re-establish cohesion of the powdery layers at the surface and subsurface levels and provide mechanical strength and abrasion resistance, without causing any discernible color alteration [13].

Recent studies [14–25] have demonstrated a considerable potential for improving consolidation methods of degraded calcium-carbonate matrices of a technique consisting in bio-mimicking the growth of hydroxyapatite (HAP, with the formula $Ca_5(PO_4)_3(OH)$ but usually written as $Ca_{10}(PO_4)_6(OH)_2$ to denote that the crystal unit cell comprises two formula units), the main mineralogical component of teeth and bones [15,26,27]. HAP was proved to be effective in binding grain boundaries and improving the mechanical properties of limestone, such as tensile strength, ultrasonic pulse velocity, resistance to abrasion [16,19,20].

HAP is formed in situ by activating reactions between Ca in calcium carbonate ($CaCO_3$)-rich layers and ammonium phosphate precursors. The theoretical chemical pathway of HAP formation using diammonium hydrogen phosphate (DAP) as the precursor is presented below (Reaction 1) [28]. The resulting hydroxyapatite network is expected to improve the cohesion between loose particles at the surface and subsurface of a wall painting [21,29].

$$10\,CaCO_3 + 6\,(NH_4)_2HPO_4 \rightarrow Ca_{10}(PO_4)_6(OH)_2 + 10\,CO_2\uparrow + 12\,NH_3\uparrow + 8\,H_2O$$

Reaction 1. Theoretical pathway of the formation of hydroxyapatite (HAP) using diammonium hydrogen phosphate as a precursor.

The superior qualities of HAP as a consolidating agent for calcium carbonate matrices lie in the fact that it has a much lower solubility ($K_{sp} = 1.6 \times 10^{-117}$ at 25 °C [30]) than calcite ($K_{sp} = 3.4 \times 10^{-9}$ at 25 °C [31]). The lattice parameters of hydroxyapatite and calcite are relatively close, respectively, a = b = 9.43 Å and c = 6.88 Å for HAP [32], and a = b = 9.96 Å and c = 17.07 Å for calcite, considering two molecules per unit cell [33]. This indicates compatibility in the nucleation of the phosphate layer onto the surface of carbonate stones and strong bonding of the newly formed layer onto the substrate [21]. The other advantage is that hydroxyapatite is the least soluble and the most stable calcium phosphate phase in aqueous solutions at pH values higher than 4.2 [34,35]. Also, it has a dissolution rate about 4–5 orders of magnitude lower than that of calcite: $R_{diss, HAP} = 1 \times 10^{-14}$ moles·cm^{-2}·s^{-1}, and $R_{diss, calcite} = 2 \times 10^{-10}$ moles·cm^{-2}·s^{-1} at pH = 5.6; $R_{diss, HAP} = 3.7 \times 10^{-14}$ moles·cm^{-2}·s^{-1}, and $R_{diss, calcite} = 5.4 \times 10^{-9}$ moles·cm^{-2}·s^{-1} at pH = 4 [36,37]. It is therefore more stable in a range of pH and it is expected to provide additional protection against acid dissolution. In addition, the precursor ammonium phosphate is non-toxic, and a good penetration depth could be obtained in the consolidation treatment [21].

However, despite successful results for the consolidation of decohesive plaster layers as substrates/surface layers of fresco wall paintings [21] and regardless of the fact that some other consolidants, such as a nano calcium hydroxide suspension, have been tested on fresco wall painting mock-ups [38], this DAP-based method has not yet been tested on archaeological wall paintings nor has any thorough assessment been performed on pigments. This research aims to fill this gap of knowledge. Following from our previous research on the application of DAP for the consolidation of fresco plaster layers, here, as a first step, we systematically investigate and evaluate in laboratory-controlled conditions the optical, physical, and chemical effects of the ammonium phosphate precursor of HAP on selected pigments (mainly those compatible with fresco application). The aim is to have a fundamental understanding of the effects (mainly on color change and phase transformations) of this inorganic 'consolidant' precursor on pigments, prior to any testing of the consolidating effect on the paint layer (both fresco and secco) in wall painting mockups and archaeological/historic wall paintings and other polychrome monuments. More specifically, this research investigates the interactions between DAP solutions and seven pigments commonly found in wall paintings and other polychrome surfaces and focuses on answering the following questions:

- Can DAP be considered as a potential precursor for a surface treatment of wall paintings (mainly fresco) and other monumental painted architectural surfaces?
- Is there any obvious color change of pigments after contact with DAP solutions?
- Are any chemical or morphological changes occurring?
- What are the possible mechanisms leading to color change and/or other forms of physical and chemical phase transformation?

2. Materials and Methods

2.1. Materials

A 1M DAP solution was prepared by adding the appropriate amount of DAP (Fisher Scientific, Hampton, NH, USA, purity: 99 + %, used as received) to deionized (DI) water. To study the chemical reaction between the DAP solution and the pigments, seven commercial inorganic pigments purchased from Kremer Pigments Inc. were tested, including six pigments commonly used for fresco application, such as cinnabar (Kremer No.10620), lapis lazuli (Kremer No. 10562), white chalk (Kremer No. 58000), French ochre (Kremer No. 40090), burnt umber (Kremer No. 40710), raw sienna (Kremer No. 40400), and one pigment, red lead (Kremer No. 42500), frequently encountered in secco paintings.

2.2. Characterization of Pigment–DAP Interaction

For the experimental application, 10 g of each pigment were dispersed in 100 mL of 1M DAP solution or in 100 mL of DI water, which was used as a control sample, and the dispersions were subsequently sealed in a glass bottle. The bottles were kept in the dark to avoid any photochemical reaction. The room temperature (T) was maintained at ~22 °C, and the relative humidity (RH) at ~50%. Using an Oakton EcoTestr® pH2 Waterproof pH Tester (standard error: ± 0.1), pH measurements were taken of the 1M DAP solution and of each pigment dispersion on day 0, a few minutes after the pigments were dispersed in the DAP solution, and subsequently at regular intervals: every 24 h between day 1 and 7 and then on day 14, 21, and 28. Monitoring of color/phase change of those pigments was carried out in the first 28 days. Red lead and chalk, however, showed phase and color change after two months of immersion in the DAP solution. For these two pigments, further monitoring will be required.

Prior to subjecting the samples to the measurements, all the powders were rinsed using DI water and left to dry overnight on filter paper. The powders were analyzed every 24 h between day 1 and 7, and then on day 14, 21, and 28, following the dispersion into 1M DAP solution. The samples listed were named using the abbreviation of the pigment name and the immersion time. For instance, CIN-raw stands for cinnabar pigment prior to the analysis, whereas CIN-d28 stands for cinnabar precipitate collected 28 days after dispersion in 1M DAP solution.

All powders were first examined using a Keyence VHX-1000 Digital Optical Microscope, using a magnification between 20× and 200×.

XRD measurements on the pigment powders were performed using a Bruker D8 diffractometer with the following measurement parameters: Cu-Kα radiation, $\lambda = 1.5404$ Å, voltage 40 kV, beam current 40 mA, and a 2–80° 2θ exploration range with a step size of 0.014° 2θ. The mineral phases were identified by using the ICDD database (International Center for Diffraction Data, Newtown Square, PA, USA).

TGA analysis was performed on selected pigment powders using a Perkin Elmer Pyris Diamond TG/DTA (Thermogravimetric/Differential Thermal Analyzer). The temperatures were scanned in the range between 40 °C to 900 °C, at the heating rate of 20 °C/min, in a flowing Ar atmosphere.

Microstructural and elemental analyses of the powders were performed on a FEI Nova NanoSEM™ 230 scanning electron microscope (SEM) with field emission gun (FEG) and variable pressure (VP) capabilities, equipped with a Thermo Scientific™ NORAN™ System 7 X-ray energy dispersive spectrometer (EDS). Gold (Au) coating to improve the electrical conductivity was applied using a Hummer® 6.2 sputtering system (Anatech Ltd., Battle Creek, MI, USA). Secondary electron (SE) imaging was performed in vacuum using the Everhart–Thornley detector (ETD). The elemental composition of single spots and area elemental maps were acquired using EDS.

FTIR spectroscopy was performed on a JASCO FT/IR-420 Fourier-Transform Infrared Spectrometer using the KBr pellet method. Pigment powders were ground and dispersed in a KBr matrix at a concentration around 0.5 wt % and then pressed into a pellet. All spectra were collected at 64 scans with a spectral resolution of 4 cm^{-1}, from 4000 to 400 cm^{-1}. The spectra were matched against the spectral database of the Infrared and Raman Users Group (IRUG, Philadelphia, PA, USA) and published literature data.

FORS (Fiber Optic Reflectance Spectroscopy) was conducted using an Ocean Optics USB 2000+ fiber optical spectrophotometer and the FieldSpec3® Spectroradiometer (Analytical Spectral Devices Inc., Boulder, CO, USA). The spectro-colorimetric measurements allowed for the quantification of incident and reflected radiation intensities, which roughly equal human color perception. During the measurement, a white diffuse reference standard was measured every 30 min. Color values were recorded in the L*a*b* color space defined in 1976 by CIE (Commission Internationale de l'Eclairage, Vienna, Austria) [39]. Changes in color/color difference (ΔE^*) were calculated with the following formula (Equation (1)) as recommended by the CIE:

$$\Delta E^* = \left[(\Delta L^*)^2 + (\Delta a^*)^2 + (\Delta b^*)^2\right]^{1/2} \quad (1)$$

where ΔL^*, Δa^*, and Δb^* are the differences in L*, a*, and b* values before and after immersion in the DAP solution. ΔL^* describes the change in luminance, Δa^* the change in red/green components, and Δb^* the change in yellow/blue components. While generally $\Delta E^* \leq 2$ is widely acceptable as the value detectable by the human eye [40], a color difference of $\Delta E^* \leq 5$ has been established as the threshold in the field of cultural heritage to evaluate color change after a conservation intervention such as consolidation treatment [16,41–48].

3. Results and Discussion

After 28 days of immersion of the pigments in the DAP solution, the pigments were assessed on the basis of phase transformations and color change. Three main groups were revealed: (1) pigments that showed no chemical and/or optical interaction (no phase or significant color change) with DAP (i.e., cinnabar, French ochre, and lapis lazuli); (2) pigments that showed phase transformation without significant color change (i.e., chalk, raw sienna, and burnt umber); and (3) pigment with strong phase and color change (i.e., red lead).

3.1. Cinnabar, French Ochre, Lapis Lazuli

The calculated ΔE* values for cinnabar, French Ochre, and lapis lazuli pigment particles before and after the 28 days of immersion in DAP were determined to be 3.5, 3.4, and 4.2, respectively (Table 1). Though these values are above the threshold of color change detected by the human eye [40], they are still below the established value (ΔE* ≤ 5) accepted for cultural heritage consolidation treatments [16,41–48].

3.1.1. Cinnabar

Cinnabar has a deep red color with angular particles of various sizes up to 100 μm (Figure 1a–d). Its identification was based on XRD analysis (Figure 1e) and FORS (Figure 1f) which showed consistently the characteristic sigmoid-shaped spectrum with an inflection point (maximum at its first derivative, Figure 1g) at ~614 nm corresponding to the bandgap of cinnabar [49]. No obvious change in shape or size of the cinnabar pigment particles (inferred by SEM–EDS analysis) was observed (Figure 1a–d).

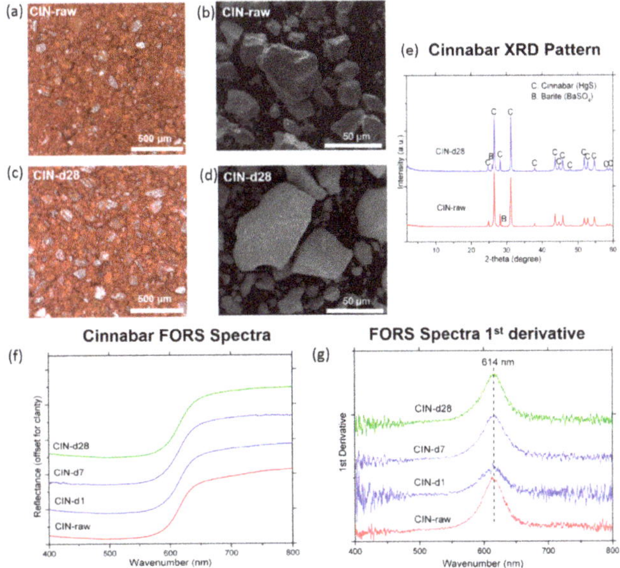

Figure 1. (**a**) Photomicrograph of the cinnabar (CIN)-raw sample; (**b**) secondary electron (SE) micrographs of the CIN-raw sample; (**c**) DM (Digital Micrograph) photomicrograph of the sample CIN-d28; (**d**) SE micrographs of the sample CIN-d28; (**e**) XRD pattern of CIN-raw and CIN-d28 ; (**f**) FORS spectra of cinnabar: CIN-raw, CIN-d1, CIN-d7, CIN-d28; (**g**) first derivative of the FORS spectra in (**f**). The intensity values of each XRD pattern, FORS spectra, and its first derivative plots were normalized and offset for comparison purposes.

3.1.2. French Ochre

Based on the XRD, FTIR, and FORS analysis (Figure 2), no detectable phase transformations were observed in the pigment particles subjected to the immersion in DAP solution (FRE-d28) as compared to the untreated powders (FRE-raw).

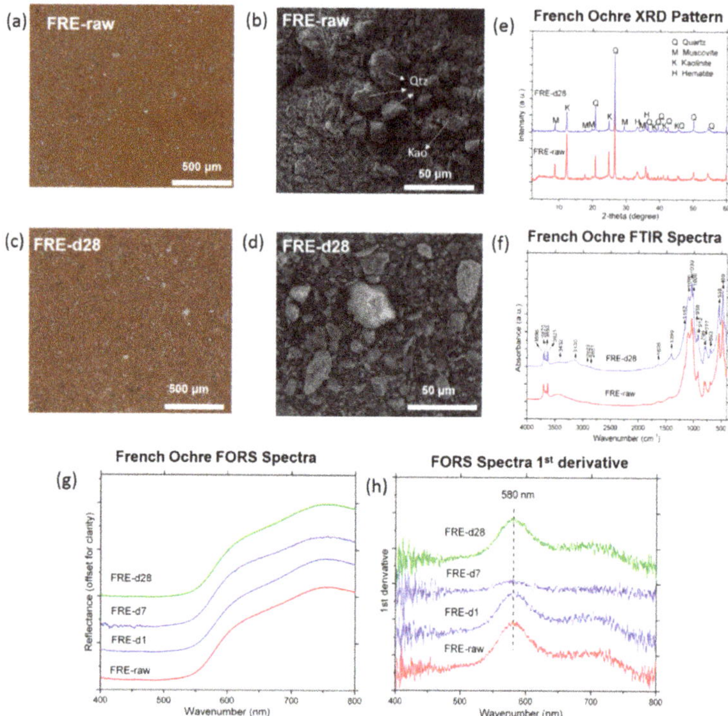

Figure 2. (**a**) Photomicrograph of the French ochre (FRE)-raw sample; (**b**) micrographs of the FRE-raw. Qtz stands for quartz and Kao stands for kaolinite; (**c**) photomicrograph of the sample FRE-d28; (**d**) micrographs of the FRE-d28; (**e**) XRD pattern of the samples FRE-raw and FRE-d28; (**f**) FTIR spectra of the FRE-raw, FRE-d1, and FRE-d28 samples; (**g**) FORS spectra of the French ochre samples FRE-raw, FRE-d1, FRE-d7, FRE-d28; (**h**) first derivative of the FORS spectra in (**g**). The intensity values of each XRD pattern, FORS spectra, and its first derivative plots were normalized and offset for comparison purposes.

XRD analysis (Figure 2e) revealed the presence of quartz, muscovite, kaolinite, and hematite. FTIR spectroscopy (Figure 2f) further corroborated the results. Kaolinite ($Al_2Si_2O_5(OH)_4$) disclosed vibrational bands at 3696, 3670, 3653 cm^{-1} (surface hydroxyl groups stretching vibration), 3621 cm^{-1} (inner hydroxyl groups stretching vibration), 1030 cm^{-1} (Si–O–Si stretching vibration), 1008 cm^{-1} (Si–O–Al stretching vibration), 938 and 912 cm^{-1} (Al–OH deformation vibration), 693 cm^{-1} (Si–O–Si symmetrical bending vibration), 538 cm^{-1} (Si–O–Al stretching vibration), and 469 cm^{-1} (Si–O–Si asymmetrical bending vibration). Quartz (SiO_2) exhibited vibrational bands at 1162 cm^{-1}(Si–O–Si rocking vibration), 1096 cm^{-1} (Si–O–Si asymmetrical stretching vibration), doublets at 777 and 799 cm^{-1}(Si–O–Si symmetrical stretching vibration) and at 693 cm^{-1} and 469 cm^{-1} (Si–O–Si symmetrical and asymmetrical bending vibration, overlapping with kaolinite). It should be noted that the Fe–O vibration of hematite which yields vibrational bands at 538 cm^{-1} and 469 cm^{-1}, were overlapping with the Si–O–Al stretching vibration of kaolinite and the Si–O–Si bending vibration of kaolinite/quartz, respectively [50–56]. The bands at 3432 cm^{-1} and 1626 cm^{-1} corresponded to the O–H stretching and O–H bending of surface-absorbed water. The bands at 3130 cm^{-1} and 1399 cm^{-1} were present probably due to the ν_3 stretching vibration and the ν_4 bending vibration of surface-adsorbed NH_4^+ [57–59].

FORS showed the characteristic inflection point (maximum at its first derivative (Figure 2g) of hematite at around 580 nm (Figure 2h). The broad absorption at ~875 nm also characteristic of hematite, could not be seen in this spectrum (cut off at 800 nm). These were attributed to ligand-to-metal charge transfer transitions in hematite [60].

3.1.3. Lapis Lazuli

The lapis lazuli pigment powder analyzed for this research (sample LAP-raw) was found to contain various minerals including lazurite, wollastonite, cancrinite, and feldspars (Figure 3), with particle sizes ranging from 2 to 50 μm.

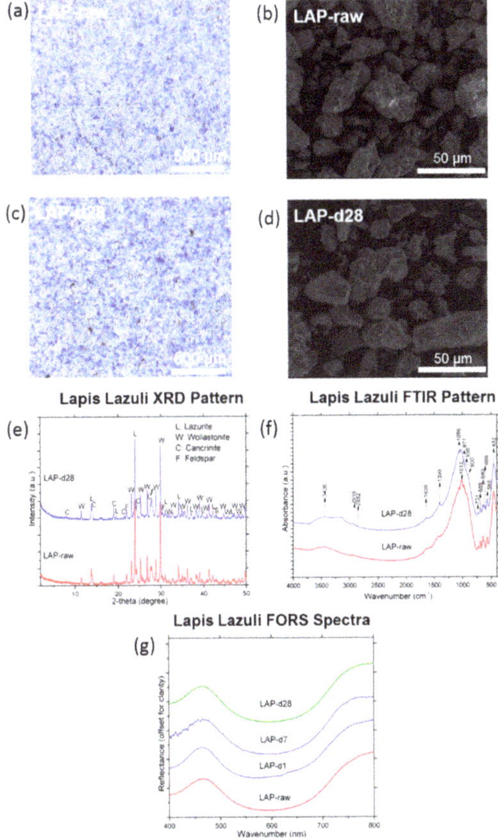

Figure 3. (**a–b**) Micrographs of the lapis lazuli pigment (LAP)-raw sample; (**c–d**) micrographs of the sample LAP-d28; (**e**) XRD pattern of the samples LAP-raw, LAP-d28; (**f**) FTIR spectra of the LAP-raw and LAP-d28 samples; (**g**) FORS spectra of the samples LAP-raw, LAP-d1, LAP-d7, LAP-d28. The intensity values of each XRD pattern and FORS spectra were normalized and offset for comparison purposes.

XRD (Figure 3e) and FTIR analysis (Figure 3f) showed no detectable phase changes resulting from the immersion in the DAP solution. The FTIR spectra showed bands in the 1100–900 cm^{-1} region that could be assigned to overlapping of Al, Si–O_4 tetrahedra asymmetric stretching vibration of lazurite and O–Si–O asymmetric stretching vibration of wollastonite, as well as bands in the 700–600 cm^{-1} region, which could be linked to an overlapping of Al, Si–O_4 tetrahedra symmetric stretching vibration of lazurite and O–Si–O symmetric stretching vibration of wollastonite [61]. The band at 568 cm^{-1} and the band at 452 cm^{-1} represent the terminal $^-$O–Si–O$^-$ bonds bending vibration and Si–O–Si bending vibration, respectively [62–64].

The visible spectrum of the lapis lazuli was dominated by an absorption band around 600 nm, corresponding to the electronic transitions for S_3^- (see Figure 3g).

3.2. *Chalk, Sienna, Burnt Umber*

The calculated ΔE* values for the chalk, raw sienna, and burnt umber pigment particles before and after 28 days of immersion in DAP were 4.9, 2.6, and 1.7, respectively (Table 1). Though the value of chalk was above the threshold of color change detected by the human eye [40], it was lower than the established value (ΔE* ≤ 5) accepted for consolidation applications in cultural heritage [16,41–48]. The color change of burnt umber pigment remained below the detection limit of human eye.

3.2.1. Chalk

The pigment (CHA-raw) used for this research was a fine powder consisting of natural white calcium carbonate ($CaCO_3$) (Figure 4a–c) and was prepared from pure microcrystalline chalk with particle sizes less than 5 µm. After 28 days of immersion in 1M DAP solution, the chalk ($CaCO_3$) particles showed evident transformation into HAP ($Ca_{10}(PO_4)_6(OH)_2$) and octacalcium phosphate (OCP, $Ca_8H_2(PO_4)_6{\cdot}5H_2O$). The habit of the original calcium carbonate crystals had changed into "plate-like" crystals (Figure 4d–f). EDS mapping of the sample CHA-d28 showed that the major phases detected consisted of Ca, O, and P elements. This transformation continued even after a period of two months with more calcium carbonate crystals been transformed into calcium phosphate.

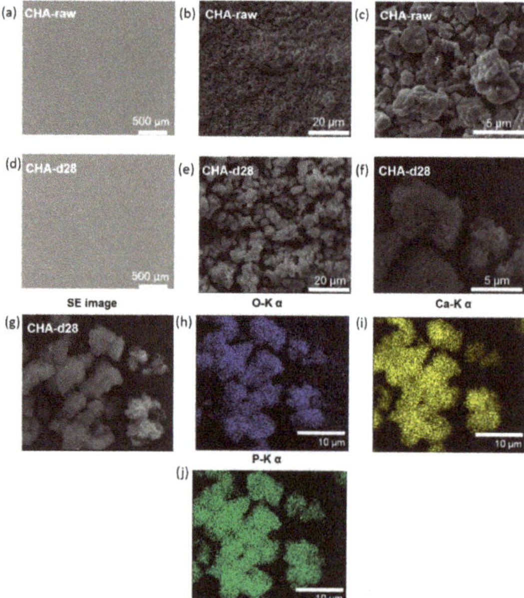

Figure 4. (**a**) Photomicrograph of the chalk (CHA)-raw sample; (**b–c**) micrographs of the CHA-raw sample; (**d**) photomicrograph of the sample CHA-d28; (**e–f**) micrographs of the CHA-d28 sample; (**g–j**) SE image and elemental mapping of the sample CHA-d28.

XRD analysis (Figure 5a) showed that the raw chalk pigment solely consisted of calcium carbonate (or calcite), while after 1 day and 28 days of immersion in DAP, some unreacted calcite, hydroxyapatite, and OCP were found to coexist. The consumption of calcite was not complete. FTIR analysis (Figure 5b) further confirmed the XRD results [65]. $CaCO_3$ yielded bands at 712 cm^{-1} (ν_4 in-plane bending vibration of CO_3^{2-}), 873 cm^{-1} (ν_2 out-of-plane bending vibration of CO_3^{2-}), 1420 cm^{-1} (ν_3 asymmetric stretching vibration of CO_3^{2-}), and combination bands at 2513 cm^{-1} and 1798 cm^{-1}. In the sample CHA-d28, bands appeared at 468 cm^{-1} (ν_2 bending mode of O–P–O bond), 562 cm^{-1} (ν_4 bending mode of O–P–O bond), 601 cm^{-1} (ν_4 bending mode of O–P–O bond), 957 cm^{-1} (ν_1 symmetric stretching mode of P–O bond), and 1033 cm^{-1} (ν_3 asymmetric stretching mode of P–O). These are the vibration modes associated with the phosphate group present in HAP and OCP. [66]. The fraction of unreacted calcite was further estimated through TGA analysis, with a weight loss recorded between 600 and 860 °C linked to the decomposition of calcite. Weight losses of 43.2 wt %, 16.9 wt %, and 11.55 wt % were observed for the samples CHA-raw, CHA-d1, and CHA-d28, respectively (Figure 5c–e). This roughly corresponded to calcite fractions of 98.2 wt %, 38.4 wt %, and 26.25 wt %, respectively. Most of the calcite was consumed on the first day of reaction with DAP.

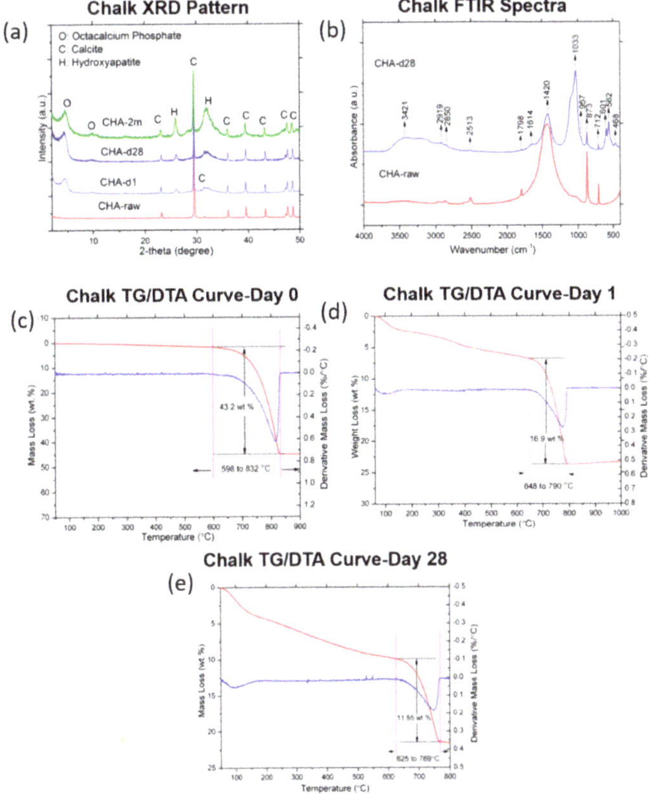

Figure 5. (**a**) XRD pattern of the samples CHA-raw, CHA-d1, CHA-d28, and CHA-2m; (**b**) FTIR spectra of the CHA-raw, CHA-d1, and CHA-d28 samples; (**c**–**e**) TGA of the sample CHA-raw, CHA-d1, and CHA-d28. The intensity of each XRD pattern was normalized and offset for comparison purposes.

OCP is commonly found to be present as an intermediate phase in the conversion process from amorphous calcium phosphates (ACP) to HAP (hydroxyapatite) [67]. This transition could explain the co-existence of HAP and OCP within the mixtures. While the formation of these phases and the kinetics of transformation largely depend on the reaction conditions such as pH and presence of foreign ions, ultimately—i.e., at thermodynamic equilibrium—they are all expected to transform to HAP, which is thermodynamically the most stable phase [67,68].

3.2.2. Raw Sienna

Microscopic examination of the sample SIE-raw (Figure 6a–b) showed that the pigment consists of different particles sizes ranging from sub-micron to 50 μm. XRD analysis of the SIE-raw and SIE-d28 (Figure 6e) samples indicated that raw sienna consisted of goethite (α-FeOOH), gypsum ($CaSO_4{\bullet}2H_2O$), calcite, quartz, and montmorillonite/clay. In the sample SIE-d1, gypsum was absent from the XRD pattern, whereas calcite could still be detected. This was due to the dissolution of gypsum into the DAP solution, while the transformation of calcite into HAP and/or OCP was not complete. For SIE-d28, however, the calcite peaks were absent, indicating that the amount of remaining calcite was probably below the detection limit (~2–3 wt %).

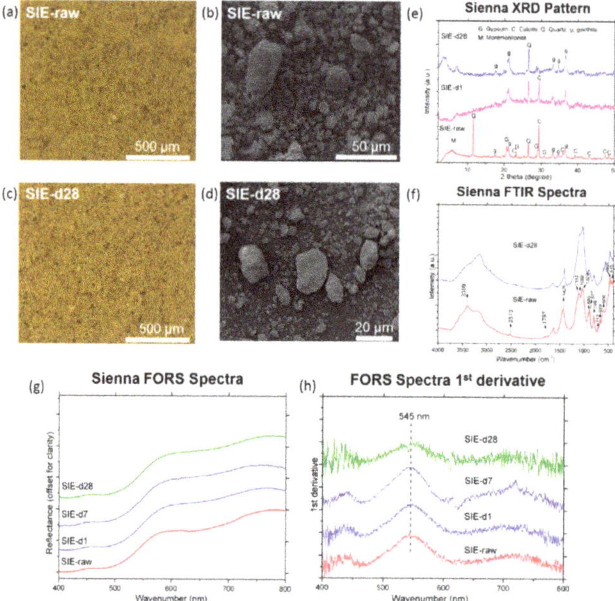

Figure 6. (**a**–**b**) Micrographs of the raw sienna pigment (SIE)-raw sample; (**c**–**d**) micrographs of the sample SIE-d28; (**e**) XRD pattern of the sample SIE-raw, SIE-d1, SIE-d28; (**f**) FTIR spectra of the SIE-raw and SIE-d28 samples; (**g**) FORS spectra of the samples SIE-raw, SIE-d1, SIE-d7, SIE-d28; (**h**) first derivative of the FORS spectra in (**g**). The intensity values of each XRD pattern, FORS spectra, and its first derivative plots were normalized and offset for comparison purposes.

After 28 days of immersion in DAP solution, newly formed phosphate-bearing phases were identified through the FTIR (Figure 6f) and SEM–EDS techniques. In the spectra of SIE-raw, gypsum yielded bands at 3541 and 3399 cm^{-1} (ν_3 asymmetric stretching vibration and ν_1 symmetric stretching vibration of water molecule, respectively), 1685 and 1621 cm^{-1} (O–H bending vibration), 1112 cm^{-1} (ν_3 asymmetric stretching vibration of SO_4^{2-} tetrahedron), 669 cm^{-1} (ν_4 asymmetric bending vibration of SO_4^{2-} tetrahedron), and 600 cm^{-1} (ν_4 asymmetric bending vibration of SO_4^{2-}tetrahedron). Quartz displayed characteristic bands at 1096 cm^{-1} (Si–O–Si asymmetrical stretching vibration), 798 cm^{-1} (Si–O–Si symmetrical stretching vibration), and 469 cm^{-1} (Si–O–Si asymmetrical bending vibration), while silicate clay had bands at 3621 cm^{-1} (O–H stretching vibration of structural hydroxyl group) and 1029 cm^{-1} (Si–O–Si asymmetrical stretching vibration) and shared (with quartz) bands at 798 cm^{-1}

and 469 cm^{-1}. $CaCO_3$ produced bands at 712 cm^{-1}, 877 cm^{-1}, 1425 cm^{-1}, 1797 cm^{-1}, and 2513 cm^{-1}. Goethite gave a broad band centered at 3141 cm^{-1} (broad, ν_2 stretching vibration of O–H) and bands at 899 cm^{-1} (δO–H bending vibration) and 798 cm^{-1} (γO–H bending vibration, overlapping with quartz and silicate clay) [69–76]. After 28 days of reaction with DAP, in the FTIR spectrum of SIE-d28, the calcite and gypsum bands disappeared with the appearance of the bands at 468 cm^{-1}, 562 cm^{-1}, 601 cm^{-1}, and 1034 cm^{-1}, corresponding to the vibration mode of the newly formed phosphate group. The bands of goethite, quartz, and silicate clay remained unchanged. The appearance of the phosphate group and the disappearance of gypsum and calcite in the FTIR spectrum further indicated that calcite and gypsum were converted into calcium phosphates.

Reflectance spectra of the yellow iron hydroxide pigment (goethite) showed the characteristic inflection point (maximum at its first derivative, Figure 6g) at around 545 nm and absorptions at 640 and ~900 nm (the latter was not visible in the spectrum) (Figure 6h).

3.2.3. Burnt Umber

The burnt umber pigment powder analyzed for this research (sample BUR-raw) contained hematite and manganese oxide (inferred by EDS point analysis) and minor phases of calcite and quartz (Figure 7). It exhibited particle sizes ranging from sub-micron to 20 μm (Figure 7b).

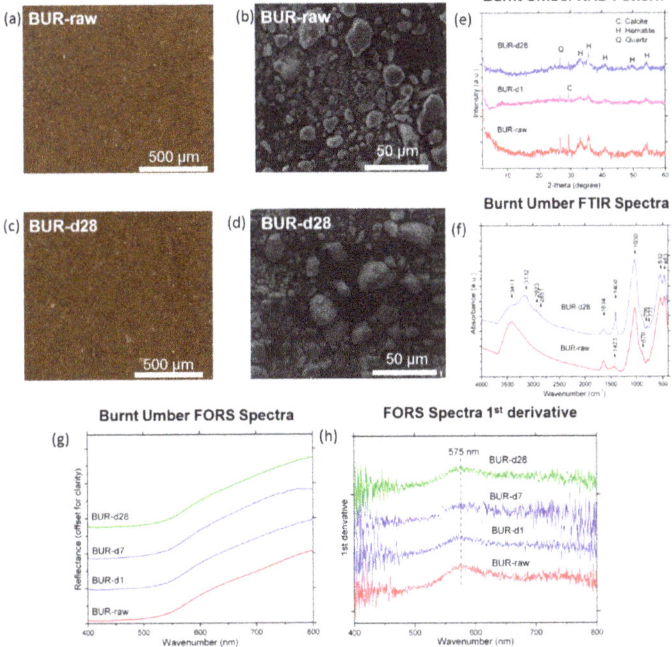

Figure 7. (**a–b**) Micrographs of the burnt umber pigment (BUR)-raw sample; (**c–d**) micrographs of the sample BUR-d28; (**e**) XRD pattern of the sample BUR-raw, BUR-d1, BUR-d28; (**f**) FTIR spectra of the BUR-raw and BUR-d28 samples; (**g**) FORS spectra of the samples BUR-raw, BUR-d1, BUR-d7, BUR-d28; (**h**) first derivative of the FORS spectra in (**g**). The intensity values of each XRD pattern, FORS spectra, and its first derivative plots were normalized and offset for comparison purposes.

After 28 days of immersion in DAP solution, the formation of calcium phosphates was first estimated from the microstructural changes revealed by SEM–EDS analysis. XRD analysis (Figure 7e) of the sample BUR-raw showed that the raw burnt umber pigment consisted of hematite, quartz, and calcite; the latter was no longer detectable after 28 days in DAP solution (sample BUR-d28). FTIR analysis (Figure 7f) showed bands at 1423 and 879 cm^{-1}, corresponding to the vibration of $CaCO_3$, and bands at 1030, 778, 797, and 463 cm^{-1} corresponding to the vibration of the silicate (possibly silicate clay and SiO_2) group. The bands at 532 and 463 cm^{-1} were indicative of the Fe–O vibration produced by hematite. After 28 days, no calcite could be detected by FTIR.

The FORS spectra of burned umber (Figure 7g) showed the same features as those collected for French ochre (Figure 7g–h), since the main component of both pigments is hematite.

3.3. *Red Lead*

The red lead pigment powder analyzed in this study was found to be pure, consisting of minium (Pb_3O_4) with small and irregular particles (Figure 8a–b) ranging in size from 2 μm to 20 μm.

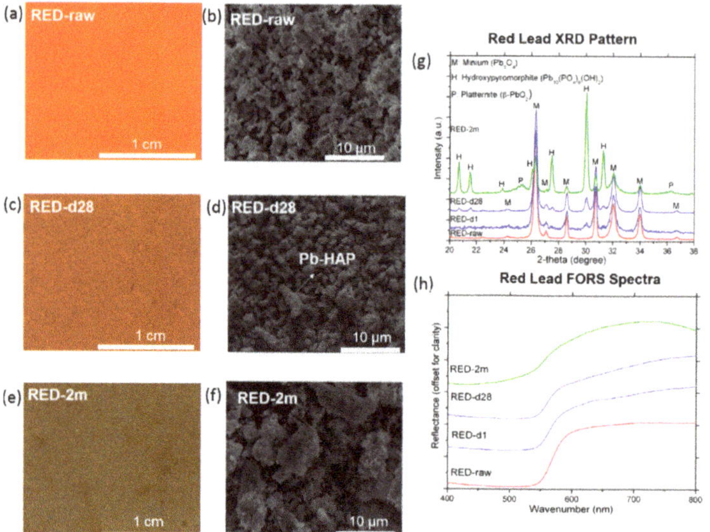

Figure 8. (**a**) Photomicrograph of the red lead (RED)-raw sample; (**b**) micrographs of the RED-raw; (**c**) photomicrograph of the sample RED-d28; (**d**) micrographs of the sample RED-d28, the elongated particle was identified as lead hydroxyapatite (Pb–HAP) by EDS point analysis; (**e**) photomicrograph of the sample RED-2m; (**f**) micrographs of the sample RED-2m; (**g**) XRD pattern of the samples RED-raw to RED-2m between 2θ of 20–38°; (**h**) FORS spectra of the samples RED-raw, RED-d1, RED-d7, RED-d28. The intensity values of each XRD pattern and FORS spectra were normalized and offset for comparison purposes.

After dispersing pigment particles in 1M DAP for 28 days, part of the minium pigment was found to be converted into hydroxypyromorphite (also known as lead hydroxyapatite, $Pb_{10}(PO_4)_6(OH)_2$, JCPDS PDF No. 01-087-2477). The color of the pigment changed from orange red to brownish red after 28 days (Figure 8c). After two months, the color was further altered to dark brown (Figure 8e). The calculated ΔE* value for the red lead pigment particles before and after the 28 days of immersion in DAP was found to be 30.6 (Table 1). This color change is significant and far beyond the threshold accepted in the field of conservation treatment (ΔE* ≤ 5).

Microscopic observations of the sample RED-d28 showed that most particles remained the same, while some new elongated crystals could be detected (Figure 8d). EDS analysis on point 1 (see arrow in Figure 8d) confirmed the presence of Pb (24.65 at %), P (13.73 at %), and O (61.62 at %). The Pb/P/O atomic ratio was close to 5:3:13, indicating the presence of hydroxypyromorphite. After two months, a significant amount of the original pigment particles was transformed into hydroxypyromorphite (Figure 8f), which are believed to be responsible for the color change from originally red to brown.

XRD analysis (Figure 8g) indicated that the raw red lead pigment (sample RED-raw) solely consisted of minium (JCPDS PDF No. 41-1493). The formation of hydroxypyromorphite (JCPDS PDF No. 8–259) was observed to begin only one day after dispersing the pigment in 1M DAP solution (sample RED-d1). After two months, the lead hydroxyapatite became a dominant phase and was identified along with the precipitates of platternite (β-PbO_2) and unreacted residual minium (RED-2m in Figure 8g). While phase transformations between the first day of reaction and after 28 days appeared similar, a more significant phase development was observed over a longer period (two months).

A similar dissolution–precipitation mechanism was reported elsewhere [77,78]. The dissolution reaction of Pb_3O_4 begins to occur at the surface of Pb_3O_4 (Reaction 2). Pb_3O_4 first releases Pb^{2+} species from a tetrahedrally coordinated Pb_3(II,IV)O_4 site through a ligand substitution, leaving unstable octahedral PbO_2 fragments (octahedral arrangement hosting Pb^{4+} ions in the crystalline structure of Pb_3O_4) in the solid. The precipitation of lead hydroxyapatite ($Pb_{10}(PO_4)_6(OH)_2$) observed after the immersion of Pb_3O_4 in DAP suggests a reaction between the supersaturated Pb^{2+} ions released during Pb_3O_4 dissolution and the phosphate ($PO_4{}^{3-}$) ions delivered through the DAP solution.

$$Pb_3O_4 + 2H_2O_{(aq)} \rightarrow 2\,Pb^{2+}_{(aq)} + PbO_{2(s)} + 4OH^-_{(aq)}$$

Reaction 2. Dissolution reaction of Pb_3O_4 in aqueous solution.

Following this step, two processes occur simultaneously:

(1) The unstable PbO_2 fragments formed from the dissolution of Pb_3O_4 are reduced to Pb^{2+}, as suggested by the Reaction 4.

$$PbO_{2(unstable)} + 2e^- + 2H_2O_{(aq)} \rightarrow Pb^{2+}_{(aq)} + 4OH^-_{(aq)}$$

Reaction 3. Reduction of unstable PbO_2 fragment into Pb^{2+}.

(2) The nucleation of a newly stable β-PbO_2 from Pb^{2+} ions, as suggested by the Reaction 5:

$$Pb^{2+}_{(aq)} + 4OH^-_{(aq)} \rightarrow PbO_{2(stable)} + 2e^- + 2H_2O_{(aq)}$$

Reaction 4. Oxidation of Pb^{2+} to stable β-PbO_2 (platternite).

Since both Pb_3O_4 and β-PbO_2 are semiconductors, the electrons can transfer between the solid phases. The driving force for the process described is provided by the decrease in both surface and lattice free energy, which results from the dissolution of the octahedral fragment of PbO_2 (labelled as $PbO_{2(unstable)}$ above) in Pb_3O_4 and the precipitation of β-PbO_2 [77].

During the dissolution reactions (Reaction 2 to Reaction 4) that occur on the surface of Pb_3O_4, a layer of very fine particles/precipitates of PbO_2 forms during the earliest dissolution stages. Once formed, PbO_2 can either remain as a spectator species or be reduced, as suggested by Reaction 5 [77,79,80], releasing more Pb^{2+}.

$$PbO_{2(s)} + H_2O_{\ (aq)} \rightarrow Pb^{2+}_{(aq)} + 2OH^-_{(aq)} + 0.5\ O_{2(aq)}$$

Reaction 5. Reductive dissolution of platternite (β-PbO_2) in aqueous solution.

However, PbO_2 formed on the surface of Pb_3O_4 is likely to passivate the substrate's surface, inhibiting further dissolution of Pb_3O_4. Still, no such particles/precipitates were detected using XRD during the first month, suggesting that the formation of PbO_2 might have been limited to an amount below the detection limit of XRD. In addition, owing to the very small porosity of the newly formed PbO_2 layer on the Pb_3O_4 surface, the $(NH_4)_2HPO_4$ solution required longer time to diffuse into the Pb_3O_4 substrate. On the basis of the XRD analysis, the β-PbO_2 phase only became detectable after two months of reaction, which suggests that the dissolution of minium continued along with the constant formation of Pb–HAP and β-PbO_2. However, the sudden increase in the precipitation of Pb–HAP and the kinetics of its precipitation rate between 28 days and two months will require further investigation. No previous research into the formation mechanism of lead hydroxyapatite in a comparable system has ever been published, and therefore future research is pivotal to understanding the reaction kinetics of that system.

3.4. *pH Value of the Supernantant Solutions*

The change of pH value of the DAP solutions as a function of time is shown in Figure 9.

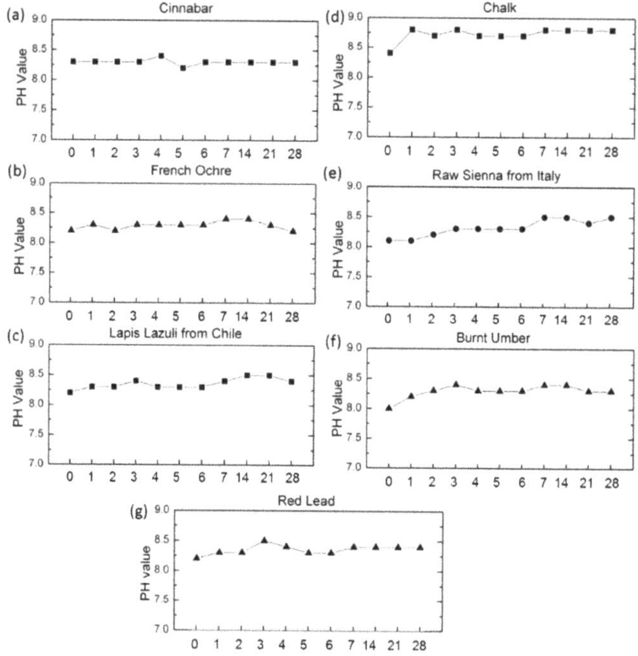

Figure 9. The change of the pH value of the DAP solutions with time (the number on the x-axis means the number of days of measurement) with a standard error of ± 0.1.

The pH value of the French ochre, lapis lazuli, and cinnabar remained almost constant (~8.3) during the first 28 days of reaction. This is consistent with the fact that no significant color, phase, or morphological changes could be observed in these pigments upon exposure to the DAP solution.

By comparing the pH values of the solution at day 0 and day 28, an increase in the pH was observed for calcium carbonate-containing pigments, including chalk, bunt umber, raw sienna (for the latter two, as accessory mineral). This was due to the chemical reaction of calcium carbonate with DAP and the formation of phosphate phases that caused the increase in the pH value of the solution. A slight elevation was also observed in the pH value of the solution containing red lead after 28 days of reaction with DAP. This change is believed to be associated with the reaction of minium (Pb_3O_4) with diammonium hydrogen phosphate, which leads to the formation of hydroxypyromorphite and, hence, to the corresponding increase in the pH value.

3.5. Summary of Color and Phase Changes in the Pigments

The color values of the pigments before immersion into the DAP solutions and after 28 days of reaction with DAP, as well as the ΔE* values, are listed in Table 1. In this research, it was demonstrated that, while the color difference ΔE* of most pigments tested, including cinnabar (deep red), French ochre (yellow), lapis lazuli (blue), chalk (white), and raw sienna (yellow), were above the threshold detected by the human eye (ΔE* > 2), with the exception of burnt umber (brown) which showed no detectable color change (ΔE* < 2), they all showed ΔE* values below the accepted threshold (ΔE* ≤ 5) for cultural heritage studies [16,41–48]. Slightly darkening (−ΔL*) was observed for most pigments, except raw sienna. Red lead, however, showed a significant color change, with ΔE* = 30.643, which is well above the accepted level.

Table 1. Changes in color values of the examined pigments before and after 28 days of immersion in DAP solutions.

Pigment Name	Value Change after 28 Days Reaction with DAP			Change in Color
	ΔL	Δa*	Δb*	ΔE*
Cinnabar	−1.759	−1.874	−2.429	3.5
French Ochre	−1.043	−0.399	−3.201	3.4
Lapis Lazuli	−2.844	1.246	2.773	4.2
Chalk	−4.424	0.508	−2.028	4.9
Raw Sienna	1.989	−1.48	−0.629	2.6
Burnt Umber	−0.865	1.224	0.709	1.7
Red lead	−10.492	−18.057	−22.425	30.6

Pigments such as chalk and calcite, found as impurity or accessory mineral in some of the colored pigments, also underwent evident phase changes from calcium carbonate into calcium phosphates such as hydroxyapatite. In this case, however, these mineralogical phase changes could be considered as 'favorable', given that they provide an additional binding mechanism which is beneficial to the overall consolidation effect.

Conversely, the changes that occurred in the red lead (Pb_3O_4) pigment can be characterized as 'non-favorable', resulting in significant color alteration from bright orange to brown (with a ΔE* = 30.6). Associated phase transformation from lead tetroxide into lead hydroxyapatite possibly occurred via the dissolution–precipitation mechanism described above. As a result, the exposure to DAP caused irreversible color damage in the red lead pigment. The phase transformation and significant color change of red lead caused by the DAP precursor poses significant concerns regarding this consolidation treatment for artifacts painted with this pigment, and therefore DAP-based consolidation would not be recommended.

4. Conclusions

The optical, physical, and chemical interactions between DAP and six pigments commonly employed in fresco applications (cinnabar, French ochre, chalk, lapis lazuli, raw sienna, and burnt umber) and one additional pigment (red lead) often used for secco applications in wall paintings and other polychrome paintings, were investigated. To study the effects of the application of the DAP precursor on the pigments' color, morphology, and mineralogy, the raw pigments (before treatment) and the reaction products after 28 days of exposure to DAP were evaluated using different and complementary characterization techniques including DM, XRD, FTIR, TGA, SEM–EDS, and FORS.

While color changes seemed to occur for most of the pigments analyzed, the majority of these were below the accepted color change threshold established for cultural heritage surface treatments. Evident phase transformations into HAP were identified only in the pigments containing calcium carbonate (calcite), such as the chalk pigment (main coloring phase of white pigment) and the pigments raw sienna and burnt umber, where calcite was identified as an accessory mineral. The formation of the HAP network in this context did not affect the overall color of these pigments. A significant color and phase change were only observed in the red lead pigment with the transformation of red lead (lead tetroxide) into hydroxypyromorphite. The DAP treatment on painted surfaces pigmented with red lead could therefore cause serious and irreversible damage to the artwork, both chromatically and chemically. For this reason, surface treatments using DAP solutions should be avoided when red lead is present. As demonstrated, measurable color differences and phase transformations of pigments, occurring immediately after the application of the DAP solution and after two months under controlled environmental exposure conditions, allowed for the assessment of the direct impact of the DAP solution on the color and mineralogy of pigments commonly encountered in archaeological and historic materials of cultural importance.

While this research did not directly evaluate the consolidation effect of DAP for wall paintings and other polychrome paintings, from our previous research evaluating the effects of DAP on calcium hydroxide-rich plaster layers [21] and the current research investigating the interactions between DAP and pigments, it can be inferred that for fresco wall paintings, where pigments are applied with water on the surface of a moist calcium hydroxide-rich plaster layer and are 'fixed' in place by the newly formed calcium carbonate crystals 'embedding' them into the 'surface skin' of the plaster layer, DAP precursors could also have a consolidating effect, without causing any phase or significant color change. As a proof of concept, further research, testing, and long-term monitoring will be conducted on mockups of fresco paintings and on site, where some other steps such as cleaning [81] and de-salination might be necessary prior to consolidation. Additional investigations will also be carried out on the effect of DAP on different organic binding media, a larger number of pigments, and secco wall paintings mockups to assess the extent of the use of DAP as a surface treatment for polychrome surfaces.

Author Contributions: Conceptualization, design of experiment, I.K. and X.M.; supervision and project administration: I.K., writing-original draft: X.M. and H.P., writing-review and editing: X.M., H.P., M.B., and I.K., investigation and formal analysis: X.M. and H.P., funding acquisition: I.K.

Funding: The work was supported financially by the National Science Foundation (NSF) (Award # 1139227, Solid State and Materials Chemistry program, Division of Materials Research).

Acknowledgments: The authors would like to acknowledge the support provided by the Molecular and Nano Archaeology Laboratory at UCLA. The co-author, Hélène Pasco, was a visiting student to the Archaeomaterials Group at UCLA during 2016. The authors would acknowledge the assistance of Christian Fischer and Roxanne Radpour from the department of materials science and engineering of UCLA in FORS measurement and interpretation of the spectra.

Conflicts of Interest: The authors declare no conflict of interest.

References

1. Neiman, M.K.; Balonis, M.; Kakoulli, I. Cinnabar alteration in archaeological wall paintings: An experimental and theoretical approach. *Appl. Phys. A* **2015**, *121*, 915–938. [CrossRef]
2. Al-Emam, E.; El-Gohary, M.; El Hady, M.A. The paint layers of mural paintings at Abydos Temples-Egypt: A closer look at the materials used. *Mediterr. Archaeol. Archaeom.* **2015**, *15*, 113–121.
3. Kamel, A.M.A. Dehydration of gypsum component of plasters and stuccos in some Egyptian archaeological buildings and evaluation of K_2SO_4 activator as a consolidant. *Sci. Cult.* **2019**, *5*, 49–59.
4. Ferroni, E.; Malaguzzi-Valerj, V.; Rovida, G. Experimental study by diffraction of heterogeneous systems as a preliminary to the proposal of a technique for the restoration of gypsum polluted murals. In Proceedings of the Reports, Plenary Meeting, Amsterdam, Netherlands, 1969-ICOM Committee for Conservation, Amsterdam, The Netherlands, 15–19 September 1969.
5. Matteini, M. In Review: An Assessment of Florentine Methods of Wall Painting Conservation Based. In *The Conservation of Wall Paintings*; Getty Conservation Institute: Los Angeles, CA, USA, 1991; pp. 137–148.
6. Giorgi, R.; Dei, L.; Baglioni, P. A New Method for Consolidating Wall Paintings Based on Dispersions of Lime in Alcohol. *Stud. Conserv.* **2000**, *45*, 154–161.

7. Baglioni, P.; Carretti, E.; Dei, L.; Giorgi, R. *Nanotechnology in Wall Painting Conservation*; IOS Press Ohmsha: Amsterdam, The Netherlands, 2003.
8. Baglioni, P.; Giorgi, R. Soft and hard nanomaterials for restoration and conservation of cultural heritage. *Soft Matter* **2006**, *2*, 293–303. [CrossRef]
9. Giorgi, R.; Ambrosi, M.; Toccafondi, N.; Baglioni, P. Nanoparticles for cultural heritage conservation: Calcium and barium hydroxide nanoparticles for wall painting consolidation. *Chem. A Eur. J.* **2010**, *16*, 9374–9382. [CrossRef] [PubMed]
10. Giorgi, R.; Baglioni, M.; Berti, D.; Baglioni, P. New Methodologies for the Conservation of Cultural Heritage: Micellar Solutions, Microemulsions, and Hydroxide Nanoparticles. *Acc. Chem. Res.* **2010**, *43*, 695–704. [CrossRef] [PubMed]
11. Chelazzi, D.; Poggi, G.; Jaidar, Y.; Toccafondi, N.; Giorgi, R.; Baglioni, P. Hydroxide nanoparticles for cultural heritage: Consolidation and protection of wall paintings and carbonate materials. *J. Colloid Interface Sci.* **2013**, *392*, 42–49. [CrossRef] [PubMed]
12. Balonis, M. Mineral Inorganic Treatments. In *The Encyclopedia of Archaeological Sciences*; Varela, S.L.L., Ed.; Wiley-Blackwell: Hoboken, NJ, USA, 2018.
13. Borgia, G.C.; Camaiti, M.; Cerri, F.; Fantazzini, P.; Piacenti, F. Hydrophobic Treatments for Stone Conservation: Influence of the Application Method on Penetration, Distribution and Efficiency. *Stud. Conserv.* **2003**, *48*, 217–226. [CrossRef]
14. Liu, Q.; Zhang, B.J. Synthesis and Characterization of a Novel Biomaterial for the Conservation of Historic Stone Buildings and Sculptures. *Adv. Mater. Sci. Technol.* **2011**, *675–677*, 317–320. [CrossRef]
15. Yang, F.; Zhang, B.; Liu, Y.; Wei, G.; Zhang, H.; Chen, W.; Xu, Z. Biomimic conservation of weathered calcareous stones by apatite. *New J. Chem.* **2011**, *35*, 887–892. [CrossRef]
16. Sassoni, E.; Naidu, S.; Scherer, G.W. The use of hydroxyapatite as a new inorganic consolidant for damaged carbonate stones. *J. Cult. Herit.* **2011**, *12*, 346–355. [CrossRef]
17. Sassoni, E.; Naidu, S.; Scherer, G.W. Preliminary results of the use of hydroxyapatite as a consolidant for carbonate stones. In Proceedings of the Materials Research Society Fall Meeting 2010, Boston, MA, USA, 29 November–2 December 2010.
18. Matteini, M.; Rescic, S.; Fratini, F.; Botticelli, G. Ammonium Phosphates as Consolidating Agents for Carbonatic Stone Materials Used in Architecture and Cultural Heritage: Preliminary Research. *Int. J. Archit. Herit.* **2011**, *5*, 717–736. [CrossRef]
19. Sassoni, E.; Franzoni, E.; Scherer, G.W.; Naidu, S. Consolidation of a porous limestone by means of a new treatment based on hydroxyapatite. In Proceedings of the 12th International Congress on Deterioration and Conservation of Stone, New York, NY, USA, 21–25 October 2012.
20. Sassoni, E.; Franzoni, E.; Pigino, B.; Scherer, G.W.; Naidu, S. Consolidation of calcareous and siliceous sandstones by hydroxyapatite: Comparison with a TEOS-based consolidant. *J. Cult. Herit.* **2013**, *14*, e103–e108. [CrossRef]
21. Balonis-Sant, M.; Ma, X.; Kakoulli, I. Preliminary Results on Biomimetic Methods Based on Soluble Ammonium Phosphate Precursors for the Consolidation of Archaeological Wall Paintings. In *Archaeological Chemistry VIII*; ACS Symposium Series; American Chemical Society: Washington, DC, USA, 2013; pp. 420–447.
22. Sassoni, E.; Franzoni, E. Sugaring marble in the Monumental Cemetery in Bologna (Italy): Characterization of naturally and artificially weathered samples and first results of consolidation by hydroxyapatite. *Appl. Phys. A* **2014**, *117*, 1893–1906. [CrossRef]
23. Sassoni, E.; Graziani, G.; Franzoni, E. Repair of sugaring marble by ammonium phosphate: Comparison with ethyl silicate and ammonium oxalate and pilot application to historic artifact. *Mater. Des.* **2015**, *88*, 1145–1157. [CrossRef]
24. Ma, X.; Balonis, M.; Pasco, H.; Toumazou, M.; Counts, D.; Kakoulli, I. Evaluation of hydroxyapatite effects for the consolidation of a Hellenistic-Roman rock-cut chamber tomb at Athienou-Malloura in Cyprus. *Constr. Build. Mater.* **2017**, *150*, 333–344. [CrossRef]
25. Sassoni, E. Hydroxyapatite and Other Calcium Phosphates for the Conservation of Cultural Heritage: A Review. *Materials* **2018**, *11*, 557. [CrossRef]
26. Kamiya, M.; Hatta, J.; Shimida, E.; Ikuma, Y.; Yoshimura, M.; Monma, H. AFM analysis of initial stage of reaction between calcite and phosphate. *Mater. Sci. Eng. B* **2004**, *111*, 226–231. [CrossRef]

27. Matteini, M. Inorganic treatments for cleaning, consolidation and protection of mural paintings: Tradition and innovation. In Proceedings of the Works of Art&Conservation Science Today, Thessaloniki, Greece, 26–28 November 2010.
28. Scherer, G.W.; Flatt, R.; Wheeler, G. Materials science research for the conservation of sculpture and monuments. *Mrs Bull.* **2001**, *26*, 44–50. [CrossRef]
29. Wang, Q.; Dove, S.; Shearman, F.; Smirniou, M. Evaluation of methods of chloride ion concentration determination and effectiveness of desalination treatments using sodium hydroxide and alkaline sulphite solutions. *Conservator* **2008**, *31*, 67–74. [CrossRef]
30. Dorozhkin, S.V. Calcium orthophosphates in nature, biology and medicine. *Materials* **2009**, *2*, 399–498. [CrossRef]
31. Lide, D.R. *CRC Handbook of Chemistry and Physics*, 85th ed.; CRC Press: Boca Raton, FL, USA, 2004.
32. Mathew, M.; Takagi, S. Structures of Biological Minerals in Dental Research. *J. Res. Natl. Inst. Stand. Technol.* **2001**, *106*, 1035–1044. [CrossRef]
33. Maslen, E.; Streltsov, V.; Streltsova, N.; Ishizawa, N. Electron density and optical anisotropy in rhombohedral carbonates. III. Synchrotron X-ray studies of $CaCO_3$, $MgCO_3$ and $MnCO_3$. *Acta Crystallogr. Sect. B Struct. Sci.* **1995**, *51*, 929–939. [CrossRef]
34. Chow, L. Solubility of calcium phosphates. In *Octacalcium Phosphate*; Karger Publishers: Berlin, Germany, 2001; Volume 18, pp. 94–111.
35. Eighmy, T.T.; Eusden, J.D. Phosphate stabilization of municipal solid waste combustion residues: Geochemical principles. *Geol. Soc. Lond. Spec. Publ.* **2004**, *236*, 435–473. [CrossRef]
36. Valsami-Jones, E.; Ragnarsdottir, K.; Putnis, A.; Bosbach, D.; Kemp, A.; Cressey, G. The dissolution of apatite in the presence of aqueous metal cations at pH 2–7. *Chem. Geol.* **1998**, *151*, 215–233. [CrossRef]
37. Chou, L.; Garrels, R.M.; Wollast, R. Comparative study of the kinetics and mechanisms of dissolution of carbonate minerals. *Chem. Geol.* **1989**, *78*, 269–282. [CrossRef]
38. Salama, K.K.; Ali, M.F.; El-Sheikh, S.M. The different influence on nano materials on pigments. *Sci. Cult.* **2018**, *4*, 1–7.
39. McLaren, K. XIII—The development of the CIE 1976 (L* a* b*) uniform colour space and colour-difference formula. *J. Soc. Dye. Colour.* **1976**, *92*, 338–341. [CrossRef]
40. Wojciech Mokrzycki, M.T. Color difference Delta E—A survey. *Mach. Graph. Vis.* **2011**, *20*, 383–411.
41. Miliani, C.; Velo-Simpson, M.L.; Scherer, G.W. Particle-modified consolidants: A study on the effect of particles on sol–gel properties and consolidation effectiveness. *J. Cult. Herit.* **2007**, *8*, 1–6. [CrossRef]
42. Sasse, H.R.; Snethlage, R. Methods for the Evaluation of Stone Conservation Treatments. In *Report of Dahlem Workshop on Saving our Architectural Heritage-Conservation of Historic Stone Structures*; Baer, N.S., Snethlage, R., Eds.; John Wiley & Sons: Berlin, Germany, 1996; pp. 223–227.
43. Duchêne, S.; Detalle, V.; Favaro, M.; Ossola, F.; Tomasin, P.; De Zorzi, C.; El Habra, N. Nanomaterials for consolidation of marble and wall paintings. In Proceedings of the 12th International Congress on the Deterioration and Conservation of Stone, New York, NY, USA, 22–26 October 2012.
44. Natali, I.; Tomasin, P.; Becherini, F.; Bernardi, A.; Ciantelli, C.; Favaro, M.; Favoni, O.; Pérez, V.J.F.; Olteanu, I.D.; Sanchez, M.D.R. Innovative consolidating products for stone materials: Field exposure tests as a valid approach for assessing durability. *Herit. Sci.* **2015**, *3*, 6. [CrossRef]
45. Martinez, P.; Soto, M.; Zunino, F.; Stuckrath, C.; Lopez, M. Effectiveness of tetra-ethyl-ortho-silicate (TEOS) consolidation of fired-clay bricks manufactured with different calcination temperatures. *Constr. Build. Mater.* **2016**, *106*, 209–217. [CrossRef]
46. Becherini, F.; Durante, C.; Bourguignon, E.; Vigni, M.L.; Detalle, V.; Bernardi, A.; Tomasin, P. Aesthetic compatibility assessment of consolidants for wall paintings by means of multivariate analysis of colorimetric data. *Chem. Cent. J.* **2018**, *12*, 98. [CrossRef] [PubMed]
47. Bourguignon, E.; Tomasin, P.; Detalle, V.; Vallet, J.-M.; Labouré, M.; Olteanu, I.; Favaro, M.; Chiurato, M.A.; Bernardi, A.; Becherini, F. Calcium alkoxides as alternative consolidants for wall paintings: Evaluation of their performance in laboratory and on site, on model and original samples, in comparison to conventional products. *J. Cult. Herit.* **2018**, *29*, 54–66. [CrossRef]
48. García, O.; Malaga, K. Definition of the procedure to determine the suitability and durability of an anti-graffiti product for application on cultural heritage porous materials. *J. Cult. Herit.* **2012**, *13*, 77–82. [CrossRef]

49. Aceto, M.; Agostino, A.; Fenoglio, G.; Idone, A.; Gulmini, M.; Picollo, M.; Ricciardi, P.; Delaney, J.K. Characterisation of colourants on illuminated manuscripts by portable fibre optic UV-visible-NIR reflectance spectrophotometry. *Anal. Methods* **2014**, *6*, 1488–1500. [CrossRef]
50. Bougeard, D.; Smirnov, K.S.; Geidel, E. Vibrational spectra and structure of kaolinite: A computer simulation study. *J. Phys. Chem. B* **2000**, *104*, 9210–9217. [CrossRef]
51. Genestar, C.; Pons, C. Earth pigments in painting: Characterisation and differentiation by means of FTIR spectroscopy and SEM-EDS microanalysis. *Anal. Bioanal. Chem.* **2005**, *382*, 269–274. [CrossRef]
52. Spence, A.; Kelleher, B.P. FT-IR spectroscopic analysis of kaolinite–microbial interactions. *Vib. Spectrosc.* **2012**, *61*, 151–155. [CrossRef]
53. Muüller, C.M.; Molinelli, A.; Karlowatz, M.; Aleksandrov, A.; Orlando, T.; Mizaikoff, B. Infrared attenuated total reflection spectroscopy of quartz and silica micro-and nanoparticulate films. *J. Phys. Chem. C* **2011**, *116*, 37–43. [CrossRef]
54. Müller, C.M.; Pejcic, B.; Esteban, L.; Delle Piane, C.; Raven, M.; Mizaikoff, B. Infrared attenuated total reflectance spectroscopy: An innovative strategy for analyzing mineral components in energy relevant systems. *Sci. Rep.* **2014**, *4*, 6764. [CrossRef]
55. Saikia, B.J.; Parthasarathy, G.; Sarmah, N. Fourier transform infrared spectroscopic estimation of crystallinity in SiO_2 based rocks. *Bull. Mater. Sci.* **2008**, *31*, 775–779. [CrossRef]
56. Bikiaris, D.; Daniilia, S.; Sotiropoulou, S.; Katsimbiri, O.; Pavlidou, E.; Moutsatsou, A.; Chryssoulakis, Y. Ochre-differentiation through micro-Raman and micro-FTIR spectroscopies: Application on wall paintings at Meteora and Mount Athos, Greece. *Spectrochim. Acta Part A Mol. Biomol. Spectrosc.* **2000**, *56*, 3–18. [CrossRef]
57. Gautier, M.; Muller, F.; Le Forestier, L.; Beny, J.-M.; Guégan, R. NH_4-smectite: Characterization, hydration properties and hydro mechanical behaviour. *Appl. Clay Sci.* **2010**, *49*, 247–254. [CrossRef]
58. Pironon, J.; Pelletier, M.; De Donato, P.; Mosser-Ruck, R. Characterization of smectite and illite by FTIR spectroscopy of interlayer NH_4^+ cations. *Clay Miner.* **2003**, *38*, 201–211. [CrossRef]
59. Mortland, M.; Fripiat, J.; Chaussidon, J.; Uytterhoeven, J. Interaction between ammonia and the expanding lattices of montmorillonite and vermiculite. *J. Phys. Chem.* **1963**, *67*, 248–258. [CrossRef]
60. Elias, M.; Chartier, C.; Prévot, G.; Garay, H.; Vignaud, C. The colour of ochres explained by their composition. *Mater. Sci. Eng. B* **2006**, *127*, 70–80. [CrossRef]
61. Dai, Q.; Dong, F.; Zhao, Y.; Deng, J.; Lu, J. Reciprocity effect between silicate bacterium and wollastonite. In Proceedings of the 11th International Congress for Applied Mineralogy (ICAM), Mianyang, China, 5–10 July 2013; pp. 59–69.
62. Sitarz, M.; Mozgawa, W.; Handke, M. Vibrational spectra of complex ring silicate anions—Method of recognition. *J. Mol. Struct.* **1997**, *404*, 193–197. [CrossRef]
63. Paluszkiewicz, C.; Blażewicz, M.; Podporska, J.; Gumuła, T. Nucleation of hydroxyapatite layer on wollastonite material surface: FTIR studies. *Vib. Spectrosc.* **2008**, *48*, 263–268. [CrossRef]
64. Salvadó, N.; Butí, S.; Aranda, M.A.; Pradell, T. New insights on blue pigments used in 15th century paintings by synchrotron radiation-based micro-FTIR and XRD. *Anal. Methods* **2014**, *6*, 3610–3621. [CrossRef]
65. Gunasekaran, S.; Anbalagan, G.; Pandi, S. Raman and infrared spectra of carbonates of calcite structure. *J. Raman Spectrosc.* **2006**, *37*, 892–899. [CrossRef]
66. Koutsopoulos, S. Synthesis and characterization of hydroxyapatite crystals: A review study on the analytical methods. *J. Biomed. Mater. Res.* **2002**, *62*, 600–612. [CrossRef] [PubMed]
67. Wang, L.; Nancollas, G.H. Calcium orthophosphates: Crystallization and dissolution. *Chem. Rev.* **2008**, *108*, 4628–4669. [CrossRef] [PubMed]
68. Miers, H.A. *Mineralogy: An Introduction to the Scientific Study of Minerals*; Macmillan and Company: London, UK, 1902.
69. Liu, Y.; Wang, A.; Freemen, J. Raman, MIR, and NIR spectroscopic study of calcium sulfates: Gypsum, bassanite, and anhydrite. In Proceedings of the Lunar and Planetary Science Conference, Woodlands, TX, USA, 23–27 March 2009; p. 2128.
70. Takahashi, H.; Maehara, I.; Kaneko, N. Infrared reflection spectra of gypsum. *Spectrochim. Acta Part A Mol. Spectrosc.* **1983**, *39*, 449–455. [CrossRef]
71. Martin, M.A.; Childers, J.W.; Palmer, R.A. Fourier Transform Infrared Photoacoustic Spectroscopy characterization of sulfur-oxygen species resulting from the reaction of SO_2 with CaO and $CaCO_3$. *Appl. Spectrosc.* **1987**, *41*, 120–126. [CrossRef]

72. Madejova, J.; Komadel, P. Baseline studies of the clay minerals society source clays: Infrared methods. *Clays Clay Miner.* **2001**, *49*, 410–432. [CrossRef]
73. Ruan, H.; Frost, R.L.; Kloprogge, J.T.; Duong, L. Infrared spectroscopy of goethite dehydroxylation: III. FT-IR microscopy of in situ study of the thermal transformation of goethite to hematite. *Spectrochim. Acta Part A Mol. Biomol. Spectrosc.* **2002**, *58*, 967–981. [CrossRef]
74. Blanchard, M.; Balan, E.; Giura, P.; Béneut, K.; Yi, H.; Morin, G.; Pinilla, C.; Lazzeri, M.; Floris, A. Infrared spectroscopic properties of goethite: Anharmonic broadening, long-range electrostatic effects and Al substitution. *Phys. Chem. Miner.* **2014**, *41*, 289–302. [CrossRef]
75. Parfitt, R.L.; Russell, J.D.; Farmer, V.C. Confirmation of the surface structures of goethite (α-FeOOH) and phosphated goethite by infrared spectroscopy. *J. Chem. Soc. Faraday Trans. 1 Phys. Chem. Condens. Phases* **1976**, *72*, 1082–1087. [CrossRef]
76. Salama, W.; El Aref, M.; Gaupp, R. Spectroscopic characterization of iron ores formed in different geological environments using FTIR, XPS, Mössbauer spectroscopy and thermoanalyses. *Spectrochim. Acta Part A Mol. Biomol. Spectrosc.* **2015**, *136*, 1816–1826. [CrossRef] [PubMed]
77. Guo, D.; Robinson, C.; Herrera, J.E. Mechanism of dissolution of minium (Pb_3O_4) in water under depleting chlorine conditions. *Corros. Sci.* **2016**, *103*, 42–49. [CrossRef]
78. Kang, Z.; Machesky, L.; Eick, H.; Eyring, L. The solvolytic disproportionation of mixed-valence compounds: III. Pb_3O_4. *J. Solid State Chem.* **1988**, *75*, 73–89. [CrossRef]
79. Xie, Y. *Dissolution, Formation, and Transformation of the Lead Corrosion Product PbO2: Rates and Mechanisms of Reactions that Control Lead Release in Drinking Water Distribution Systems*; Washington University in St. Louis: St. Louis, MO, USA, 2010.
80. Lin, Y.-P.; Valentine, R.L. The release of lead from the reduction of lead oxide (PbO_2) by natural organic matter. *Environ. Sci. Technol.* **2008**, *42*, 760–765. [CrossRef] [PubMed]
81. El-Sheikh, S.M.; Ali, M.F.; Salama, K.K. Low cost pulps with microemulsions for cleaning of *fresco* painting surfaces. *Sci. Cult.* **2017**, *3*, 41–46.

© 2019 by the authors. Licensee MDPI, Basel, Switzerland. This article is an open access article distributed under the terms and conditions of the Creative Commons Attribution (CC BY) license (http://creativecommons.org/licenses/by/4.0/).

MDPI
St. Alban-Anlage 66
4052 Basel
Switzerland
Tel. +41 61 683 77 34
Fax +41 61 302 89 18
www.mdpi.com

Sustainability Editorial Office
E-mail: sustainability@mdpi.com
www.mdpi.com/journal/sustainability

www.ingramcontent.com/pod-product-compliance
Lightning Source LLC
LaVergne TN
LVHW071256170726
843515LV00006BA/1414

* 9 7 8 3 0 3 6 5 3 1 2 0 5 *